MDPI

St. Alban-Anlage 66

4052 Basel

Switzerland

Tel. +41 61 683 77 34

Fax +41 61 302 89 18

www.mdpi.com

Horticulturae Editorial Office

E-mail: horticulturae@mdpi.com

www.mdpi.com/journal/horticulturae

33. Gambuti, A.; Strollo, D.; Ugliano, M.; Lecce, L.; Moio, L. trans-Resveratrol, quercetin, (+)-catechin, and (−)-epicatechin content in south Italian monovarietal wines: Relationship with maceration time and marc pressing during winemaking. *J. Agric. Food Chem.* **2004**, *52*, 5747–5751. [CrossRef] [PubMed]

34. Moio, L.; Ugliano, M.; Gambuti, A.; Genovese, A.; Piombino, P. Influence of Clarification Treatment on Concentrations of Selected Free Varietal Aroma Compounds and Glycoconjugates in Falanghina (*Vitis vinifera* L.) Must and Wine. *Am. J. Enol. Vitic.* **2004**, *55*, 7–12.

35. Zhang, Y.; Jayaprakasam, B.; Seeram, N.P.; Olson, L.K.; DeWitt, D.; Nair, M.G. Insulin Secretion and Cyclooxygenase Enzyme Inhibition by Cabernet Sauvignon Grape Skin Compounds. *J. Agric. Food Chem.* **2004**, *52*, 228–233. [CrossRef] [PubMed]

36. Boido, E.; García-Marino, M.; Dellacassa, E.; Carrau, F.; Rivas-Gonzalo, J.C.; Escribano-Bailón, M.T. Characterisation and evolution of grape polyphenol profiles of Vitis vinifera L. cv. Tannat during ripening and vinification. *Aust. J. Grape Wine Res.* **2011**, *17*, 383–393. [CrossRef]

37. Caccavello, G.; Giaccone, M.; Scognamiglio, P.; Forlani, M.; Basile, B. Influence of intensity of post-veraison defoliation or shoot trimming on vine physiology, yield components, berry and wine composition in Aglianico grapevines. *Aust. J. Grape Wine Res.* **2017**, *23*, 226–239. [CrossRef]

38. Caccavello, G.; Giaccone, M.; Scognamiglio, P.; Mataffo, A.; Teobaldelli, M.; Basile, B. Vegetative, Yield, and Berry Quality Response of Aglianico to Shoot-Trimming Applied at Three Stages of Berry Ripening. *Am. J. Enol. Vitic.* **2019**, *70*, 351–359. [CrossRef]

39. Boulton, R. The Copigmentation of Anthocyanins and Its Role in the Color of Red Wine: A Critical Review. *Am. J. Enol. Vitic.* **2001**, *52*, 67–87.

40. Zhang, X.-K.; Lan, Y.-B.; Huang, Y.; Zhao, X.; Duan, C.-Q. Targeted metabolomics of anthocyanin derivatives during prolonged wine aging: Evolution, color contribution and aging prediction. *Food Chem.* **2021**, *339*, 127795. [CrossRef]

41. de Freitas, V.A.P.; Fernandes, A.; Oliveira, J.; Teixeira, N.; Mateus, N. A review of the current knowledge of red wine colour. *Oeno One* **2017**, *51*. [CrossRef]

42. Mateus, N.; Silva, A.M.S.; Rivas-Gonzalo, J.C.; Santos-Buelga, C.; de Freitas, V. A New Class of Blue Anthocyanin-Derived Pigments Isolated from Red Wines. *J. Agric. Food Chem.* **2003**, *51*, 1919–1923. [CrossRef]

43. Waterhouse, A.L.; Zhu, J. A quarter century of wine pigment discovery. *J. Sci. Food Agric.* **2020**, *100*, 5093–5101. [CrossRef]

44. Li, E.; Mira de Orduña, R. Acetaldehyde kinetics of enological yeast during alcoholic fermentation in grape must. *J. Ind. Microbiol. Biotechnol.* **2017**, *44*, 229–236. [CrossRef]

45. Jeandet, P.; Bessis, R.; Gautheron, B. The Production of Resveratrol (3,5,4'-trihydroxystilbene) by Grape Berries in Different Developmental Stages. *Am. J. Enol. Vitic.* **1991**, *42*, 41–46.

46. Vrhovsek, U.; Wendelin, S.; Eder, R. Effects of Various Vinification Techniques on the Concentration of cis- and trans-Resveratrol and Resveratrol Glucoside Isomers in Wine. *Am. J. Enol. Vitic.* **1997**, *48*, 214–219.

47. Peña-Neira, A.; Dueñas, M.; Duarte, A.; Hernandez, T.; Estrella, I.; Loyola, E. Effects of ripening stages and of plant vegetative vigor on the phenolic composition of grapes (*Vitis vinifera* L.) cv. Cabernet Sauvignon in the Maipo Valley (Chile). *Vitis* **2004**, *43*, 51. [CrossRef]

48. González-Manzano, S.; Rivas-Gonzalo, J.C.; Santos-Buelga, C. Extraction of flavan-3-ols from grape seed and skin into wine using simulated maceration. *Anal. Chim. Acta* **2004**, *513*, 283–289. [CrossRef]

49. Kantz, K.; Singleton, V.L. Isolation and Determination of Polymeric Polyphenols in Wines Using Sephadex LH-20. *Am. J. Enol. Vitic.* **1991**, *42*, 309–316.

50. Forino, M.; Picariello, L.; Lopatriello, A.; Moio, L.; Gambuti, A. New insights into the chemical bases of wine color evolution and stability: The key role of acetaldehyde. *Eur. Food Res. Technol.* **2020**, *246*, 733–743. [CrossRef]

51. Jensen, J.S.; Demiray, S.; Egebo, M.; Meyer, A.S. Prediction of wine color attributes from the phenolic profiles of red grapes (Vitis vinifera). *J. Agric. Food Chem.* **2008**, *56*, 1105–1115. [CrossRef] [PubMed]

52. Dennis, E.G.; Keyzers, R.A.; Kalua, C.M.; Maffei, S.M.; Nicholson, E.L.; Boss, P.K. Grape contribution to wine aroma: Production of hexyl acetate, octyl acetate, and benzyl acetate during yeast fermentation is dependent upon precursors in the must. *J. Agric. Food Chem.* **2012**, *60*, 2638–2646. [CrossRef]

53. Kalua, C.M.; Boss, P.K. Evolution of volatile compounds during the development of Cabernet Sauvignon grapes (*Vitis vinifera* L.). *J. Agric. Food Chem.* **2009**, *57*, 3818–3830. [CrossRef] [PubMed]

54. Brander, C.F.; Kepner, R.E.; Webb, A.D. Identification of some Volatile Compounds of Wine of Vitis Vinifera Cultivar Pinot Noir. *Am. J. Enol. Vitic.* **1980**, *31*, 69–75.

55. Etiévant, P.X. Wine. In *Volatile Compounds in Foods and Beverages*; Routledge: Milton Park, UK, 2017; pp. 483–546. ISBN 0-203-73428-9.

56. Lee, S.-J.; Rathbone, D.; Asimont, S.; Adden, R.; Ebeler, S.E. Dynamic Changes in Ester Formation during Chardonnay Juice Fermentations with Different Yeast Inoculation and Initial Brix Conditions. *Am. J. Enol. Vitic.* **2004**, *55*, 346–354.

57. Di Stefano, R. Free and glycoside terpenes and actinidols changes during must and wine storage at various pH. *Riv. Vitic. Enol.* **1989**, *42*, 11–23.

58. Genovese, A.; Lisanti, M.T.; Gambuti, A.; Piombino, P.; Moio, L. Relationship between sensory perception and aroma compounds of monovarietal red wines. In Proceedings of the International Workshop on Advances in Grapevine and Wine Research 754, Venosa, Italy, 15–17 September 2005; pp. 549–556.

Nutritive Value, Polyphenolic Content, and Bioactive Constitution of Green, Red and Flowering Plants

Editors

Christophe El-Nakhel
Leo Sabatino

MDPI • Basel • Beijing • Wuhan • Barcelona • Belgrade • Manchester • Tokyo • Cluj • Tianjin

Editors
Christophe El-Nakhel
University of Naples Federico II
Italy

Leo Sabatino
University of Palermo
Italy

Editorial Office
MDPI
St. Alban-Anlage 66
4052 Basel, Switzerland

This is a reprint of articles from the Special Issue published online in the open access journal *Horticulturae* (ISSN 2311-7524) (available at: https://www.mdpi.com/journal/horticulturae/special_issues/Flowering_Plants).

For citation purposes, cite each article independently as indicated on the article page online and as indicated below:

LastName, A.A.; LastName, B.B.; LastName, C.C. Article Title. *Journal Name* **Year**, *Volume Number*, Page Range.

ISBN 978-3-0365-4759-6 (Hbk)
ISBN 978-3-0365-4760-2 (PDF)

Cover image courtesy of Christophe El-Nakhel

Contents

About the Editors

Christophe El-Nakhel

Dr. Christophe El-Nakhel is a Postdoctoral Researcher at the University of Naples Federico II, Italy. He has served as a Guest Editor of international journals such as Agriculture (MDPI) and Horticulturae (MDPI). He is a member of the Italian Horticulture Society and International Horticulture Society. His research is focused on horticultural science (soilless cultivation, protected cultivation, leafy vegetables, space farming, microgreens, the application of urine derivatives, etc.) and biostimulants.

Leo Sabatino

Dr. Leo Sabatino is a Researcher at the University of Palermo, Italy. He has served as a Guest Editor of international journals such as Applied Sciences (MDPI), Agriculture (MDPI), Horticulturae (MDPI), and Plants (MDPI). Dr. Sabatino is a member of the Editorial Boards of Agronomy (MDPI) and Frontiers in Plant Sciences. He is a member of the Italian Horticulture Society and International Horticulture Society. His research area covers vegetable grafting, biofortification, soilless cultivation, the propagation of ornamental plants, and the application of microbial and nonmicrobial biostimulants in horticulture.

horticulturae

MDPI

Editorial

Nutritive Value, Polyphenolic Content, and Bioactive Constitution of Green, Red and Flowering Plants

Christophe El-Nakhel

Department of Agricultural Sciences, University of Naples Federico II, 80055 Portici, Italy; christophe.elnakhel@unina.it

Citation: El-Nakhel, C. Nutritive Value, Polyphenolic Content, and Bioactive Constitution of Green, Red and Flowering Plants. *Horticulturae* **2022**, *8*, 461. https://doi.org/10.3390/horticulturae8050461

Received: 13 May 2022
Accepted: 16 May 2022
Published: 20 May 2022

Publisher's Note: MDPI stays neutral with regard to jurisdictional claims in published maps and institutional affiliations.

Plants, including vegetables are a well-known source of health-promoting phytochemicals (plant secondary metabolites) that take part in several physiological processes and play a major role in plant defense and adaptation, in particular plant–environment interactions. The accumulation of these health-promoting phytochemicals depends predominantly on genetic factors and the phenological stage; nonetheless, preharvest factors, e.g., eustress, fertilization, irrigation, light, biostimulant application and other agronomic practices, interfere in modulating and shuffling these phytochemicals. Nowadays, healthier lifestyles are strictly related to plant consumption, especially functional foods rich in bioactive phytochemicals or "ecochemicals", knowing that low occurrence of chronic diseases are well correlated with a vegetable-rich diet. Such vegetables are nutrient dense, endowed with bioactive content that boosts the nutritional quality of food and food security particularly.

The current Special Issue, "Nutritive Value, Polyphenolic Content, and Bioactive Constitution of Green, Red and Flowering Plants", compiles 11 original research articles focusing on the quality of seeds, microgreens, leafy vegetables, herbs, flowers, berries, fruits, and by-products. This Special Issue gathers scientific papers from several research groups around the world, where preharvest and postharvest factors were assessed regarding their effect on the qualitative aspects of the different plants tested.

The quality of leafy vegetables is largely dictated by the level of nitrate present in the leaves, notwithstanding the high nutritional components presented by this commodity, especially given that leafy vegetables tend to accumulate nitrate under adverse conditions (low light or high fertilization rate). This accumulation in plant tissues is also related to N type, uptake, and metabolism. For this purpose, Di Mola and coworkers [1] assessed the quality and yield of *Spinacia oleracea* L. in relation to light and fertilization in greenhouse conditions in a winter season under Mediterranean conditions. The authors tested two greenhouse plastic films with different optical characteristics (clear and diffused light films) in combination with different nitrogen fertilization regime (optimal, sub-optimal, and unfertilized) to determine the effect on quality and yield of spinach. Fertilization treatments and diffused light film decreased the lightness of spinach leaves. Although this film increased the yield by 22.3%, it decreased total ascorbic acid content and dry matter percentage by 61.2 and 8.5%, respectively, though it increased nitrate by 6.3-fold compared to the clear plastic film treatment. As for fertilization, the sub-optimal treatment did not decrease the yield significantly, it increased total phenols by 16.1% and decreased nitrate by 38.6% when compared to optimal fertilization. The authors suggested the feasibility of using the diffused film in winter to boost spinach yield, but with marginal quality depression since nitrate content in leaves did not reach the legal limit fixed by the European Commission. Petropoulos et al. [2] studied nutrient solution as well on the yield, mineral profile, and phytochemical composition of spinach microgreens, but in a controlled growth chamber. Nutrient deficiency stress could generate a positive effect on crop quality, based on its level, thus reducing the production cost and increasing the concentration of secondary metabolites. Based on this background, the authors applied four different nutrient solution (Hoagland) feeding regimes of 0, 5, 10 and 20 days to *Spinacia oleracea* L. microgreens grown

in a peat-based substrate. The increase in feeding days increased linearly the fresh yield and nitrate content of the produced microgreens. The 20-day feeding regime resulted in the highest fresh yield (1.59 kg m^{-2}), ABTS antioxidant activity, total chlorophylls, lutein, β-carotene, quercetin-3-sophoroside-7-glucoside, and patuletin derivative. The 10-day feeding regime did not significantly reduce total phenols compared to 20 days, in contrast to 0 and 5 days, which decreased these secondary metabolites in a significant way and, concomitantly, increased P, K, Ca, and Mg, based on a fresh weight basis. The authors concluded that the 10-day feeding regime proved to be cost-effective and a quality booster, since the content of nitrate was reduced by 70.7% and total ascorbic acid increased by 7.0%, and there was no significant decrease in total phenols but higher dry matter content when compared to the 20-day feeding regime; however, it was compromised due to a 12.6% yield decrease.

El-Nakhel et al. [3] grew *Daucus carota* L. and *Anethum graveolens* L. as microgreens in a floating raft technique under greenhouse conditions. The authors opted for this innovative technique for microgreens, to apply biostimulants (vegetal-based protein hydrolysate) in direct contact with the root system, thus testing how this legume-derived biostimulant would improve the yield, colorimetric parameters, minerals, carotenoids, free amino acids, and other secondary metabolites of these two species belonging to the Apiaceae botanical family. Microgreens are known to be a rich asset of minerals and bioactive metabolites, in addition to offering a range of particular tastes and alluring colors. The addition of protein hydrolysate at a dose of 0.3 mL L^{-1} in this experiment engendered an increase in dill fresh yield (13.5%) and an increase in carrot dry matter % and led to a significant improvement in the canopy colorimetric indices (a*, b*, and Chroma). The biostimulant treatment caused an increase in Ca, S, total chlorophylls, carotenoids, soluble proteins (20.6%), and free amino acids (18.5%) for both species, an increase in anthocyanins (461.7%) and total phenols (12.4%) for carrot, and an increase in total ascorbic acid (17.2%) and nitrate for dill microgreens. Seeds of the same botanical family (Apiaceae), were assessed for their nutritional value by Golubkina and coworkers [4]. Such seeds are highly appreciated for their value in traditional medicine and for their utilization as spices. The authors aimed at evaluating the biochemical profile of 11 species and 43 cultivars grown under similar conditions. Lovage and anise seeds were characterized by the highest total antioxidant activity, total phenolics, and water-soluble proteins. Moreover, some celery cultivars demonstrated high total phenolics, while fennel and coriander revealed high water-soluble proteins as well. Regarding total dissolved solids, fennel and stem celery (cv. Atlant) were the richest. The authors concluded that lovage, anise, parsley, and celery seeds contain the highest levels of antioxidants, and through this comparative study it is possible to orient the choice and implement selected Apiaceae seeds as natural food preservatives and dietary supplements.

Protected cultivation, such as growing modules, facilitate the modulation and management of plant growing conditions to improve qualitative attributes. In this situation, vapor pressure deficit (VPD) is a crucial microclimate factor influencing plant transpiration rate and subsequently physiological and biochemical responses associated with transpiration. With this in mind, Amitrano and collaborators [5] cultivated two differentially pigmented butterhead lettuce cultivars under different VPDs (0.69 and 1.76 kPa) to evaluate any potential shuffling of minerals, phytochemicals, antioxidant capacity, growth, and morpho-physiological parameters. Low VPD caused an increase in both lettuce cultivars fresh and dry biomass, leaf number and area, and a higher Fv/Fm ratio. In addition, lettuce cultivars under low VPD accumulated less nitrate. Under the same conditions, the red cultivar accumulated more calcium, magnesium, and malate, whereas the green cultivar accumulated more phosphorus. On the other hand, a high VPD boosted total ascorbic acid in both cultivars, whereas it boosted phenols and antioxidant activity in the green cultivar by 16.1 and 8.1%. Such results shed light on high VPD as a mild stress, aiming to enhance leafy greens quality. Additionally in protected cultivation, Chowdury et al. [6] cultivated kale, but in a plant factory under different environmental conditions. Kale is

considered a nutrient dense leafy vegetable, endowed with medicinal properties. The aim of the authors was to estimate glucosinolates and anthocyanins accumulated in this crop based on diffuse spectral reflectance using regression methods, especially that reflectance spectroscopic techniques are alternative non-destructive techniques. In this study, the applied wavelengths ranged from 300 to 1050 nm, and the used regression procedures to relate the spectral data to the functional components were: (i) principal component regression; (ii) partial least squares regression; and (iii) stepwise multiple linear regression. The authors found that the last model performed better than the others, and they identified wavelengths in the early near-infrared region to estimate glucosinolates and anthocyanins. However, progoitrin and glucobrassicin were the most detected glucosinolates, whereas cyanidin and malvidin were the most detected anthocyanins in the laboratory analysis.

In the same category of greens, curly endive grown in open field was assessed under iodine biofortification, with the knowledge that approximately 45% of European inhabitants are distressed by iodine deficiency, as declared by the World Health Organization. Sabatino et al. [7] aimed to evaluate the outcomes of four levels of iodine foliar application (0, 50, 250, and 500 mg L^{-1}) on yield, mineral profile, sugars, and bioactive compounds of curly endive (*Cichorium endivia* L., var. *crispum* Hegi). An increasing dose of I decreased head fresh weight and soluble solid content gradually, but did boost calcium, total phenolics, and ascorbic acid. In addition, fructose and glucose were boosted until the dose of 250 mg L^{-1}. Biofortification was mostly accentuated at 250 mg L^{-1}, particularly in fall. Bioactive phenols are highly present in medicinal plants, and numerous plants around the globe are implicated in traditional remedies against type 2 diabetes. In this context, Bljajic and collaborators [8] assessed the activity of hydroethanolic and aqueous extracts of traditional antidiabetics present in Croatian ethnomedicine. For this purpose, extracts from *Achillea millefolium, Artemisia absintium, Centaurium erythrae, Morus alba, Phaseolus vulgaris, Sambucus nigra* and *Salvia officinalis* were assessed for their chemical composition, antioxidant ability, and α-glucosidase inhibiting activity. Rutin, ferulic, and chlorogenic acid were well present in the extracts, as revealed by HPLC analysis. The content of phenolics was correlated with ABTS and DPPH radical scavenging activity, enzyme inhibiting properties, and reducing properties (towards Fe^{3+} and Mo^{6+}), with better efficacy from ethanolic extracts. *S. officinalis* (leaves) and *A. millefolium* (areal parts) ethanolic extracts were characterized by having notable antioxidant activity and being inhibitors of α-glucosidase. Chromium, a mineral that boosts the action of insulin was only detected in *A. absinthium*. The authors suggested that the investigated plants embody a potential alternative for complementary treatment of diabetes and its complications.

The introduction of plant cultivation in space includes an investigation of cosmic radiation effects on plant growth and development. For this reason, Dzhos et al. [9] studied tomato seeds after half a year of storage in the International Space Station, by depicting the quality and biochemical characteristics of tomato fruits originating from the cultivated seeds (*Solanum lycopersicum* L. cv. Podmoskovny ranny dwarf type). Space-stored seeds generated higher plants, yield, and fruit weight, but lower fruit dry matter percentage. The same seeds also induced lower nitrate accumulation and higher β-carotene in both cultivations, whereas lutein and lycopene were significantly higher, though only in field conditions; moreover, the ascorbic acid and antioxidant activity of fruits (space-stored) were higher under greenhouse conditions, while total phenolics were higher in both conditions from the same seeds. However, space-stored seeds induced a decrease in fruit iron and copper, total sugar, titratable acidity, and taste index. Numerous processing techniques are convenient to assure shelf-life extension, as well as the quality and safety of plant food products. In this context, Araujo-Rodrigues et al. [10] investigated the impact of freezing, hot air drying, and storage time on the antioxidant capacity and bioactive compounds of pulps and powders of baby carrot and cherry tomato by-products. This study revealed the high nutritional and functional value of these by-products when converted into pulps or powders, showing a high content of phenolic compounds, carotenoids, and tocopherols. Nonetheless, the drying process decreased polyphenolic content, carotenoids, and antioxidant capacity,

but dry by-products guaranteed stability during the selected storage. However, α-, β-, γ- and δ-tocopherol in tomato and γ-tocopherol in carrot increased significantly during freezing storage and after the drying process. The authors in this study demonstrated that these processing methods engender value-added products with high nutritional profile and microbiologically safe, and pose an interesting economic and environmental impact. An appreciated plant beverage product is wine, where the quality relies on detecting the optimal maturity of grapes at harvest, which determines the main biochemical parameters of the berry. For this reason, Genovese and coworkers [11] tested the effects of four berry ripening stages (total soluble solids of 18, 20, 22 and 25 ^{0}Brix) on aged "Aglianico" wine, where they assessed key secondary metabolites, phenolics, and volatile compounds. The grape maturity degree increased wine color intensity, the level of anthocyanins, and total *trans*-resveratrol. Grapes issued from late harvest of Vitis *vinifera* L. cv. "Aglianico" produced wines richer in aliphatic alcohols, esters, acetates, benzyl alcohol, and α-terpineol. Eventually, grapes of 25 ^{0}Brix soluble solids content produced wines with more biologically active phenolic compounds and a stable color, as well as being richer in aroma compounds.

Conflicts of Interest: The author declares no conflict of interest.

References

1. Di Mola, I.; Ottaiano, L.; Cozzolino, E.; Sabatino, L.; Sifola, M.I.; Mormile, P.; El-Nakhel, C.; Rouphael, Y.; Mori, M. Optical Characteristics of Greenhouse Plastic Films Affect Yield and Some Quality Traits of Spinach (*Spinacia oleracea* L.) Subjected to Different Nitrogen Doses. *Horticulturae* **2021**, *7*, 200. [CrossRef]
2. Petropoulos, S.A.; El-Nakhel, C.; Graziani, G.; Kyriacou, M.C.; Rouphael, Y. The Effects of Nutrient Solution Feeding Regime on Yield, Mineral Profile, and Phytochemical Composition of Spinach Microgreens. *Horticulturae* **2021**, *7*, 162. [CrossRef]
3. El-Nakhel, C.; Ciriello, M.; Formisano, L.; Pannico, A.; Giordano, M.; Gentile, B.R.; Fusco, G.M.; Kyriacou, M.C.; Carillo, P.; Rouphael, Y. Protein Hydrolysate Combined with Hydroponics Divergently Modifies Growth and Shuffles Pigments and Free Amino Acids of Carrot and Dill Microgreens. *Horticulturae* **2021**, *7*, 279. [CrossRef]
4. Golubkina, N.; Kharchenko, V.; Moldovan, A.; Zayachkovsky, V.; Stepanov, V.; Pivovarov, V.; Sekara, A.; Tallarita, A.; Caruso, G. Nutritional Value of Apiaceae Seeds as Affected by 11 Species and 43 Cultivars. *Horticulturae* **2021**, *7*, 57. [CrossRef]
5. Amitrano, C.; Rouphael, Y.; De Pascale, S.; De Micco, V. Modulating vapor pressure deficit in the plant micro-environment may enhance the bioactive value of lettuce. *Horticulturae* **2021**, *7*, 32. [CrossRef]
6. Chowdhury, M.; Ngo, V.-D.; Islam, M.N.; Ali, M.; Islam, S.; Rasool, K.; Park, S.-U.; Chung, S.-O. Estimation of glucosinolates and anthocyanins in kale leaves grown in a plant factory using spectral reflectance. *Horticulturae* **2021**, *7*, 56. [CrossRef]
7. Sabatino, L.; Di Gaudio, F.; Consentino, B.B.; Rouphael, Y.; El-Nakhel, C.; La Bella, S.; Vasto, S.; Mauro, R.P.; D'Anna, F.; Iapichino, G. Iodine biofortification counters micronutrient deficiency and improve functional quality of open field grown curly endive. *Horticulturae* **2021**, *7*, 58. [CrossRef]
8. Bljajić, K.; Brajković, A.; Čačić, A.; Vujić, L.; Jablan, J.; Saraiva de Carvalho, I.; Zovko Končić, M. Chemical composition, antioxidant, and α-glucosidase-inhibiting activity of aqueous and hydroethanolic extracts of traditional antidiabetics from Croatian ethnomedicine. *Horticulturae* **2021**, *7*, 15. [CrossRef]
9. Dzhos, E.; Golubkina, N.; Antoshkina, M.; Kondratyeva, I.; Koshevarov, A.; Shkaplerov, A.; Zavarykina, T.; Nechitailo, G.; Caruso, G. Effect of Spaceflight on Tomato Seed Quality and Biochemical Characteristics of Mature Plants. *Horticulturae* **2021**, *7*, 89. [CrossRef]
10. Araújo-Rodrigues, H.; Santos, D.; Campos, D.A.; Ratinho, M.; Rodrigues, I.M.; Pintado, M.E. Development of frozen pulps and powders from carrot and tomato by-products: Impact of processing and storage time on bioactive and biological properties. *Horticulturae* **2021**, *7*, 185. [CrossRef]
11. Genovese, A.; Basile, B.; Lamorte, S.A.; Lisanti, M.T.; Corrado, G.; Lecce, L.; Strollo, D.; Moio, L.; Gambuti, A. Influence of Berry Ripening Stages over Phenolics and Volatile Compounds in Aged Aglianico Wine. *Horticulturae* **2021**, *7*, 184. [CrossRef]

Article

Optical Characteristics of Greenhouse Plastic Films Affect Yield and Some Quality Traits of Spinach (*Spinacia oleracea* L.) Subjected to Different Nitrogen Doses

Ida Di Mola [1], Lucia Ottaiano [1,*], Eugenio Cozzolino [2], Leo Sabatino [3], Maria Isabella Sifola [1], Pasquale Mormile [4], Christophe El-Nakhel [1,*], Youssef Rouphael [1] and Mauro Mori [1]

[1] Department of Agricultural Sciences, University of Naples Federico II, 80055 Portici, Italy; ida.dimola@unina.it (I.D.M.); sifola@unina.it (M.I.S.); youssef.rouphael@unina.it (Y.R.); mori@unina.it (M.M.)
[2] Council for Agricultural Research and Economics (CREA)—Research Center for Cereal and Industrial Crops, 81100 Caserta, Italy; eugenio.cozzolino@crea.gov.it
[3] Department of Agricultural, Food and Forest Sciences, University of Palermo, Viale delle Scienze 4, 90128 Palermo, Italy; leo.sabatino@unipa.it
[4] Institute of Applied Science and Intelligent Systems of National Research Council (ISASI-CNR), 80078 Pozzuoli, Italy; p.mormile@isasi.cnr.it
* Correspondence: lucia.ottaiano@unina.it (L.O.); christophe.elnakhel@unina.it (C.E.-N.)

Citation: Mola, I.D.; Ottaiano, L.; Cozzolino, E.; Sabatino, L.; Sifola, M.I.; Mormile, P.; El-Nakhel, C.; Rouphael, Y.; Mori, M. Optical Characteristics of Greenhouse Plastic Films Affect Yield and Some Quality Traits of Spinach (*Spinacia oleracea* L.) Subjected to Different Nitrogen Doses. *Horticulturae* **2021**, *7*, 200. https://doi.org/10.3390/horticulturae7070200

Academic Editor: Nikolaos Katsoulas

Received: 25 May 2021
Accepted: 13 July 2021
Published: 18 July 2021

Publisher's Note: MDPI stays neutral with regard to jurisdictional claims in published maps and institutional affiliations.

Abstract: Light and nitrogen strongly affect the growth, yield, and quality of food crops, with greater importance in green leafy vegetables for their tendency to accumulate nitrate in leaves. The purpose of this research was to explore the effect of two greenhouse films (Film A and B) on yield, and quality of spinach grown under different nitrogen regimes (not fertilized—N0%; sub-optimal N dose—N50%; optimal N dose—N100%). Film A and Film B were used as clear and diffused light films, with 75% and 87% thermicity, and 85% and 90% total transmittivity, respectively, where only Film B had a UV-B window. Film B elicited an increase in yield (22%) and soil–plant analysis development (SPAD) index (4.6%) compared to the clear film, but did not affect chlorophyll a, b, and total chlorophyll content. In addition, the diffuse film significantly decreased ascorbic acid in the crop but had no effect on lipophilic antioxidant activity and phenols content, but decreased ascorbic acid content. Finally, nitrate content was strongly increased both by nitrogen dose (about 50-fold more than N0%) and greenhouse films (about six-fold higher under diffuse light film), but within the legal limit fixed by European Commission. Therefore, irrespective of N levels, the use of diffuse-light film in winter boosts spinach yield without depressing quality.

Keywords: greenhouse clear film; greenhouse diffuse-light film; spinach yield; nitrate content; antioxidant activity; ascorbic acid

1. Introduction

Spinach (*Spinacia oleracea* L.) is a major nutrient-dense leafy vegetable. It contains high amounts of Fe, K, Mg, vitamins (A, B6, C, K, E), antioxidants, chlorophylls [1,2] and it is rich in fiber, but very low in calories. Due to its special nutritional properties, about 1.7 million hectares are dedicated to spinach cultivation over the world [3]. Europe accounts for about 39,000 ha [3], mainly concentrated in the Mediterranean basin countries where the environmental conditions (light and temperature) are optimal to achieve the best response in terms of both yield and quality. In Italy, about 6300 ha are dedicated to spinach cultivation [4], of which 8.0% are under controlled conditions (greenhouses or tunnels).

One of the major factors that affects the high nutritional quality of spinach, and other green leafy vegetables is the nitrate content of the leaves, which shows adverse effects on human health [5–7]. Nitrate accumulation in plant tissues directly depends on (i) nitrogen (N) fertilization doses [6,8,9], (ii) type of N fertilizers [6,9–12], (iii) root N uptake [5] and (iv) N metabolism [13,14]; the latter includes both N mobilization from

the root to the shoot or N assimilation. This assimilation depends on the activity of the nitrate-reductase enzyme that is reported to be regulated by light through activation of the gene codifying it [15], and it is regulated by nitrate on the transcription and post-translation level. Therefore, there is a clear interactive light * N fertilization effect on the nitrate content of plants [5,6,16,17]. Regardless of N fertilization, light intensity is inversely correlated with the nitrate content of plant tissues [5,6,13,17–21], therefore nitrate content can be usefully reduced by increasing light intensity. Nevertheless, light is positively related to several other quality components like minerals, vitamins and antioxidants [17,21–23], including chlorophylls. Chlorophylls are responsible for leaf greenness that contributes to the esthetical value of several green-leafy vegetables including spinach [24]; in addition, they are involved in preventing mutant DNA from proliferating, that is responsible for some forms of cancer [25].

Light intensity changes across the seasons, and it is higher during spring/summer (maximum values in June/July) than autumn/winter period (minimum values in December) [26]. The nitrate content of green leafy vegetables is generally higher when they are grown during winter than during spring [19,20]. In addition, the Rd/Rg ratio (Diffuse radiation/Global radiation) also changes across the seasons. The Rd/Rg ratio is higher during autumn/winter than spring/summer period, due to the increasing cloud cover and precipitations [27]. Marpaung and Hirano [27] reported maximum and minimum Rd/Rg of 0.68 in October and 0.51–0.52 in June and July. It is well known that diffuse radiation determines beneficial effects on plant productivity [26], which are frequently reported as diffuse radiation fertilization effects [28] since they enhance the radiation use efficiency of plants. Global radiation (short-wave radiation) consists of direct solar radiation and diffuse radiation resulting from reflected or scattered sunlight due to several factors/compounds, such as air molecules, water vapor, clouds, dust, pollutants, forest fires, and volcanoes; overall, atmospheric conditions can reduce direct radiation by 10% on clear and dry days and by 100% on thick andcloudy days [29]. Similarly, the glass or plastic cover of a greenhouse intercepts a part of solar radiation, creating light conditions different from open field conditions [30]. The amount of daylight received may be reduced around 30% by the glasshouse structure [31,32]. On the other hand, cultivation under a protected environment allows off-season production, resulting in a better price for farmers; in addition, the adoption of greenhouses or tunnels, combined with intensive production methods, makes it possible to reach higher yield than the same vegetable crops grown in open field conditions [33].

About half of greenhouse spinach cultivation areas in Italy are concentrated in the Campania region (Southern Italy), where radiation is not a limiting factor. Nevertheless, also in this optimal condition, in a context of energy sustainability, it is important to optimize light penetration into the greenhouses by using photo-selective plastic films as covering material [34], particularly during the winter period to balance both light and temperature. Various commercial products are available in the market, that differ by several factors like thickness, light diffusion, solar radiation transmission, mist control, thermicity, anti-drip, and anti-dust effect. However, the optical properties of the plastic cover film of greenhouses/tunnels diversely affect yield and quality traits of vegetable crops.

We previously tested two of these products on lamb's lettuce grown during spring, a season with a generally high light intensity but a potentially low Rd/Rg, and fertilized with increasing rates of N [35]. By contrast, the present study was conducted on spinach grown during winter in two greenhouses covered each with a different plastic film that has distinct optical properties (low and high Rd/Rg ratio). The scope of the current study was to explore the influence of light conditions on the productivity and quality of spinach cultivated under diverse nitrogen levels.

2. Materials and Methods

2.1. Experimental Set-Up and Design, and Plastic Films Properties

The "Platypus RZ" F_1 spinach (RIJK ZWAAN) was cultivated during winter 2019 in large pots (0.38 m^2 area and 60 cm height) placed in plastic tunnels, at the Department of Agricultural Science (Portici, Naples, Italy; N40° 48.870'; E14° 20.821'; 70 m a.s.l.). Pots were filled with sandy soil (91% sand, 4.5% silt, and 4.5% clay), with good fertility (P_2O_5 253 ppm, K_2O 490 ppm, organic matter 2.5% and total N 0.09%), pH 7.4 and electrical conductivity 0.151 dS m^{-1}.

The experimental design provided a split-plot factorial combination between two greenhouses covered each with a different film (Film A and Film B) and three N levels: not fertilized control (N0%), sub-optimal N dose (25 kg ha^{-1}—N50%), and optimal N dose (50 kg ha^{-1} —N100%) under each film. Each of the three fertilization treatments was replicated three times and completely randomized in the plastic tunnels, for a total of nine experimental units greenhouse-^{1} (18 in total).

The plastic films used to cover the two tunnels, have different optical characteristics. Film A, supplied by Lirsa srl (Ottaviano, Naples, Italy; commercial name LIRSALUX), is a thermal and clear plastic film, 150 microns thick, with anti-drip effect (an additive is added to the film formulation, which flattens the water droplets into a layer of water that runs down the sides of the greenhouse). Film B, manufactured by Ginegar Plastic Products and supplied by Polyeur srl (Benevento, Italy; commercial name SUNSAVER), is a thermal and diffused light film, 150 microns thick, with a light diffusivity of 58% and anti-drip effect. Transmission measurements on the two plastic films were realized using an UV/VIS Spectrophotometer (JASCO V-650 (JASCO Corporation, Tokyo Japan), accuracy 0.5 nm, range 190–900 nm) with an Integrating sphere (JASCO ISN-722, inside diameter 60 mm, range 200–870 nm), which allows the estimation of the total light transmission.

The total transmission UV-Vis spectrum of the plastic Film A and Film B are reported in Figures S1 and S2, respectively. Film A does not transmit UV radiation, identically to most traditional films present in the market and used for greenhouse covering. The thermicity of this film is 75% and the total transmittivity (direct plus diffused components of light transmitted) in photosynthetically active radiation (PAR) is about 85%.

In Figure S2, the spectrum of film B in the same range, is reported and it notes a UV-B "window". This means that from 270 to 330 nm (UV-B range: 280–320 nm) the film partially transmits. The value of the thermicity of this film is 87% while its total transmittivity in PAR is 90%. The optical characteristics of the two plastic films (Transmittivity [%], Radiation [Lux], Light intensity [μmol m^{-2} s^{-1}]) were monitored constantly during the trial.

2.2. Plant Management

Spinach was sown on 17 January with a plant density of 340 seeds m^{-2}, resulting in 130 seeds pot^{-1}. N was added as calcium nitrate (26%) and it was given in a single solution on 13 February due to the short cycle length. The irrigation was managed accurately to avoid any leaching risk. Experimental pots were irrigated with an amount of water equal to the evapotranspiration, which was calculated by the Hargreaves formula. The harvesting occurred on 12 March.

2.3. Photosynthetically Active Radiation and Temperature Measurements

During the growing period, the light intensity in the photosynthetically active radiation (PAR-400 to 700 nm wavelength range) and temperatures were monitored continuously. Light intensity was recorded by a WatchDog A150 data logger (Spectrum Technologies Inc., Aurora, IL, USA) placed at canopy level, and was expressed as micro-moles of light energy $m^{-2}s^{-1}$. Temperatures were measured with probes (Vantage Pro2, Davis Instruments) placed in the pots at canopy level and distributed randomly across the greenhouses. Both data were reported as hourly day mean divided into 15-day intervals over the growing period (60 days) starting from sowing up to final harvest.

2.4. Yield Measurements, SPAD Index and Color Parameters

At harvest, the yield was determined by cutting the whole pot surface, and was expressed as t ha^{-1}. For each treatment replicate, a representative leaf sample of each pot/replicate was collected and oven-dried at 70 °C for 72 h in order to assess dry matter percentage, and subsequently utilize the dry material for the assessment of nitrate concentration. SPAD index measurements were conducted on a representative vegetable sample (ten undamaged young fully expanded leaves) per replicate, via a chlorophyll meter SPAD-502 (Konica Minolta, Tokyo, Japan). The CIELAB (Commission international de l'eclairage) color parameters (L*; a*; b*) were measured by a Minolta CR-300 Chroma Meter (Minolta Camera Co. Ltd., Osaka, Japan).

2.5. Qualitative Parameters Assessments

On each dried sample of leaves, nitrate concentration was assessed by Foss FIAstar 5000 (FOSS Italia S.r.l., Padova, Italy) continuous flow Analyzer. On the fresh sample (leaves), after freezing and lyophilizing, ABTS antioxidant activity (ABTS) and total phenols were assessed, whereas total ascorbic acid content (AsA) and chlorophyll a, and b measurements were assessed on fresh samples (leaves) before harvest.

ABTS antioxidant activity was determined by 2,2-azinobis 3-ethylbenzothiazoline-6-sulfonic acid (ABTS) method [36]. Antioxidant compounds inhibit ABTS$^+$ radical, proportionally to their concentration; therefore they are indirectly measured by UV-Vis spectrophotometry (ONDA V-10 Plus (Giorgio Bormac s.r.l., Carpi, Italy) at 734 nm. The data are presented in mmol of Trolox eq. 100 g^{-1} dw.

Total ascorbic acid concentration was evaluated following the Kampfenkel et al. method [37], which is based on the reduction of Fe3$^+$ to Fe^{2+} by ascorbic acid; Fe $^{2+}$ with 2,2-dipyridyl form a complex, of which the quantitation was performed at 525 nm ONDA V-10 Plus (UV-Vis spectrophotometry, ONDA V-10 Plus, Giorgio Bormac s.r.l., Carpi, Italy). Results were expressed as mg 100 g^{-1} fresh weight (fw).

Total phenols concentration was determined by Folin–Ciocalteau method [38]; absorption was appraised at 765 nm through a UV-Vis spectrophotometer (ONDA V-10 Plus, Giorgio Bormac s.r.l., Carpi, Italy). Data of total phenol were expressed in mg gallic acid eq. 100 g^{-1} dw.

As for chlorophyll a, and b, after extraction in 99% acetone and centrifugation at 3000× *g* for 5 min, pigments content was determined by a Hach DR 2000 spectrophotometer (Hach Co., Loveland, CO, USA) at 662 and 647 nm, respectively. The extinction coefficients, used for the determination of chlorophyll a, b were described by Lichtenhaler and Wellburn [39].

2.6. Statistical Analysis

All data were examined with the SPSS software package (SPSS version 22, Chicago, IL, USA), using a general linear model (GLM) for the Analysis Of Variance (ANOVA). The means were separated using the Duncan's Multiple Range Test (DMRT) test at $p \leq 0.05$.

3. Results

3.1. Environmental Conditions

The hourly day mean values of light intensity in the PAR (photosynthetic active radiation) and temperature during the growing period of spinach (15-day intervals) from sowing up to final harvest, are reported in Figures 1 and 2, respectively.

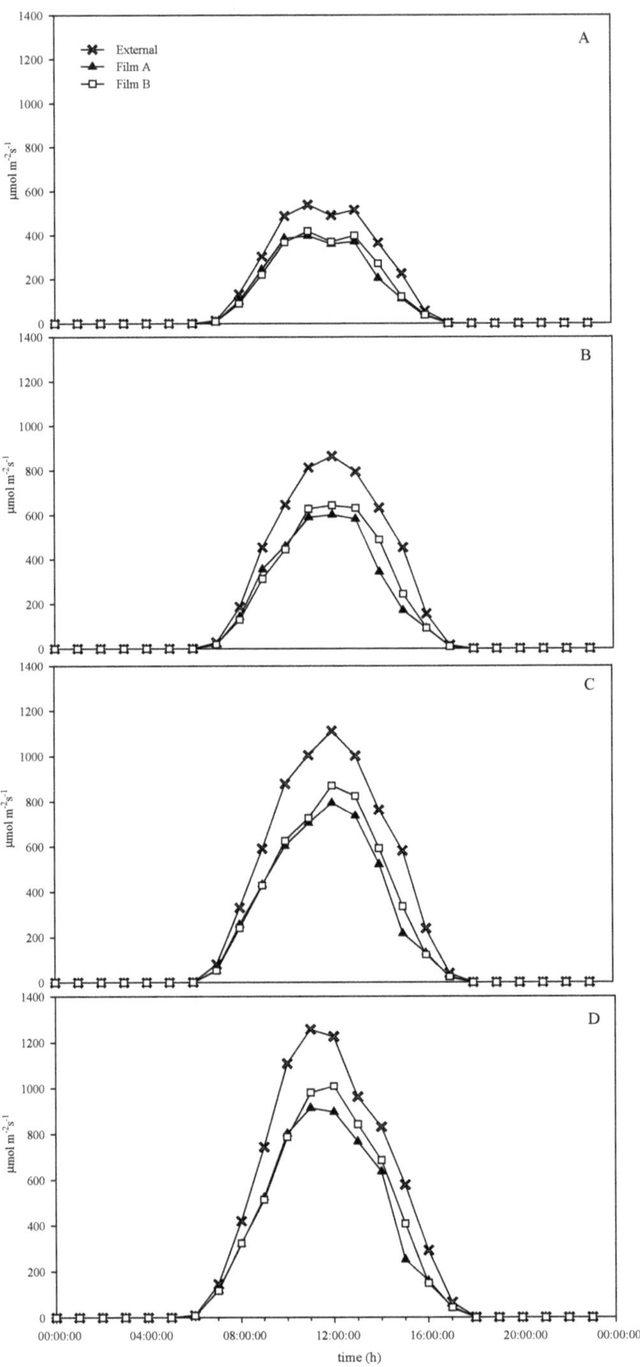

Figure 1. Hourly day mean value of light intensity in the PAR (photosynthetic active radiation) during the growing period of spinach referred to 15-day intervals starting from sowing (A = 1–15 days; B = 16–30 days; C = 31–45 days; D = 46–60 days).

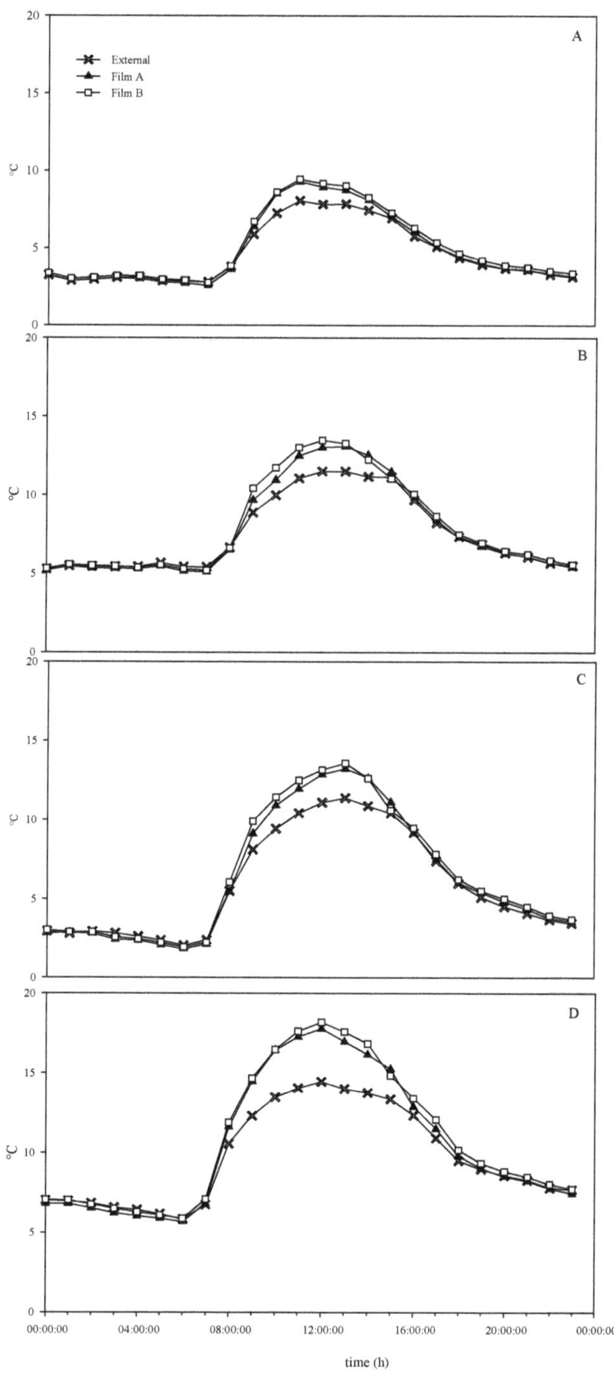

Figure 2. Hourly day mean value of temperature during the growing period of spinach referred to 15-day intervals starting from sowing (A = 1–15 days; B = 16–30 days; C = 31–45 days; D = 46–60 days).

The light intensity had an increasing trend in the four intervals, ranging on average from ~230 μmol m^{-2} s^{-1} in the first interval to ~570 μmol m^{-2} s^{-1} in the last interval; concomitantly it overcame 500 and 1200 μmol m^{-2} s^{-1} under external conditions, in the first and fourth interval, respectively (Figure 1). Film A and Film B decreased external light intensity by 31 and 26%, respectively. Differences between the two films were particularly evident starting from the second interval (16–30; Figure 1B), with a mean value of 7.7% (Figure 1B–D), with respect to 3.8% of the 1–15 day interval (Figure 1A).

The mean temperature inside the greenhouses was higher than that measured outside, increasing by 7.6% on average over the entire growing period (Figure 2). Notably, under Film B the mean temperature was always higher than that under Film A (7.59 °C vs. 7.37 °C); with the highest increase in the first 15 days interval (+4.1%) (Figure 2).

3.2. Yield and SPAD Index

The main effect of the two experimental factors (greenhouse cover film and N doses) on yield is reported in Figure 3. Plants yielded significantly more (+22.3%) under plastic Film B than under plastic Film A. The yield also had an increasing trend when N ranged between 0 kg ha^{-1} (N0%) to 50 kg ha^{-1} (N100%), but without a significant difference between the sub-optimal N dose (N50%) and optimal N dose (N100%) (Figure 3).

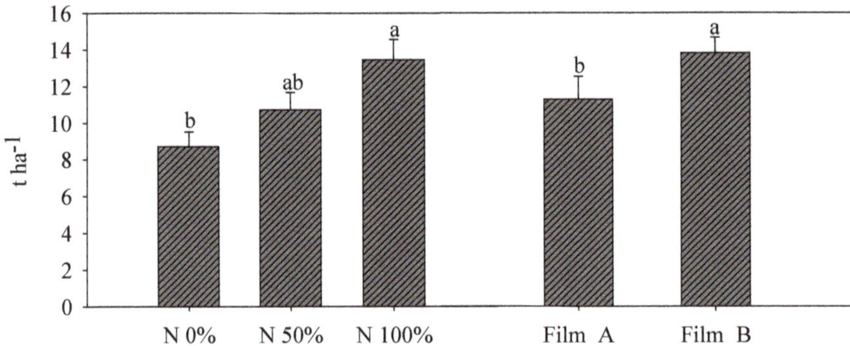

Figure 3. Effect of N fertilization (not fertilized = N0%; fertilized with 25 kg N ha^{-1} = N50%; fertilized with 50 kg N ha^{-1} = N100%) and plastic film (clear plastic film = Film A; light diffusion plastic film = Film B) on spinach yield (tons hectare^{-1}: t ha^{-1}). Diverse letters on the bars indicate significant differences according to Duncan's test ($p = 0.05$). Vertical bars designate $\pm$ SE (standard error) of means.

Furthermore, in the case of SPAD index investigation, Film B equally elicited an increase of 4.6% over Film A (Figure 4). Similarly, the effect of N fertilization treatments on SPAD was less strong than the effect on yield, though significant, with a 6.4% and 11.3 % increase, for sub-optimal and optimal dose compared to N0% fertilization, respectively.

3.3. Spinach Colorimetric Indices, Bioactive Qualities and Dry Matter Percentage

Regarding the CIELAB color parameters, the statistical analysis showed that only lightness (L*) was significantly affected by both N doses and plastic films (Table 1). Notably, the two fertilized treatments were not different and they were significantly lower than control (−4.1%). Instead, Film B showed the lowest value of the L* parameter (−3.1% as compared to Film A).

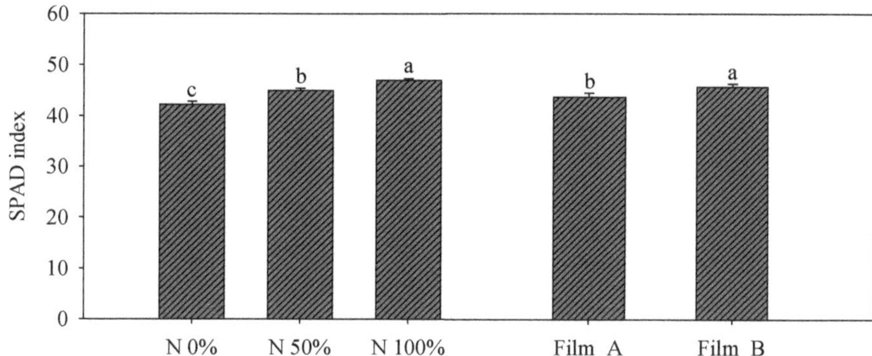

Figure 4. Effect of N fertilization (not fertilized = N0%; fertilized with 25 kg N ha^{-1} = N50%; fertilized with 50 kg N ha^{-1} = N100%) and plastic film (clear plastic film = Film A; light diffusion plastic film = Film B) on spinach SPAD (soil and plant development) index. Diverse letters on the bars indicate. Significant differences according to Duncan's test (p = 0.05). Vertical bars designate ± SE (standard error) of means.

Table 1. Effect of N fertilization (not fertilized = N0%; fertilized with 25 kg N ha^{-1} = N50%; fertilized with 50 kg N ha^{-1} = N100%) and plastic film (clear plastic film = Film A; light diffusion plastic film = Film B) on color parameters (L* = lightness, a* = green/red coordinate, and b* = blue/yellow coordinate) of spinach leaves.

Treatments	L*	a*	b*
Fertilization			
N0%	38.99 ± 0.63 a	−14.33 ± 0493	19.87 ± 0.65
N50%	37.48 ± 0.55 b	−13.52 ± 0.35	18.44 ± 0.44
N100%	37.31 ± 0.76 b	−13.40 ± 0.67	18.60 ± 0.82
Film			
Film A	38.53 ± 0.67 a	−13.92 ± 0.47	19.42 ± 0.82
Film B	37.32 ± 0.51 b	−13.58 ± 0.42	18.51 ± 0.69
Significance			
Fertilization (N)	*	ns	ns
Film (F)	*	ns	ns
N × F	ns	ns	ns

ns, *: non-significant or significant at $p \leq 0.05$, respectively. All data are expressed as mean ± SE (standard error), n = 3. Diverse letters within each column indicate significant differences according to Duncan's test (p = 0.05).

There was no effect of N fertilization on ABTS, and AsA, whereas total phenols decreased significantly with increasing N fertilization rates, but only the N100% treatment was significantly different from the other two treatments (−18.8% with respect to their mean value) (Table 2). Greenhouse cover film affected only AsA, which was significantly lower (−61.2%) in plants grown under Film B than plants grown under Film A (Table 2).There was no change in ABTS antioxidant activity and total phenols content due to different films used (Table 2). Finally, a significant higher value of leaves dry matter percentage was recorded in N0% treatment, with about a 15% increase over the mean value of the two fertilized treatments; also, Film A elicited a 9.3% leaves DM increase over the Film B value (Table 2).

Table 2. Effect of N fertilization (not fertilized = N0%; fertilized with 25 kg N ha^{-1} = N50%; fertilized with 50 kg N ha^{-1} = N100%) and plastic film (clear plastic film = Film A; light diffusion plastic film = Film B) on ABTS antioxidant activity (ABTS), total phenols, Total Ascorbic Acid (AsA) and dry matter (DM) of spinach leaves.

Treatments	ABTS	Total Phenols	AsA	DM
	mM Trolox eq. 100g^{-1} dw	mg Gallic Acid g^{-1} dw	mg 100 g^{-1} fw	%
Fertilization				
N0%	22.77 ± 1.70	3.22 ± 0.18 a	33.49 ± 4.93	9.9 ± 0.6 a
N50%	21.90 ± 1.29	2.88 ± 0.17 a	27.45 ± 7.33	8.8 ± 0.2 b
N100%	20.55 ± 1.72	2.48 ± 0.15 b	23.62 ± 3.81	8.3 ± 0.3 b
Film				
Film A	22.10 ± 1.91	2.95 ± 0.17	40.62 ± 6.71 a	9.4 ± 0.5 a
Film B	21.42 ± 1.23	2.78 ± 0.16	15.75 ± 5.33 b	8.6 ± 0.2 b
Significance				
Fertilization (N)	ns	**	ns	*
Film (F)	ns	ns	**	*
N × F	ns	ns	ns	ns

ns, *, **: non-significant or significant at $p \leq 0.05$ and 0.01, respectively. All data are expressed as mean ± SE (standard error), $n = 3$. Diverse letters within each column indicate significant differences according to Duncan's test ($p = 0.05$). dw: dry weight, fw: fresh weight.

Chlorophyll a, chlorophyll b and total chlorophylls significantly increased with N fertilization up to 50 kg ha^{-1} (Table 3). Furthermore, nitrate content increased linearly (y = 43.348x − 11.784, R^2 = 0,9974) with N dose (Table 3):

Table 3. Effect of N fertilization (not fertilized = N0%; fertilized with 25 kg N ha^{-1} = N50%; fertilized with 50 kg N ha^{-1} = N100%) and plastic film (clear plastic film = Film A; light diffusion plastic film = Film B) on chlorophyll a, chlorophyll b, and total chlorophylls, and nitrate of spinach leaves.

Treatments	Chlorophyll a	Chlorophyll b	Total Chlorophylls	Nitrate
	mg g^{-1} fw	mg g^{-1} fw	mg g^{-1} fw	mg kg^{-1} fw
Fertilization				
N0%	0.905 ± 0.034 b	0.547 ± 0.045 b	1.452 ± 0.079 b	52.3 ± 20.8 c
N50%	0.976 ± 0.044 ab	0.716 ± 0.105 a	1.692 ± 0.146 a	1968.4 ± 650.7 b
N100%	1.015 ± 0.025 a	0.786 ± 0.075 a	1.800 ± 0.099 a	3205.5 ± 537.5 a
Film				
Film A	0.976 ± 0.029	0.689 ± 0.075	1.665 ± 0.103	476.4 ± 134.4 b
Film B	0.954 ± 0.040	0.677 ± 0.076	1.632 ± 0.114	3007.7 ± 671.6 a
Significance				
Fertilization (N)	*	*	*	**
Film (F)	ns	ns	ns	**
N x F	ns	ns	ns	ns

ns, *, **: non-significant or significant at $p \leq 0.05$ and 0.01, respectively. All data are expressed as mean ± SE (standard error), n = 3. Diverse letters within each column indicate significant differences according to Duncan's test ($p = 0.05$). fw: fresh weight.

There was no effect of different films on leaf content of chlorophyll a, chlorophyll b, and total chlorophyll (Table 3). By contrast, nitrate content was significantly affected by higher levels in spinach grown under Film B compared to Film A (Table 3).

4. Discussion

In the current research, the two plastic films with different properties, clear (Film A) and diffuse-light film (Film B), resulted in a different micro-climate. However, under Film B, the light intensity was higher than that recorded under Film A. Under both films conditions, the light intensity was lower than that of open-air conditions by 26% and 31%, for Film B and Film A, respectively.

Data reported in Figure 3 showed that the differences in light and temperature conditions resulted in differences in the yield of spinach, which was higher for plants grown under diffuse light film (Film B) than that of plants grown under clear film (Film A), specifically +22.3%. Our findings are consistent with the results of our previous research [34] on lamb's lettuce, where we found a similar increase in yield (by 22%) with a light diffused cover film. Diffuse light is reported as well to increase biomass yield of other vegetables like *Solanum lycopersicum* [40], and *Capsicum annuum* [41]. Kanniah et al. [28] reported that in conditions of high diffuse radiation, plant production increases as a result of a more efficient yield per unit of PAR, and this phenomenon is also known as the "diffuse fertilization effect". Roderick et al. [42] explained that on clear days, direct sunlight mainly reaches the upper canopy; whereas in the presence of diffuse radiation, the sub-canopy vegetation, usually shaded, is uniformly illuminated, since radiation comes from all directions and can penetrate deeper into the canopy. Hence, this produces higher light use efficiency and a higher yield, as photosynthesis is the primary factor driving plant productivity [43]. Additionally, some authors argued that plants cultivated under diffuse light suffer fewer stress events related to water and heat [44,45].

The yield increase reflects the trend of the SPAD index, which is a good indicator of the physiological and biochemical status of plants, and indirectly of chlorophyll content, which plays a key role in photosynthesis. The SPAD index of spinach plants grown under Film B increased by 4.6% with respect to that of spinach grown under Film A, although chlorophyll a, b, and total were not significantly affected by light conditions. Riga and Benedicto [46] found that plastic films affected leaf pigments (chlorophyll a, total, and carotenoids) of lettuce only in the harvests made from April to September, and no effects were recorded for winter harvest. The effect of Film B on the yield of the present experiment could be also ascribed to its partial UV-B transmission. Heuberger et al. [30] studied spinach subjected to different light conditions (no additional lighting, corresponding to standard growing practice; additional photosynthetically active radiation; and three different UV-B intensity, corresponding to field-grown condition; UV intensive, and UV permanent), and they found an increase in fresh and dry weight of spinach under additional PAR and intensive UV-B. The UV radiation (UV-A and UV-B) can modulate the growth and development of plants, but they vary depending on plant species [47,48]. In fact, growth was reduced by solar UV-B in lettuce [49], wheat and cotton [50], boosted in basil [51] and unaltered in other species like maize [52].

The yield and SPAD index were also affected by the N rate, with an increasing trend when the N dose increased from 0 to 50 kg ha^{-1}. Nonetheless, despite the stronger effect recorded for yield (+23.0% and +54.0%, for N50% and N100%, respectively), no significant difference was found between the sub-optimal and optimal doses. Instead, the increase in the SPAD index was only 6.4% and 11.3 %, for N50%, and N100%, respectively. Cozzolino et al. [34] reported an increase in SPAD index when the lamb's lettuce plants are grown under the diffuse light cover film and with increased N dose. Furthermore, the chlorophyll was positively affected by increased N doses, according to our previous research on spinach and lamb's lettuce [14,35,53,54].

Although there are few studies concerning the effect of light (clear and diffuse, high or low presence of UV-B) on crop quality, it is indisputable the importance that product quality assumes in the consumers' choices. The concept of quality differs with customer groups (consumers, food industry, participants markets, etc.) [33] and with the crop choice. Overall, size, color, consistency, shape, and freshness are the main components of external

quality; vitamins, minerals, bioactive compounds, carbohydrates, nitrate, residues, etc., are a part of the internal value of a product constituting its nutritional and health values [33].

In the present research, we investigated the effect of two experimental factors (light and N) on some traits of spinach quality, such as color, antioxidant activity (ABTS), total ascorbic acid, total phenols, and nitrate content. In regards to colorimetric parameters, only L* was influenced by the tested factors. L* decreased by 3.1% under Film B compared to A, and decreased equally in both fertilized treatments by 4.1% compared to unfertilized treatment. These outcomes contrast with those of Cozzolino et al. [34] on lamb's lettuce, where they observed an increase in a* (greenness) and b* values. It is likely that the different responses are due to different natural light conditions during the two growing periods: winter (low light intensity) in our research vs. spring (higher light intensity). Additionally, Kittas et al. [48] did not find differences in L* and a* when they cultivated eggplants at three different levels of UV transmission (5%, 3%, and 0%); they concluded that fruit color is primarily genetically determined, and then, it is influenced by environmental factors such as nutrients, temperature and light conditions.

Regarding nutritional and health benefits in the current research, no influence of cover film was recorded for ABTS antioxidant activity and total phenols content (Table 2). The results regarding the antioxidant activities did not confirm what was reported by Cozzolino et al. [34], who found a significant decrease of ABTS (−33.9% in lamb's lettuce) under the diffuse light film. These contrasting results could be due to different light conditions, as explained for color parameters. It seems that at a high light intensity, growth conditions are more favorable, therefore, plants do not induce the increase of antioxidant activity in contrast to what occurs under diffuse light conditions. In addition, Colonna et al. [55] reported a decrease of ABTS in several baby leaf vegetables, including spinach, when the plants were exposed to high light intensity.

Total phenols were not affected by greenhouse cover films, according to previously reported results in lamb's lettuce [34], spinach and rocket [55]. On the other hand, some authors [56,57] reported that UV-B induces the accumulation of secondary metabolites, which influence several physiological processes of plants. Irrespective of the greenhouse cover films, ABTS and AsA were not affected by N fertilization, however, total phenols significantly decreased at 50 kg ha^{-1} of N. This is probably due to the absence of nutritional stress conditions. Indeed, low availability of N may elicit the antioxidant system resulting in an increase of total phenolics, hydrophilic antioxidant activity and total antioxidant power in mustard [58].

Regarding the synthesis of ascorbic acid, it seems that it is influenced by the amount and intensity of light during the growing season, but it also is a cultivar-dependent trait [33]. Ascorbic acid concentration usually increases when the exposition to light increases, especially in leafy greens [59,60]. Our results are in line with the abovementioned ones, since total ascorbic acid concentration was significantly higher under clear than under light diffusion film, where the light intensity was higher. Heuberger et al. [30] suggested that slight oxidative stress due to UV-B exposure (1–2 kJ m^{-2}d^{-1} UV-B), as well as the addition of photosynthetically active radiation, increased the leaf ascorbate content, whereas the high oxidative stress leads to a reduction of the leaf ascorbate pool in spinach leaves. Instead, Colonna et al. [55] did not record any effect of light on ascorbic acid, probably because they evaluated the effect of different PAR only at harvest (two harvest times, corresponding to low and high PAR).

A deep discussion must be carried out on nitrate content in spinach. Indeed, this species has the characteristic to accumulate nitrate in its leaves. The nitrate accumulation depends on several factors: genetics, N fertilization, light conditions, and growing conditions (open field or protected environment, and autumn–winter or spring–summer periods) [6]. The interest in nitrate content in vegetables is strongly linked to their effects on human health. Bruning-Fann and Kaneene [61] reported that 5–7% of the total nitrate intake is converted to nitrite by oral bacteria and salivary enzymes, with a higher conversion rate in infants or patients with gastroenteritis. Although nitrate is relatively non-toxic, nitrite

and other reaction products of nitrate (nitric oxide and N-nitrous compounds) generally cause concern due to their adverse health effects. Nevertheless, results in this area are contrasting. In particular, several epidemiological studies did not confirm any direct correlation between nitrate concentration in food and the incidence of cancer [62,63]. However, the European Commission (EC; Regulation No. 1258/2011) [64] fixed a limit for nitrate content in leafy vegetables, which for fresh spinach is 3500 mg kg^{-1} fw.

In the current research, nitrate content was about six-fold higher in spinach leaves grown under diffuse light film than that under clear film, though it was within the limit fixed by the EC, and the highest nitrate content was recorded in N100% treatment. The effects of light and N fertilization on nitrate accumulation are well known. In the present experiment, the high light intensity under the clear film presumably allowed to stimulate the nitrate reductase activity, since the nitrate accumulation was reduced. In addition, the N rate of 50 kg ha^{-1} (N100%) was probably excessive since it caused an accumulation of nitrate in leaves, and also manifested no difference in yield when compared to N50% treatments. Proietti et al. [65] found that spinach grown at a photon flux density of 200 μmol m^{-2} s^{-1} (10/14 h photoperiod; light/dark), at five weeks stage showed more nitrate compared to plants grown at 800 μmol m^{-2} s^{-1}. The activity of the main enzymes involved in the metabolism of nitrate and vitamin C increased under higher light intensity [66,67]; additionally, high light provides more energy to fix carbon dioxide to boost vitamin C synthesis and nitrate assimilation in plant leaves [68].

Irrespective of the greenhouse cover films, ABTS and AsA were not affected by N fertilization, however, total phenols significantly decreased at 50 kg ha^{-1} of N. This is probably due to the absence of nutritional stress conditions. Indeed, low availability of N may elicit the antioxidant system resulting in an increase of total phenolics, and total antioxidant power in mustard [58].

5. Conclusions

Light and nitrogen are among the limiting factors for obtaining an optimal growth of plants, and they have greater importance in green leafy vegetables, such as spinach, as they strongly affect some quality traits such as nitrate content. Additionally, the optical properties of greenhouse cover films can have a great effect on light intensity.

Our findings, recorded in the Mediterranean area during the winter season, demonstrated that the cultivation of spinach under diffuse-light film resulted in an increase in yield and SPAD index, which is indicative of a better physiological and biochemical status of plants. The external quality of spinach, specifically color, as well as chlorophyll content, ABTS, and total phenols, were not influenced by the cover; instead, total ascorbic acid content decreased and nitrate content increased under the diffused light film. The optimal N dose boosted yield and increased pigments content but without a significant increase over sub-optimal dose; finally, phenols content decreased at optimal N dose whereas nitrate content increased. Therefore, it would seem that irrespective of the N levels, the use of diffuse-light film boosts spinach yield without depressing quality.

Supplementary Materials: The following are available online at https://www.mdpi.com/article/10.3390/horticulturae7070200/s1 Figure S1: UV-Vis transmission as a function of the wavelength of clear plastic film A (Lirsalux by Lirsa). Figure S2: UV-Vis transmission as a function of the wavelength of the light diffusion plastic film B (Sunsaver by Ginegar Plastic Products).

Author Contributions: Conceptualization, M.M., E.C., I.D.M. and Y.R.; methodology, M.M., I.D.M., E.C. and P.M.; software, L.O., C.E.-N. and L.S.; validation, M.M. and Y.R.; formal analysis, I.D.M., C.E.-N. and L.O.; investigation, I.D.M. and M.I.S.; resources, E.C.; data curation, M.M. and P.M.; writing—original draft preparation, I.D.M., M.I.S. and Y.R.; writing—review and editing, I.D.M., L.S., Y.R. and C.E.-N.; visualization, L.O., E.C. and M.I.S.; supervision, Y.R. and P.M.; project administration, M.M.; funding acquisition, M.M. All authors have read and agreed to the published version of the manuscript.

Funding: This research received no external funding.

Institutional Review Board Statement: Not applicable.

Informed Consent Statement: Not applicable.

Data Availability Statement: The datasets generated for this study are available on request to the corresponding author.

Acknowledgments: We would like to thank Sabrina Nocerino and Roberto Maiello for their support in laboratory work, and Luigi Formisano for his formatting support.

Conflicts of Interest: The authors declare no conflict of interest.

References

1. Maeda, N.; Yoshida, H.; Mizushina, Y. Chapter 26-Spinach and Health: Anticancer Effect. In *Bioactive Foods in Promoting Health*; Fru Vege; Academic Press: Cambridge, MA, USA, 2010; pp. 393–405. [CrossRef]
2. Hanif, R.; Iqbal, Z.; Iqbal, M.; Hanif, S.; Rasheed, M. Use of vegetables as nutritional food: Role in human health. *J. Agric. Biol. Sci.* **2006**, *1*, 18–22.
3. FAOSTAT 2018. Available online: http://www.fao.org/faostat/en/#data/QC (accessed on 2 March 2021).
4. ISTAT 2019. Available online: https://www.istat.it/it/agricoltura?dati (accessed on 2 March 2021).
5. Non Renseigné, U.A.; Iqbal, M. Nitrate accumulation in plants, factors affecting the process, and human health implications. A review. *Agron. Sustain. Dev.* **2007**, *27*, 45–57.
6. Colla, G.; Kim, H.J.; Myriacou, M.C.; Rouphael, Y. Nitrate in fruits and vegetables. *Sci. Hortic.* **2018**, *237*, 221–238. [CrossRef]
7. Salehzadeh, H.; Maleki, A.; Rezaee, R.; Shahmoradi, B.; Ponnet, K. The nitrate content of fresh and cooked vegetables and their health-related risks. *PLoS ONE* **2020**, *15*, e0227551. [CrossRef]
8. Barker, A.V.; Maynard, D.N. Nutritional factors affecting nitrate accumulation in spinach. *Commun. Soil Sci. Plant Anal.* **1971**, *2*, 471–478. [CrossRef]
9. Stagnari, F.; Di Bitetto, V.; Pisante, M. Effects of N fertilizers and rates on yield, safety and nutrients in processing spinach genotypes. *Sci. Hortic.* **2007**, *114*, 225–233. [CrossRef]
10. Goh, K.M.; Vity Akon, P. Effects of fertilizers on vegetable production 2. Effects of nitrogen fertilizers on nitrogen content and nitrate accumulation of spinach and beetroot. *N. Zeal. J. Agric. Res.* **1986**, *29*, 485–494. [CrossRef]
11. Porto, M.L.; Alves, J.C.; Souza, A.P.; Araujo, R.C.; Arruda, J.A. Nitrate production and accumulation in lettuce as affected by mineral Nitrogensupply and organic fertilization. *Hortic. Bras.* **2008**, *26*, 227–230. [CrossRef]
12. Abubaker, S.M.; Abu-Zahra, T.R.; Alzubi, Y.A.; Tahboub, A.B. Nitrate accumulation in spinach (*Spinacia oleracea* L.) tissues under different fertilization regimes. *J. Food Agric. Environ.* **2010**, *8*, 778–780.
13. Breimer, T. Environmental factors and cultural measures affecting the nitrate content in spinach. *Fert. Res.* **1982**, *3*, 191–292. [CrossRef]
14. Di Mola, I.; Cozzolino, E.; Ottaiano, L.; Nocerino, S.; Rouphael, Y.; Colla, G.; El-Nakhel, C.; Mori, M. Nitrogen Use and Uptake Efficiency and Crop Performance of Baby Spinach (*Spinacia oleracea* L.) and Lamb's Lettuce (*Valerianella locusta* L.) Grown under Variable Sub-Optimal N Regimes Combined with Plant-Based Biostimulant Application. *Agronomy* **2020**, *10*, 278. [CrossRef]
15. Bian, Z.; Wang, Y.; Zhang, X.; Li, T.; Grundy, S.; Yang, Q.; Cheng, R. A Review of Environment Effects on Nitrate Accumulation in Leafy Vegetables Grown in Controlled Environments. *Foods* **2020**, *9*, 732. [CrossRef] [PubMed]
16. Stagnari, F.; Galieni, A.; Pisante, M. Shading and nitrogen management affect quality, safety and yield of greenhouse-grown leaf lettuce. *Sci. Hort.* **2015**, *192*, 70–79. [CrossRef]
17. Fu, Y.; Li, H.Y.; Yu, J.; Liu, H.; Cao, Z.Y.; Manukovsky, N.S.; Liu, H. Interaction effects of light intensity and nitrogen concentration on growth, photosynthetic characteristics and quality of lettuce (*Lactuca sativa* L. Var. youmaicai). *Sci. Hort.* **2017**, *214*, 51–57. [CrossRef]
18. Santamaria, P. Nitrate in vegetables: Toxicity, content, intake and EC regulation (review). *J. Sci. Food Agric.* **2006**, *86*, 10–17. [CrossRef]
19. Di Mola, I.; Rouphael, Y.; Colla, G.; Fagnano, M.; Paradiso, R.; Mori, M. Morph physiological traits and nitrate content of greenhouse lettuce as affected by irrigation with saline water. *HortScience* **2017**, *52*, 1716–1721. [CrossRef]
20. Di Mola, I.; Rouphael, Y.; Ottaiano, L.; Duri, L.G.; Mori, M.; De Pascale, S. Assessing the effects of salinity on yield, leaf gas exchange and nutritional quality of spring greenhouse lettuce. *Acta Hortic.* **2018**, *1227*, 479–484. [CrossRef]
21. Caruso, G.; Formisano, L.; Cozzolino, E.; Pannico, A.; El-Nakhel, C.; Rouphael, Y.; Tallarita, A.; Cenvinzo, V.; De Pascale, S. Shading affects Yield, Elemental Composition and Antioxidants of Perennial Wall Rocket Crops Grown from Spring to Summer in Southern Italy. *Plants* **2020**, *9*, 933. [CrossRef] [PubMed]
22. Cometti, N.N.; Martins, M.Q.; Bremenkamp, C.A.; Nunes, J.A. Nitrate concentration in lettuce leaves depending on photosynthetic photon flux and nitrate concentration in the nutrient solution. *Hortic. Bras.* **2011**, *29*, 548–553. [CrossRef]
23. Song, J.; Huang, H.; Hao, Y.; Song, S.; Zhang, Y.; Su, W.; Liu, H. Nutritional quality, mineral and antioxidant content in lettuce affected by interaction of light intensity and nutrient solution concentration. *Sci. Rep.* **2020**, *10*, 2796. [CrossRef] [PubMed]
24. Limantara, L.; Dettling, M.; Indrawati, R.; Brotosudarmo, T.H.P. Analysis on the Chlorophyll Content of Commercial Green Leafy Vegetables. *Procedia Chem.* **2015**, *14*, 225–231. [CrossRef]

25. Hedges, J.; Lister, C.E. Nutritional attributes of spinach, silver beet and eggplant. Crop & Food Research Confidential Report No. 1928. *N. Zeal. J. Agric. Res.* **2007**, *29*, 1–30.
26. Jiang, H.; Yang, Y.; Wang, H.; Bai, Y.; Bai, Y. Surface Diffuse Solar Radiation Determined by Reanalysis and Satellite over East Asia: Evaluation and Comparison. *Remote Sens.* **2020**, *12*, 1387. [CrossRef]
27. Marpaung, F.; Hirano, T. Environmental dependence and seasonal variation of diffuse solar radiation in tropical peatland. *J. Agric. Meteorol.* **2014**, *70*, 223–232. [CrossRef]
28. Kanniah, K.D.; Beringer, J.; North, P.; Hutley, L. Control of atmospheric particles on diffuse radiation and terrestrial plant productivity: A review. *Prog. Phys. Geog.* **2012**. [CrossRef]
29. Department of Energy. Available online: https://www.energy.gov/eere/solar/solar-radiation-basics (accessed on 19 March 2021).
30. Heuberger, H.; Preager, U.; Georgi, M.; Schirrmacher, G.; Graßmann, J.; Schnitzler, W.H. Precision stressing by UV-B radiation to improvequality of spinach under protected cultivation. *Acta Hort.* **2004**, *659*, 201–206. [CrossRef]
31. Peet, M.M. Greenhouse crop stress management. *Acta Hort.* **1999**, *481*, 643–654. [CrossRef]
32. Warren, W.J.; Hand, D.W.; Hannah, M.A. Light interception and photosynthetic efficiency in some glasshouse crops. *J. Exp. Bot.* **1992**, *43*, 363–373.
33. Gruda, N. Impact of environmental factors on product quality of greenhouse vegetables for fresh consumption. *CRC Crit. Rev. Plant Sci.* **2005**, *24*, 227–247. [CrossRef]
34. Cozzolino, E.; Di Mola, I.; Ottaiano, L.; El-Nakhel, C.; Mormile, P.; Rouphael, Y.; Mori, M. The potential of greenhouse diffusing cover material on yield and nutritive values of lamb's lettuce grown under diverse nitrogen regimes. *Italus Hortus* **2020**, *27*, 55–67. [CrossRef]
35. Di Mola, I.; Cozzolino, E.; Ottaiano, L.; Giordano, M.; Rouphael, Y.; Colla, G.; Mori, M. Effect of Vegetal- and Seaweed Extract-Based Biostimulants on Agronomical and Leaf Quality Traits of Plastic Tunnel-Grown Baby Lettuce under Four Regimes of Nitrogen Fertilization. *Agronomy* **2019**, *9*, 571. [CrossRef]
36. Pellegrini, N.; Re, R.; Yang, M.; Rice-Evans, C. Screening of dietary carotenoids and carotenoid-rich-fruit extracts for antioxidant activities applying 2,2-Azinobis (3-ethylenebenzothiazoline-6-sulfonic acid radical cation decoloration assay. In *Methods in Enzymology*; Academic Press: Cambridge, MA, USA, 1999; Volume 299, pp. 379–384.
37. Kampfenkel, K.; Van Montagu, M.; Inzé, D. Extraction and determination of ascorbate and dehydroascorbate from plant tissue. *Anal. Biochem.* **1995**, *225*, 165–167. [CrossRef]
38. Singleton, V.L.; Orthofer, R.; Lamuela-Raventós, R.M. Analysis of total phenols and other oxidation substratesand antioxidants by means of Folin-Ciocalteu reagent. In *Methods in Enzymology*; Academic Press: Cambridge, MA, USA, 1999; Volume 299, pp. 152–178.
39. Lichtenhaler, H.K.; Wellburn, A.R. Determinations of total carotenoids and chlorophylls a and b of leaf extracts in different solvents. In Proceedings of the Biochemical Society Transactions 603rd Meeting, Liverpool, UK, 1 October 1983; Volume 11, pp. 591–592. [CrossRef]
40. Duek, T.; Janse, J.; Li, T.; Kempkes, F.; Eveleens, B. Influence of diffuse glass on the growth and production of tomato. *Acta Hort.* **2012**, *956*, 75–82. [CrossRef]
41. Chun, H.; Yum, S.; Kang, Y.; Kim, H.; Lee, S. Environments and canopy productivity of green pepper (Capsicum annuum L.) in a greenhouse using light-diffused woven film. *Korean J. Hortic. Sci. Technol.* **2005**, *23*, 367–371.
42. Roderick, M.L.; Farquhar, G.D.; Berry, S.L.; Noble, I.R. On the direct effect of clouds and atmospheric particles on the productivity and structure of vegetation. *Oecologia* **2001**, *129*, 21–30. [CrossRef] [PubMed]
43. Mola, I.D.; Conti, S.; Cozzolino, E.; Melchionna, G.; Ottaiano, L.; Testa, A.; Sabatino, L.; Rouphael, Y.; Mori, M. Plant-Based Protein Hydrolysate Improves Salinity Tolerance in Hemp: Agronomical and Physiological Aspects. *Agronomy* **2021**, *11*, 342. [CrossRef]
44. Urban, O.; Klem, K.; Ač, A.; Havránková, K.; Holišová, P.; Navrátil, M.; Zitová, M.; Kozlová, K.; Pokorný, R.; Šprtová, M.; et al. Impact of clear and cloudy sky conditions on the vertical distribution of photosynthetic CO_2 uptake within a spruce canopy. *Funct. Ecol.* **2012**, *26*, 46–55. [CrossRef]
45. Steiner, A.L.; Chameides, W.L. Aerosol-induced thermal effects increase modelled terrestrial photosynthesis and transpiration. *Tellus B* **2005**, *57*, 404–411. [CrossRef]
46. Riga, P.; Benedicto, L. Effects of light-diffusing plastic film on lettuce production and quality attributes. *Span. J. Agric. Res.* **2017**, *15*, 085. [CrossRef]
47. Neugart, S.; Schreiner, M. UVB and UVA as eustressors in horticultural and agricultural crops. *Sci. Hort.* **2018**, *234*, 370–381. [CrossRef]
48. Kittas, C.; Tchamitchian, M.; Katsoulas, N.; Karaiskou, P.; Papaioannou, C.H. Effect of two UV-absorbing greenhouse-covering films on growth and yield of an eggplant soilless crop. *Sci. Hort.* **2006**, *110*, 30–37. [CrossRef]
49. Paul, N.D.; Moore, J.P.; McPherson, M.; Lambourne, C.; Croft, P.; Heaton, J.C.; Wargent, J.J. Ecological responses to UV radiation: Interactions between the biological effects of UV on plants and on associated organisms. *Phys. Plant.* **2012**, *145*, 565–581. [CrossRef]
50. Kataria, S.; Guruprasad, K.N.; Ahuja, S.; Singh, B. Enhancement of growth, photosynthetic performance and yield by exclusion of ambient UV components in C3 and C4 plants. *J. Phot. Phot. B Biol.* **2013**, *127*, 140–152. [CrossRef]

51. Sakalauskaite, J.; Viškelis, P.; Duchovskis, P.; Dambrauskiene, E.; Sakalauskiene, S.; Samuoliene, G.; Brazaityte, A. Supplementary UV-B irradiation effects on basil (*Ocimum basilicum* L.) growth and phytochemical properties. *J. Food Agric. Environ.* **2012**, *10*, 342–346.
52. Pal, M.A.; Sharma, A.; Abrol, Y.P.; Sengupta, U.K. Exclusion of UV-B radiation from normal solar spectrum on growth of mung bean and maize. *Agric. Ecosyst. Environ.* **1997**, *61*, 29–34. [CrossRef]
53. Di Mola, I.; Ottaiano, L.; Cozzolino, E.; Senatore, M.; Giordano, M.; El-Nakhel, C.; Sacco, A.; Rouphael, Y.; Colla, G.; Mori, M. Plant-based biostimulants influence the agronomical, physiological, and qualitative responses of baby rocket leaves under diverse nitrogen conditions. *Plants* **2019**, *8*, 522. [CrossRef] [PubMed]
54. Di Mola, I.; Cozzolino, E.; Ottaiano, L.; Giordano, M.; Rouphael, Y.; El Nakhel, C.; Leone, V.; Mori, M. Effect of seaweed (Ecklonia maxima) extract and legume-derived protein hydrolysate biostimulants on baby leaf lettuce grown on optimal doses of nitrogen under greenhouse conditions. *Aust. J. Crop Sci.* **2020**, *14*, 1456–1464. [CrossRef]
55. Colonna, E.; Rouphael, Y.; Barbieri, G.; De Pascale, S. Nutritional quality of leafy vegetables harvested at two light intensities. *Food Chem.* **2020**, *199*, 702–710. [CrossRef]
56. Nitz, G.M.; Schnitzler, W.H. Effect of PAR and UV-B radiation on the quality and quantity of the essential oil in sweet basil (*Ocimum basilicum* L.). *Acta Hort.* **2004**, *659*, 375–381. [CrossRef]
57. Jansen, M.A.K.; Gaba, V.; Greenberg, B.M. Higher plants and UV-B radiation: Balancing damage, repair and acclimation. *Trends Plant Sci.* **1998**, *3*, 131–135. [CrossRef]
58. Li, J.; Zhu, Z.; Gerendás, J. Effects of nitrogen and sulfur on total phenolics and antioxidant activity in two genotypes of leaf mustard. *J. Plant Nutr.* **2008**, *31*, 1642–1655. [CrossRef]
59. Weerakkody, W.A.P. Nutritional value of fresh leafy vegetables as affected by pre-harvest factors. *Acta Hort.* **2003**, *604*, 511–515. [CrossRef]
60. Oyama, H.; Shinohara, Y.; Ito, T. Effects of air temperature and light intensity on β-carotene concentration in spinach and lettuce. *J. Japan. Soc. Hort. Sci.* **1999**, *68*, 414–420. [CrossRef]
61. Bruning-Fann, C.S.; Kaneene, J.B. The effects of nitrate, nitrite and N-nitroso compounds on human health: A review. *Vet. Hum. Toxicol.* **1993**, *35*, 521–538.
62. Milkowski, A.; Garg, H.K.; Coughlin, J.R.; Bryan, N.S. Nutritional epidemiology in the context of nitric oxide biology: A risk–benefit evaluation for dietary nitrite and nitrate. *Nitric Oxide* **2010**, *22*, 110–119. [CrossRef]
63. Speijers, G.; Van den Brandt, P.A. Nitrite and potential endogenous formation of N-nitroso compounds; safety evaluation of certain food additives, JECFA. *WHO Food Addit. Ser.* **2003**, *50*, 49–74.
64. European Community. Reg. n° 1258 of 2 December 2011. *Off. J. Eur. Union* **2011**, *L320*, 15–17.
65. Proietti, S.; Moscatello, S.; Giacomelli, G.A.; Battistelli, A. Influence of the interaction between light intensity and CO_2 concentration on productivity and quality of spinach (*Spinacia oleracea* L.) grown in fully controlled environment. *Adv. Space Res.* **2013**, *52*, 1193–1200. [CrossRef]
66. Smirnoff, N. Ascorbate biosynthesis and function in photoprotection. *Philos. T Roy. Soc. B* **2000**, *355*, 1455–1464. [CrossRef] [PubMed]
67. Zhou, W.L.; Liu, W.K.; Yang, Q.C. Quality changes in hydroponic lettuce grown under pre-harvest short-duration continuous light of different intensities. *J. Hort. Sci. Biotech.* **2012**, *87*, 429–434. [CrossRef]
68. Demšar, J.; Osvald, J.; Vodnik, D. The effect of light-dependent application of nitrate on the growth of aeroponically grown lettuce (*Lactuca sativa* L.). *J. Am. Soc. Hortic. Sci.* **2004**, *129*, 570–575. [CrossRef]

horticulturae

Article

The Effects of Nutrient Solution Feeding Regime on Yield, Mineral Profile, and Phytochemical Composition of Spinach Microgreens

Spyridon A. Petropoulos [1,*,†], Christophe El-Nakhel [2,†], Giulia Graziani [3], Marios C. Kyriacou [4] and Youssef Rouphael [2]

[1] Department of Agriculture Crop Production and Rural Environment, University of Thessaly, Fytokou Street, N. Ionia, 38446 Magnissia, Greece

[2] Department of Agricultural Sciences, University of Naples Federico II, Via Universita 100, 80055 Portici, Italy; christophe.elnakhel@unina.it (C.E.-N.); youssef.rouphael@unina.it (Y.R.)

[3] Department of Pharmacy, University of Naples Federico II, 80131 Naples, Italy; giulia.graziani@unina.it

[4] Department of Vegetable Crops, Agricultural Research Institute, Nicosia 1516, Cyprus; m.kyriacou@ari.gov.cy

* Correspondence: spetropoulos@uth.gr; Tel.: +30-24210-93196

† These authors have contributed equally to this work.

Citation: Petropoulos, S.A.; El-Nakhel, C.; Graziani, G.; Kyriacou, M.C.; Rouphael, Y. The Effects of Nutrient Solution Feeding Regime on Yield, Mineral Profile, and Phytochemical Composition of Spinach Microgreens. *Horticulturae* **2021**, *7*, 162. https://doi.org/10.3390/horticulturae7070162

Academic Editor: Francisco Garcia-Sanchez

Received: 19 May 2021
Accepted: 23 June 2021
Published: 25 June 2021

Publisher's Note: MDPI stays neutral with regard to jurisdictional claims in published maps and institutional affiliations.

Abstract: Microgreens are receiving increasing popularity as functional and healthy foods due to their nutritional value and high content of bioactive compounds. The aim of the present study was to evaluate the effects of nutrient deprivation through the regulation of nutrient solution (NS) feeding days on the plant growth and chemical composition of spinach microgreens. For this purpose, spinach microgreens were subjected to four different fertigation treatments—namely, 0 (control), 5, 10, and 20 NS feeding days before harvesting—and harvested tissues were evaluated with regard to fresh and dry yield, color of true leaves, antioxidant activity, and chlorophyll, carotenoid, and phenolic compound contents. The results of our study revealed that prolonged NS feeding (20 NS) resulted in the highest fresh yield and photosynthetic pigment contents (chlorophylls, lutein, and β-carotene). In contrast, mineral concentrations (P, K, Ca, and Mg) were the lowest for the 20 NS, whereas the control (0 NS) and 5 NS recorded the highest concentrations. Apart from that, spinach microgreens subjected to 10 NS treatment recorded 70.7% less nitrates, better mineral concentrations, 7.0% higher total ascorbic acid, similar polyphenol contents, higher DM%, and only 12.6% yield decrease compared to 20 NS treatment. In conclusion, although the highest overall fresh yield was recorded with the 20 NS treatment, the highest nitrate concentrations and the lowest mineral concentrations may raise food safety concerns. On the other hand, 10 NS treatment seems to be the most promising, since it combined high yields with high mineral concentrations and low nitrate concentrations, without compromising bioactive compound (e.g., polyphenols) contents, presenting a cost-effective and sustainable practice for microgreen cultivation.

Keywords: macronutrients; *Spinacia oleracea* L.; carotenoids; nitrates; phenolic acids; flavonoids; *UHPLC-HRMS*; chlorophylls; vitamin C

1. Introduction

Microgreens are a novel and emerging category of food products obtained from harvesting the aerial parts of young seedlings of various species—such as vegetables, herbs, and aromatic plants—while wild edible species have recently been included in this category of food products [1,2]. They are distinct from sprouts and baby leaves since they are harvested at the cotyledon stage and before the true leaves emerge [3]. They usually contain high contents of bioactive molecules, such as polyphenols, carotenoids, vitamins, tocopherols, and other antioxidant compounds, which has raised the interest of consumers who are seeking to include new healthy and functional foods in their diets [4]. Their

high concentrations of valuable nutrients also make them perfect candidates in health-supporting diets, since the consumption of low amounts of microgreens may prevent nutrient deficiencies and chronic diseases that plague the modern world [5]. Moreover, the biofortification of microgreens through nutrient solution management is easier to facilitate than conventional growing systems, and may also help to achieve food and nutrition security [6–8]. Therefore, the farming sector, forced by consumer needs and marketing trends for newly designed healthy food products, is seeking new agronomic approaches that may improve the quality of the final product, while at the same time decreasing production costs.

Microgreens' shelf life is relatively limited, and spans 2–4 days at ambient temperatures, but may extend to up to 10–14 days at 5 °C. This is perhaps the most serious limitation encountered in their supply chain, but it mainly reflects changes in visual quality induced by dehydration and ageing, which also impact sensory quality [9,10]. The changes in the phytonutrient contents and in vitro/in vivo bioactive values of microgreens introduced by the postharvest period and conditions constitute an area that has not yet received extensive research attention; in fact, literature on this subject remains scarce, although some reports suggest the use of coating and packaging techniques to extend shelf life [11,12]. However, the general rule is that microgreens should be consumed as closely to their harvest as possible in order to ensure a full organoleptic experience. It is therefore unsurprising that they are commonly grown by chefs in upscale restaurants and by consumers at home in order to ensure immediate use after harvest [1,5,13].

Abiotic stressors such as water and nutrient deficiency stress have a significant impact on plant growth and quality and, depending on the severity of the stress, may have negative or positive effects on crops [14]. In this context, the use of nutrient deprivation techniques has been considered a cost-effective agronomic practice, which may beneficially affect plants through the increase in the secondary metabolites that plants synthesize to cope with this eustress [15,16]; Their application in hydroponic systems is easy to facilitate, also allowing the regulation of nutrient solution composition in favor of plant growth and quality [7,17]. This approach has also been studied in field crops, aiming to exploit the plasticity of plants' ionomes via the expression of specific genes that regulate root transporters [18]. However, the line between nutrient deficiency and regulated deprivation is very fine, and the critical threshold up to which nutrients should be deprived from plants in order to have beneficial effects has yet to be defined. This situation is more complex in hydroponic systems, where plants rely on constant nutrient supplementation in order to obtain the required amounts of nutrients due to limited buffering of the growing medium [19,20]. However, the cultivation of microgreens has completely different requirements due to the early harvesting of the plants, and nutrient deprivation may result in significant benefits for producers and consumers alike via the reduced cost of production and improved functional properties [3].

Spinach (*Spinacia oleracea* L.) is a very popular and highly nutritious vegetable of the Amaranthaceae family, which is widely consumed throughout the world in fresh, frozen, or canned form [9]. It is a rich source of vitamins, minerals, and trace elements, as well as various bioactive compounds including carotenoids, flavonoids, and tocopherols [9,21]. The harvesting stage has a great impact on the chemical composition of spinach microgreens, which are harvested earlier than conventionally grown spinach and, therefore, present a different chemical profile. Ghoora et al. [22] suggested that spinach microgreens contained significantly higher amounts of α-tocopherol and lower amounts of oxalic acid than mature leaves, while similar findings were reported by Lester et al. [23] for baby spinach leaves compared to mature ones with regard to carotenoid and vitamin contents. Moreover, in the study of Ghoora et al. [22], it was noted that spinach microgreens achieved a 2.5–3.0 times higher nutrient quality score (NQS) than mature spinach, which further highlights the importance of microgreens in healthy diets. Despite its high content of beneficial compounds, spinach is also considered to be a hyperaccumulator of nitrates, which may have a negative impact on human health when consumption exceeds recommended

daily intake (RDI) values [24,25]. Pre-harvest factors—such as genotype selection and cultivation practices—and post-harvest factors may regulate chemical composition and improve the quality of the final product [14]. For example, Erfani et al. [26] studied the impact of genotype minerals, vitamins, fatty acids, macronutrients, and oxalic acid, and suggested significant differences among the various genotypes tested. Moreover, according to Bergquist et al. [27], the growth stage at harvest is also essential for the nutritional value and chemical composition of edible spinach leaves, while the early harvesting of baby spinach leaves was found to increase their flavonoid contents [27].

The aim of the present study was to evaluate the effects of nutrient solution application in spinach grown for microgreen production, and to further elucidate how nutrient solution deprivation may affect plant growth and—most importantly—the chemical composition of the final product. For this purpose, spinach plants were supplemented with nutrient solution for 0, 5, 10, and 20 days, and harvested tissues were analyzed for plant growth parameters (fresh and dry weight), as well as for their chemical composition (minerals, nitrates, total ascorbic acid, chlorophylls, carotenoids, and phenolic compounds content). The results of our study could be useful, and offer cost-effective practices to improve the nutritional value of the final product without severe effects on crop performance, further increasing the added value of microgreen products.

2. Materials and Methods

2.1. Genetic Material, Growth Chamber Settings, and Nutrient Feeding

Spinacia oleracea L. var. Palco F1 (CN Seeds Ltd., Pymoor, Ely, Cambrigeshire, UK) was chosen as a nitrate-accumulating species to be grown as microgreens under diverse nutrient solution (NS) feeding days (0, 5, 10, and 20 days of NS application). Spinach microgreens were sown at a density of 60,000 seeds m^{-2} in plastic trays (19 × 14 × 6 cm) filled with a peat-based medium (pH 5.48 and EC = 282 $\mu S\ cm^{-1}$; Special Mixture, Floragard Vertriebs-GmbH, Oldenburg, Germany) mixed with vermiculite (50% v/v). The peat-based medium was characterized by the following elements: NO_3 (11 mg kg^{-1}), PO_4 (140 mg kg^{-1}), K (796 mg kg^{-1}), Ca (2402 mg kg^{-1}), Mg (303 mg kg^{-1}), SO_4 (235 mg kg^{-1}), and Na (540 mg kg^{-1}), expressed on a dry weight basis. The NS consisted of a quarter-strength modified Hoagland solution (pH = 6 ± 0.2 and EC = 500 ± 50 $\mu S\ cm^{-1}$), described in detail in the work of Kyriacou et al. [15]. The NS was replaced by osmotic water (pH = 6 ± 0.2 and EC = 100 ± 25 $\mu S\ cm^{-1}$) when the feeding treatment was over.

The experiment was conducted in a controlled growth chamber (KBP-6395F, Termaks, Bergen, Norway) at the Department of Agricultural Sciences, University of Naples Federico II, Portici, Italy. The growth chamber settings were: 24/18 ± 2 °C, day/night temperatures, 12 h photoperiod provided by an LED panel (K5 Series XL 750, Kind LED, Santa Rosa, CA, USA) delivering a mean intensity of 300 ± 15 $\mu mol\ m^{-2}\ s^{-1}$ at canopy level (optimal absorption spectrum for the photosynthesis; 400–700 nm), and a relative humidity of 65–75 ± 5%. A completely randomized design (CRD) with three replicates (e.g., trays) was used to compare four NS feeding treatments (0, 5, 10, and 20 days of NS application). A daily rotation scheme was performed during the growing cycle. Spinach microgreens were harvested when they formed the first two fully expanded true leaves, at 25 days after sowing (DAS) for 0 and 5 days of NS feeding, and at 20 DAS for the 10- and 20-day NS feeding treatments.

2.2. CIELAB Color Space Parameters Measurement of Spinach Microgreens' Canopy, Sampling, and Yield Assessment

The CIELAB color space parameters (L*, a* and b*) of spinach microgreens' canopy were measured before harvesting using a portable Minolta Chroma Meter (CM-2600d, Minolta Camera Co. Ltd., Osaka, Japan), and then the hue angle (h◦) and chroma (C*) were calculated as follows: hue angle = $\tan^{-1}$ (b*/a*), and chroma = $((a^*)^2 + (b^*)^2)^{0.5}$. Eight measurements per replicate/tray were taken into consideration, accounting for twenty-four measurements per treatment. Afterwards, the microgreens (stems and leaves) of each tray/replicate were cut with scissors at the substrate level, and the fresh weight of

each replicate was assessed and expressed as kg of fresh weight m^{-2}. One part of the batch sample of each replicate was dried at 65 °C in a forced-air oven until reaching constant dry weight, which was used to calculate dry matter percentage. These oven-dried microgreen materials (stems and leaves) were ground and used for macromineral concentration analysis. The remaining part of each batch sample was stored at −80 °C for subsequent qualitative analysis (chlorophyll pigments and total ascorbic acid), while a part of the frozen material was lyophilized in a freeze drier (Christ, Alpha 1–4, Osterode, Germany) for ABTS antioxidant activity, carotenoids (lutein and β-carotene), and polyphenol profile determination.

2.3. Determination of Minerals, Nitrates, and Total Ascorbic Acid

The oven-dried microgreen materials were used to determine the concentrations of minerals and nitrates, following a previously described methodology [28]. In brief, 250 mg of plant tissues were extracted with 50 mL of ultrapure water and put in a water bath (ShakeTemp SW 22, Julabo, Seelbach, Germany) for 10 min at 80 °C, with constant shaking. After that, the extracts were centrifuged, and the supernatant was collected and stored in a vial for chromatographic analysis with an ion chromatography instrument (ICS-3000, Dionex, CA, USA) coupled with an electrical conductivity detector. Nitrate (NO_3^-) and mineral (phosphorus (P), potassium (K), calcium (Ca), magnesium (Mg), sulfur (S), and sodium (Na)) concentrations were determined on a dry weight basis (g kg^{-1}) and then converted to mg kg^{-1} fresh weight (fw), based on the recorded dry matter content. For the determination of total ascorbic acid (TAA), 400 mg of fresh material kept at deep-freezing conditions was extracted according to the protocol of Kampfenkel et al. [29] and analyzed at 525 nm using an UV–Vis spectrophotometer (Hach DR 4000; Hach Co, Loveland, CO, USA). The results were expressed as mg 100 g^{-1} fw.

2.4. Chlorophyll Pigments, ABTS Antioxidant Activity, Carotenoid Extraction, and Quantification by HPLC-DAD

Total chlorophylls and chlorophyll *a* and *b* contents were determined according to the protocol previously described by Lichtenthaler and Buschmann [30]. In particular, 500 mg of fresh material stored at deep-freezing conditions was extracted in 10 mL of 90% acetone. The extracts were centrifuged, and then the supernatant was collected and the absorbance at 662 and 645 nm was measured via spectrophotometry (Hach DR 4000; Hach Co., Loveland, CO, USA) in order to quantify chlorophylls *a* and *b*, respectively. The total chlorophyll content was calculated as the sum of chlorophylls *a* and *b*, and expressed as mg 100 g^{-1} fw.

200 mg of freeze-dried material was extracted with methanol. The ABTS antioxidant activity of this extract was measured with the 2,20-azino-bis-3-ethylbenzothiazoline-6-sulfonic acid ABTS method [31]. The results were expressed in mol Trolox equivalents 100 g^{-1} fw.

For lutein and β-carotene determination, the method of Kim et al. [30], as modified by Kyriacou et al. [12], was implemented. In brief, 100 mg of the lyophilized microgreen materials was extracted in 6 mL ethanol + 0.1% butylated hydroxytoluene. The quantification of lutein and β-carotene followed a reversed-phase HPLC separation using a Shimadzu HPLC LC-10 (Shimadzu, Osaka, Japan). The results were expressed as µg g^{-1} fw.

2.5. Phenolic Compound Extraction and Conditions of UHPLC-HRMS Analysis

Freeze-dried material was extracted according to the method of Huang et al. [32] after modifications. In brief, 100 mg of freeze-dried samples were extracted in 2.5 mL of methanol/water (70:30, *v/v*) acidified with formic acid (0.5%), and then sonicated for 30 min at room temperature. The extracts were centrifuged at 4000 rpm for 10 min at 4 °C, and the supernatant was collected after filtering through a 0.2 µm nylon membrane syringe filter (Phenomenex, Castel Maggiore, BO, Italy); 5 µL were used for UHPLC-HRMS analysis.

The analysis of phenolic compounds was performed using a UHPLC system (Thermo Fisher Scientific, Waltham, MA, USA) equipped with a degassing system, a quaternary UH-

PLC pump, and an autosampler device (Dionex Ultimate 3000). The separation of polyphenols was performed using a thermostatted (25 °C) column (Luna Omega PS 1.6 μm; Phenomenex; 50 mm × 2.1 mm). The volume of the injected sample was 5 μL. Two mobile phases were used—namely, Phase A (water with 0.1% formic acid v/v), and Phase B (acetonitrile with 0.1% formic acid v/v) [33]. The Q Exactive Orbitrap LC-MS/MS equipment was calibrated on a daily basis before the analysis of samples using a reference standard mixture. In full scan MS and AIF modes, a 5 ppm mass tolerance window was set, while the Xcalibur software v. 3.0.63 (Xcalibur, Thermo Fisher Scientific, Waltham, MA, USA) was used to analyze and process the obtained data. All of the results were expressed as μg 100 g^{-1} fw.

2.6. Statistics

The experiment was performed according to randomized complete blocks (RCB), with three replicates per feeding treatment, and all of the data are presented as mean ± standard error (SE). The mean values of the studied parameters were subjected to analysis of variance (ANOVA), and the means were compared according to Duncan's multiple range test (DMRT) at $p \leq 0.05$ using the SPSS 20 software package (SPSS Inc., Chicago, IL, USA). Regression analysis was conducted in order to identify relationships between the measured parameters (fresh yield, dry biomass, dry matter percentage, nitrates, P, K, and total ascorbic acids) and nutrient solution feeding days. This analysis was also performed with the SPSS 20 software package (SPSS Inc., Chicago, IL, USA).

3. Results and Discussion

3.1. Spinach Microgreens' Biometric and Colorimetric Parameters

Figure 1 illustrates the yield, dry biomass, and dry matter percentage of *Spinacia oleracea* L. var. Palco F1 microgreens, subjected to the four NS feeding regimes (0, 5, 10, and 20 days). The yield of spinach increased linearly (Supplementary Figure S1) and significantly when additional days of NS were supplemented. At 20 days of NS feeding, the yield increased by 41.97% compared to 0 days of NS feeding, and registered 1.59 kg m^{-2}. Interestingly, our results demonstrated that 10 days of NS feeding caused a marginal decrease in fresh biomass of spinach microgreens (12.5%) compared to 20 days of NS feeding. As for dry biomass (g m^{-2}), 5 and 10 days of NS feeding presented the highest values (123.5 and 128.9 g m^{-2}, respectively) compared to the other two treatments (Figure 1B). Finally, the highest dry matter percentage was registered with 5 days of NS feeding (10.33%; Figure 1C). The relationship resulting from the regression analysis of the dry biomass and dry matter percentages is best described by a quadratic function (Supplementary Figure S1). These findings are of great importance, since plants that received the highest number of feeding days (20 days) not only showed a higher yield, but also were harvested 5 days earlier than the plants with 0 and 5 days of NS feeding. The low dry matter content for the 20-day NS treatment also indicates that the highest yield is attributed to high moisture content in plant tissues, probably due to better functioning of the roots. Similar results were reported by El-Nakhel et al. [17], who studied the effects of macronutrient deprivation on lettuce plants, and suggested that lower availability of nutrients resulted in reduced fresh weight, while it increased dry matter content. Fallovo et al. [34] also reported the importance of nutrient solution concentration to fresh yield in lettuce plants grown in a floating system, while Murphy and Pill [35] and Wieth et al. [36] suggested similar findings in the case of arugula and red cabbage microgreens, respectively. Moreover, the higher DM% in plants with 0, 5, and 10 days of NS feeding compared to 20-day NS is in line with other studies on lettuce [37], rocket, and Brussels sprout microgreens [33], when NS was completely replaced with water thorughout the whole growing cycle. Based on these reports, the findings of our study could be attributed to the osmotic stress and nutrient deficiency that plants experienced when nutrient solution feeding was applied for less than 20 days. This is also corroborated by the earlier harvesting of plants that received prolonged NS feeding for 10 and 20 days compared to the rest of the treatments (0 and 5 days of feeding), which indicates that plant growth was held back under the latter conditions. In contrast,

Chen et al. [38] did not find any decrease in lettuce plants' fresh weight when N, P, and K concentration in nutrient solution was reduced to 10% of the control treatment, while Murphy and Pill [35] also did not observe a significant decrease in the fresh weight of cabbage and Brussels sprouts when they were deprived of nitrogen. Therefore, it seems that plants' responses to nutrient deprivation may have genotypic implications, and different species or cultivars may respond differently to such conditions.

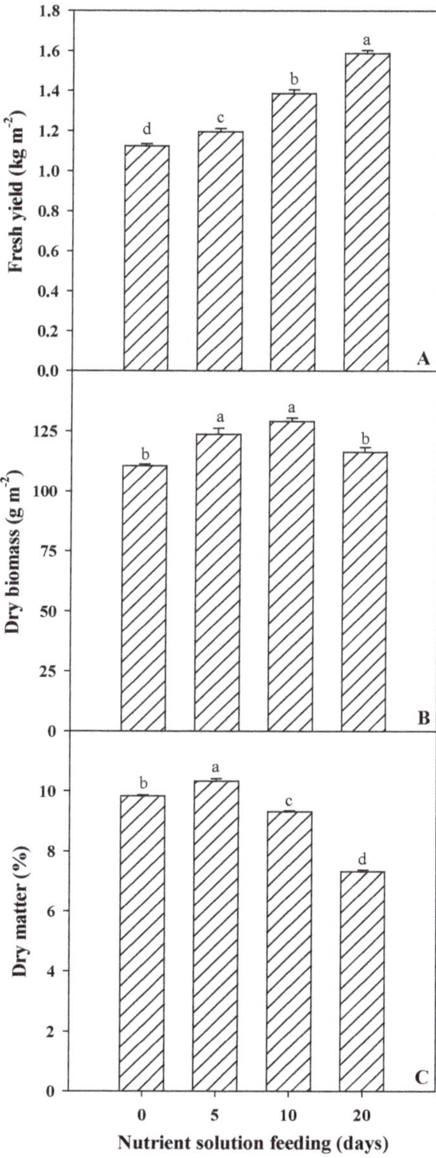

Figure 1. The effect of nutrient solution feeding (number of days) on plant growth parameters (fresh yield (**A**), dry biomass (**B**), and dry matter content (**C**)). Different letters within the above bars indicate significant differences according to Duncan's multiple range test ($p = 0.05$).

The three coordinates of CIELAB are presented in Table 1. The perceptual lightness (L*) presented the lowest value in the 10-day NS feeding treatment—8.06% lower than the other treatments. The 20-day NS feeding treatment exhibited a significantly darker green than then other treatments (+15.98%), as marked by the measured a* parameter, in addition to having a significantly higher hue angle. Concomitantly, the same treatment manifested a significant lower yellow color compared to the other treatments, as supported by a lower b* value. On the other hand, the 0- and 5-day NS feeding treatments generated spinach microgreens with lighter green canopies, as demonstrated by the lower a* and higher b* values presented in Table 1, as well as showing higher saturation of their color as indicated by significantly higher chroma values. These differences in color parameters between the various NS feeding treatments indicate that plants that received prolonged feeding presented a darker green color than the rest of the treatments, meaning better visual quality and less chlorosis. This finding could be further justified by the very low nitrate concentrations (see description in the following sections) in the control treatment and the 5-day NS feeding treatment, which also recorded the highest b* and chroma values and the lowest values for hue angle. According to the literature, low nutrient availability—and particularly that of nitrogen—is associated with low chlorophyll content in leaves and more yellowish leaves [34,39]. Similarly to our study, El-Nakhel et al. [33] reported that nutrient availability may affect the color of microgreens' leaves and, thus, improve the visual quality of the final product.

Table 1. CIELAB color space parameters, chroma, and hue angles of spinach microgreens in relation to nutrient solution feeding regime.

Nutrient Solution Feeding (Days)	L *	a *	b *	Chroma	Hue Angle
0	38.52 ± 0.35 a	-7.61 ± 0.42 a	31.21 ± 0.32 a	16.05 ± 0.16 a	103.7 ± 0.61 c
5	38.04 ± 0.71 a	-7.51 ± 0.04 a	30.60 ± 0.22 a	15.81 ± 0.09 a	103.8 ± 0.08 c
10	35.31 ± 0.61 b	-7.42 ± 0.35 a	25.67 ± 0.57 b	13.87 ± 0.22 b	106.2 ± 0.96 b
20	38.66 ± 0.95 a	-8.71 ± 0.02 b	22.41 ± 0.75 c	12.73 ± 0.29 c	111.3 ± 0.71 a
	*	*	***	***	***

* and ***: Significant at $p \leq 0.05$ and 0.001, respectively. Different letters within each column indicate significant differences according to Duncan's multiple range test $p = 0.05$). All data are expressed as mean $\pm$ standard error, $n = 3$.

3.2. Spinach Microgreens' Nitrate and Macromineral Concentrations

As illustrated in Table 2, the nitrate concentration of spinach microgreens proved to be directly correlated with NS feeding days, as it increased exponentially from 10 to 20 days of NS feeding (Supplementary Figure S1), reaching 1698 mg kg^{-1} fw at the highest feeding treatment, whereas this concentration was about 498 mg kg^{-1} fw in the 10-day NS feeding treatment (-70.67%). In addition, it was 85.87% and 90.64% less when NS feeding lasted 5 days or when it was completely replaced by osmotic water, respectively. Quadratic equations were fitted to quantify the effects of NS feeding days on P and K concentrations in spinach microgreens (Supplementary Figure S1). When irrigated constantly for 20 days with a quarter-strength NS, spinach microgreens accumulated the lowest values of phosphorus, potassium, calcium, and magnesium concentrations when expressed on a fresh weight basis and compared to the other NS feeding treatments (Table 2). These latter macroelements accumulated significantly when the NS feeding days were reduced. Indeed, the P concentration of microgreens was higher in the other three NS treatments (0, 5, and 10 days), while K and Mg accumulated the most in the 5-day NS feeding treatment, and Ca accumulated the most when microgreens were only irrigated with osmotic water. Overall, K was the most abundant macroelement detected in spinach microgreens, followed by Mg, P, and Ca. The highest concentrations of P, K, and Mg, and the second highest concentration of Ca, for 5-day NS feeding could be partly attributed to this treatment having the highest dry matter content (see Figure 1), which is also justified by the fact that the 20-day NS treatment—which had the lowest dry matter content—also

recorded the lowest values for mineral concentrations. Therefore, the dilution effect should be partially responsible for this trend.

Table 2. Nitrate and mineral concentrations (mg kg^{-1} fw) in relation to nutrient solution feeding regime.

Nutrient Solution Feeding (Days)	NO$_3$ (mg kg^{-1} fw)	P (mg kg^{-1} fw)	K (mg kg^{-1} fw)	Ca (mg kg^{-1} fw)	Mg (mg kg^{-1} fw)
0	159 ± 14 c	789 ± 19 a	8225 ± 72 c	147 ± 12 a	1038 ± 18 b
5	240 ± 8.3 c	775 ± 12 a	8897 ± 84 a	115 ± 8.4 b	1110 ± 2.8 a
10	498 ± 43 b	817 ± 15 a	8568 ± 86 b	95.6 ± 1.5 b	893 ± 28 c
20	1698 ± 24 a	694 ± 7.6 b	6115 ± 89 d	53.5 ± 3.9 c	618 ± 18 d
	***	***	***	***	***

*** significant at $p \leq 0.001$. Different letters within each column indicate significant differences according to Duncan's multiple range test ($p = 0.05$). All data are expressed as mean ± standard error, $n = 3$.

The observed highest concentration of nitrates for prolonged NS feeding is in agreement with other reports, where another explanation for the increase in nitrate concentration under prolonged NS feeding could be that nitrogen is preferably used by plants as an osmoticum [40]; therefore, considering that prolonged feeding may increase osmotic levels in growth media, plants counteract by accumulating nitrates in order to increase turgor pressure and maintain nutrient uptake. In any case, the recorded values of nitrate concentration are within the safety limits established by the European Commission Regulation (EU) No 1258/2011, although high concentrations of nitrates in food products should be avoided considering the contribution they may have to overall nitrate intakes on a daily basis [7].

The lowest concentration of minerals found in the 20-day NS treatment should be associated with the highest fresh yield observed in the same treatment, which indicates that apart from the dilution effect discussed above, plants treated with prolonged NS feeding used the absorbed minerals more efficiently for biosynthetic purposes. The same trends were observed in the study of El-Nakhel et al. [33], who also reported lower mineral concentrations in plants fed with nutrient solution compared to untreated plants—without statistically significant differences being observed, however. Similarly, El-Nakhel et al. [17] suggested that plants grown in nutrient solution with higher concentrations of minerals had lower concentrations of Ca and Mg compared to plants grown in quarter-strength nutrient solution, whereas the opposite trends were observed with regard to P concentration. Finally, apart from the dilution effect mentioned above, the high concentration of Ca in the 0-day NS treatment plants could be partially associated with the presence of stress conditions, since Ca is involved in the signaling pathways of plants' responses to stress [41]. Therefore, despite the highest yield being observed for prolonged NS feeding (20 days), the 10-day NS feeding treatment was also promising in terms of fresh yield and nitrate and mineral concentrations, which are important parameters for the nutritional value of the final product.

3.3. Spinach Microgreens' Pigments, Total Ascorbic Acid, and ABTS Antioxidant Activity

Table 3 reflected the investigation of pigments (total chlorophylls and carotenoids), ABTS antioxidant activity, and the total ascorbic acid content of spinach microgreens subjected to the tested NS feeding treatments. Total chlorophylls increased starting at 10 NS feeding days, and reached their maximum at 20 NS feeding days; total chlorophylls proved to be correlated (0.73) with the a* parameter listed previously in Table 1, and inversely correlated (−0.96) with b*. This finding also confirms the deeper green color and better visual quality of leaves for these treatments compared to 0 and 5 NS feeding days; however, it is interesting to highlight the fact that total chlorophyll content was lower in the 0- and 5-day NS feeding treatments, despite the higher concentration of Mg in plant tissues, which helps in the light harnessing process as a basic ingredient in chlorophyll molecules [42]. This contradiction could be attributed to the fact that chlorophyll-bound Mg

concentration depends on the species, and in specific species such as spinach and lettuce a significant amount of Mg is bound to chlorophylls [43]. Therefore, the high free Mg concentration detected in plants with 0 and 5 NS feeding days could be partly related to the chlorophyll content for the same treatments being the lowest. Another explanation could be associated with less nutrient availability in the 0 and 5 NS feeding day treatments, which results in reduced chlorophyll biosynthesis [34]. Moreover, the highest total chlorophyll contents in the 10- and 20-day NS treatments are associated with the highest fresh yields recorded for these treatments, since chlorophyll is the main photosynthetic pigment that allows plants to harvest energy from soil and transform it into biosynthetic products [44].

Table 3. Antioxidant activity, pigments, and total ascorbic acid content in relation to nutrient solution feeding regime (means ± SD).

Nutrient Solution Feeding (Days)	ABTS	Total Chlorophylls	Lutein	β-Carotene	Total Ascorbic Acid
	mol Trolox eq. $100g^{-1}$ fw	mg $100g^{-1}$ fw	µg g^{-1} fw	µg g^{-1} fw	mg $100g^{-1}$ fw
0	663.7 ± 22.2 b	43.0 ± 1.19 c	33.9 ± 0.59 c	19.3 ± 0.86 c	130.5 ± 1.61 d
5	725.8 ± 26.2 b	41.3 ± 0.87 c	37.4 ± 0.46 c	21.4 ± 0.87 c	145.1 ± 2.47 c
10	751.6 ± 15.9 b	56.9 ± 2.07 b	48.8 ± 1.04 b	27.2 ± 2.34 b	167.3 ± 5.03 a
20	994.8 ± 36.2 a	82.8 ± 5.36 a	54.2 ± 2.33 a	44.0 ± 0.96 a	156.4 ± 3.04 b
	***	***	***	***	***

*** Significant at $p \leq 0.001$. Different letters within each column indicate significant differences according to Duncan's multiple range test ($p = 0.05$).

Moreover, lutein and β-carotene exhibited the same trend as total chlorophylls, registering the highest values in plants with 20 NS feeding days, at 54.2 and 44 µg g^{-1} fw, respectively. These latter were positively correlated with ABTS antioxidant activity (0.87 and 0.9, respectively), which also increased when the NS feeding days increased, and proved by a significant margin to be the highest in spinach microgreens under 20-day NS feeding treatment (994.8 mmol Trolox eq. 100^{-1} fw). According to the literature, nutritional stress may result in decreased carotenoid contents, as suggested in the studies of El-Nakhel et al. [17], who subjected hydroponically grown lettuce plants under nutrient deprivation, and of Pannico et al. [37], who applied nutrient stress to lettuce microgreens. The increased contents of carotenoids in 10- and 20-day NS feeding treatments are in line with the increased fresh yields observed in the same treatments, since, along with chlorophylls, carotenoids are also very important pigments for the light-harvesting photosystem II and the light-harvesting antenna complexes of plants [45,46]. Regarding the high antioxidant activity recorded in the 10- and 20-day NS feeding treatments, this finding could be attributed to the high content of antioxidant compounds detected in these treatments, such as chlorophylls and carotenoids, as already described (Table 3), as well as phenolic compounds (described in the following section). The correlation between antioxidant compound content and antioxidant activity in the plant tissues of leafy vegetables is well established, and has been confirmed in numerous literature reports [47,48]. In contrast, total ascorbic acid accumulated more in microgreens under 10-day NS feeding treatment, being 6.97% higher than the closest 20-day NS feeding treatment. Particularly, regression analysis indicated a quadratic relationship between NS feeding days and total ascorbic acid (Supplementary Figure S1). Total ascorbic acid ranged from 130.5 to 167.3 mg 100 g^{-1} fw, and reached its maximum at 10 NS, then started to decrease with NS feeding days (Table 3; Supplementary Figure S1). From this finding, it could be assumed that although ascorbic acid is considered to be one of the major antioxidant compounds, several other compounds may also contribute to the overall antioxidant mechanism of plants, as already observed in our study and in other literature reports [49,50].

3.4. Spinach Microgreens' Polyphenol Profiles and Total Polyphenols

The phenolic compound profiles of spinach microgreens in relation to NS feeding days are presented in Table 4. Sixteen individual compounds were detected in all of the tested samples, including fourteen flavonoids and two phenolic acids. The most abundant compounds were patuletin derivative (peak 16) and 5,3′,4′-trihydroxy-3 methoxy-6,7-methylendioxyflavone 4′ glucuronide (peak 15), followed by quercetin-3-sophoroside-7-glucoside, kaempferol-3-sinapoylsophoroside-7-glucoside, and spinacetin derivative (peaks 2, 4, and 14, respectively). The two major compounds have been previously reported by Berquist et al. [9], who also identified various patuletin derivatives and 5,3′,4′-trihydroxy-3 methoxy-6,7-methylendioxyflavone 4′ glucuronide.

Table 4. Polyphenol composition (μg 100 g^{-1} fw) in relation to nutrient solution feeding regime (means $\pm$ SD).

Peak No	Phenolic Compounds	Nutrient Solution Feeding (Days)				Sig.
		0	5	10	20	
1	Km 3-hydroxyferuloylsophorotrioside-7-glucoside	46.36 $\pm$ 3.41 a	35.65 $\pm$ 1.12 b	34.98 $\pm$ 1.58 b	28.20 $\pm$ 0.57 c	***
2	Qn 3-sophoroside-7-glucoside	539.8 $\pm$ 27.0 c	529.5 $\pm$ 42.1 c	690.1 $\pm$ 15.5 b	861.9 $\pm$ 35.4 a	***
3	Kaempferol-3-diglucoside	1.34 $\pm$ 0.13 c	2.50 $\pm$ 0.04 ab	2.40 $\pm$ 0.17 b	2.91 $\pm$ 0.18 a	***
4	Km 3-sinapoylsophoroside-7-glucoside	464.8 $\pm$ 16.1	464.8 $\pm$ 9.51	489.7 $\pm$ 29.7	556.8 $\pm$ 30.8	ns
5	Km 3-sinapoylsophorotrioside-7-glucoside	36.99 $\pm$ 0.15 a	31.25 $\pm$ 2.78 b	29.00 $\pm$ 0.92 b	19.78 $\pm$ 1.11 c	***
6	Qn 3-sinapoyltriglucoside	113.7 $\pm$ 5.95 a	101.0 $\pm$ 2.34 b	87.68 $\pm$ 1.72 c	90.85 $\pm$ 1.81 bc	**
7	Synapoyl-hexose	86.54 $\pm$ 2.88 ab	90.78 $\pm$ 0.74 a	81.98 $\pm$ 3.35 b	34.63 $\pm$ 1.38 c	***
8	Caffeoylquinic acid	9.81 $\pm$ 0.33 c	14.08 $\pm$ 0.46 b	33.05 $\pm$ 1.18 a	5.37 $\pm$ 0.21 d	***
9	Rutin	1.35 $\pm$ 0.08 c	1.94 $\pm$ 0.15 bc	3.13 $\pm$ 0.58 a	2.86 $\pm$ 0.30 ab	*
10	Coumaroyl-diglucoside	9.71 $\pm$ 0.22 b	11.87 $\pm$ 1.01 a	6.73 $\pm$ 0.55 c	4.35 $\pm$ 0.17 d	***
11	Ferulic acid	141.9 $\pm$ 5.28 b	189.3 $\pm$ 7.81 a	182.0 $\pm$ 3.99 a	154.1 $\pm$ 8.40 b	**
12	Km 3-p-coumaroylsophoroside-7-glucoside	1.90 $\pm$ 0.06 b	1.30 $\pm$ 0.03 c	2.17 $\pm$ 0.10 b	2.95 $\pm$ 0.16 a	***
13	Isorhamnetin-3-gentiobioside	5.56 $\pm$ 0.20 c	14.31 $\pm$ 0.40 b	25.13 $\pm$ 1.06 a	13.93 $\pm$ 1.01 b	***
14	Spinacetin derivative	368.6 $\pm$ 5.66	369.9 $\pm$ 15.6	410.5 $\pm$ 34.5	436.2 $\pm$ 23.3	ns
15	5,3′,4′-trihydroxy-3 methoxy-6,7-methylendioxyflavone 4′ glucuronide	2015 $\pm$ 77.9 b	2314 $\pm$ 43.8 a	2335 $\pm$ 44.6 a	1933 $\pm$ 32.2 b	***
16	Patuletin derivative	2479 $\pm$ 221 c	3005 $\pm$ 232 bc	3114 $\pm$ 136 b	3874 $\pm$ 102 a	**
	Total phenols	6323 $\pm$ 321 c	7178 $\pm$ 247 b	7528 $\pm$ 78.2 ab	8021 $\pm$ 146 a	**

ns,*,**, ***: Non-significant or significant at $p \leq 0.05$, 0.01, and 0.001, respectively. Km: kaempferol; Qn: quercetin. Different letters within each row indicate significant differences according to Duncan's multiple range test ($p = 0.05$). All data are expressed as mean $\pm$ standard error, $n = 3$.

All of the studied polyphenols in this study showed a varied content when subjected to diverse NS feeding treatments, except for kaempferol-3-sinapoylsophoroside-7-glucoside and spinacetin derivative (peaks 4 and 14) which were not significantly affected by NS feeding days (Table 4). Moreover, total polyphenols increased gradually when more NS was administered to spinach microgreens—the 20-day NS feeding treatment resulted in the highest accumulation of total polyphenols (8021 μg 100 g^{-1} fw), without being significantly different from 10-day NS feeding (7528 μg 100 g^{-1} fw). In contrast, when irrigated with only osmotic water throughout the growing cycle (0 NS feeding days), spinach microgreens accumulated the least polyphenols (6323 μg 100 g^{-1} fw), and around 11.91% less than the closest treatment (5 NS feeding days). Spinach microgreens irrigated with NS throughout the growing cycle (20 NS feeding days) accumulated significantly more quercetin-3-sophoroside-7-glucoside, kaempferol-3-p-coumaroyl-sophoroside-7glucoside, and patuletin derivative. Moreover, kaempferol-3-diglucoside content was the highest in the 20-day NS treatment, without being significantly different from the 5-day NS treatment, while rutin content was the highest in the 10-day NS treatment,

without being different from the 20-day NS treatment. In contrast, spinach microgreens subjected to the 10-day NS feeding treatment accumulated the highest content of 5,3′,4′-trihydroxy-3 methoxy-6,7-methylendioxyflavone 4′ glucuronide, caffeoylquinic acid, and isorhamnetin-3-gentiobioside, and when completely deprived of NS they accumulated the highest amounts of kaempferol-3-hydroxyferuloylsophorotrioside-7-glucoside, kaempferol-3-sinapoylsophorotrioside-7-glucoside, and quercetin-3-sinapoyltriglucoside. Coumaroyl-diglucoside content was the highest in the 5-day NS treatment. In contrast to our study, El-Nakhel et al. [17] suggested that nutrient deprivation may increase the content of phenolic compounds in hydroponically grown lettuce, while El-Nakhel [33] reported an increase in total phenolic compounds in rocket microgreens that received nutrient solution compared to untreated ones. Moreover, the same authors observed a varied effect of nutrient solution supplementation on individual phenolic compounds—a finding that is consistent with our study. The contradictory results in the literature reports indicate that plants' response to nutrient stress varies according to the species and the severity of the stress. Therefore, although nutrient deprivation is expected to increase the content of phenolic compounds as part of plants' defense mechanisms against nutrient stress [17,46], this was not the case in our study—probably due to the genotype-dependent response to nutrient stress suggested in the literature [51].

4. Conclusions

Microgreens are a novel category of crop products of increasing interest to consumers and the marketing sector. The ease of facilitating cropping—even in domestic conditions—and the shortness of their growth cycle are very promising features that may contribute to addressing food security and malnutrition. Moreover, the supplementation of macro- and micronutrients through fertigation allows the manipulation of their chemical composition and production of tailor-made products with improved bioactive and functional properties. In this context, the application of nutritional eustress via the regulation of nutrient solution feeding might be a cost-effective cultivation practice to further improve the nutritional value of microgreens. Our results demonstrated that mild nutritional stress through the supplementation of nutrient solution for 10 consecutive days resulted in a slight decrease in fresh yield, without compromising quality features such as mineral and bioactive compound contents, whereas it increased total ascorbic acid content and reduced nitrates by 70.67%, and still generated the same growth cycle and days until harvest as the 20-day NS ftreatment. Therefore, with the aim being the production of not only more but also better products, the application of mild nutritional stress is a cost-effective and sustainable practice for microgreen cultivation. However, considering the species-dependent response to nutritional stress, further studies with varied species are needed in order to establish cultivation protocols and best practices guides accordingly.

Supplementary Materials: The following are available online at https://www.mdpi.com/article/10.3390/horticulturae7070162/s1: Figure S1: Relationships between the fresh yield (A), dry yield (B), dry matter percentage (C), nitrate concentrations (D), P and K concentrations (E and F), and total ascorbic acid content (TAA) (G) in relation to nutrient solution feeding days.

Author Contributions: Conceptualization, S.A.P. and C.E.-N.; methodology, C.E.-N.; software, C.E.-N.; validation, C.E.-N. and G.G.; formal analysis, C.E.-N.; investigation, C.E.-N. and G.G.; resources, S.A.P. and Y.R.; data curation, C.E.-N. and G.G.; writing—original draft preparation, S.A.P. and C.E.-N.; writing—review and editing, S.A.P., M.C.K., and Y.R.; supervision, S.A.P. and Y.R.; project administration, S.A.P.; funding acquisition, S.A.P. All authors have read and agreed to the published version of the manuscript.

Funding: This research received no external funding.

Data Availability Statement: The study did not report any data.

Acknowledgments: The authors would like to thank Francesco Cristofano, Luigi Formisano, Luigi G. Duri, and Michele Ciriello from OrtoLab UNINA for their technical assistance during the experiment, Antonio Pannico for data analysis, and Maria Giordano for laboratory analysis.

Conflicts of Interest: The authors declare no conflict of interest.

References

1. Kyriacou, M.C.; Rouphael, Y.; Di Gioia, F.; Kyratzis, A.; Serio, F.; Renna, M.; De Pascale, S.; Santamaria, P. Micro-scale vegetable production and the rise of microgreens. *Trends Food Sci. Technol.* **2016**, *57*, 103–115. [CrossRef]
2. Palmitessa, O.D.; Renna, M.; Crupi, P.; Lovece, A.; Corbo, F.; Santamaria, P. Yield and Quality Characteristics of Brassica Microgreens as A ff ected by the NH_4:NO_3 Molar Ratio and Strength of the Nutrient Solution. *Foods* **2020**, *9*, 677. [CrossRef]
3. Mir, S.A.; Shah, M.A.; Mir, M.M. Microgreens: Production, shelf life, and bioactive components. *Crit. Rev. Food Sci. Nutr.* **2017**, *57*, 2730–2736. [CrossRef]
4. Di Gioia, F.; Renna, M.; Santamaria, P. Sprouts, Microgreens and "Baby Leaf" Vegetables. In *Minimally Processed Refrigerated Fruits and Vegetablesâ*; Yildiz, F., Wiley, R.C., Eds.; Springer: Boston, MA, USA, 2017; pp. 403–432. ISBN 978-1-4939-7016-2.
5. Choe, U.; Yu, L.L.; Wang, T.T.Y. The Science behind Microgreens as an Exciting New Food for the 21st Century. *J. Agric. Food Chem.* **2018**, *66*, 11519–11530. [CrossRef]
6. Di Gioia, F.; Petropoulos, S.A.; Ozores-Hampton, M.; Morgan, K.; Rosskopf, E.N. Zinc and iron agronomic biofortification of Brassicaceae microgreens. *Agronomy* **2019**, *9*, 677. [CrossRef]
7. Rouphael, Y.; Kyriacou, M.C.; Petropoulos, S.A.; De Pascale, S.; Colla, G. Improving vegetable quality in controlled environments. *Sci. Hortic.* **2018**, *234*, 275–289. [CrossRef]
8. Giordano, M.; El-Nakhel, C.; Pannico, A.; Kyriacou, M.C.; Stazi, S.R.; De Pascale, S.; Rouphael, Y. Iron biofortification of red and green pigmented lettuce in closed soilless cultivation impacts crop performance and modulates mineral and bioactive composition. *Agronomy* **2019**, *9*, 290. [CrossRef]
9. Bergquist, S.Å.; Gertsson, U.E.; Knuthsen, P.; Olsson, M.E. Flavonoids in baby spinach (*Spinacia oleracea* L.): Changes during plant growth and storage. *J. Agric. Food Chem.* **2005**, *53*, 9459–9464. [CrossRef] [PubMed]
10. Xiao, Z.; Lester, G.E.; Park, E.; Saftner, R.A.; Luo, Y.; Wang, Q. Evaluation and correlation of sensory attributes and chemical compositions of emerging fresh produce: Microgreens. *Postharvest Biol. Technol.* **2015**, *110*, 140–148. [CrossRef]
11. Renna, M.; Paradiso, V.M. Ongoing research on microgreens: Nutritional properties, shelf-life, sustainable production, innovative growing and processing approaches. *Foods* **2020**, *9*, 862. [CrossRef]
12. Ghoora, M.; Srividya, N. Effect of Packaging and Coating Technique on Postharvest Quality and Shelf Life of *Raphanus sativus* L. and *Hibiscus sabdariffa* L. Microgreens. *Foods* **2020**, *9*, 653. [CrossRef] [PubMed]
13. Sharma, A.; Rasane, P.; Dey, A.; Singh, J.; Kaur, S.; Dhawan, K.; Kumar, A.; Joshi, H.S. Optimization of a Process for a Microgreen and Fruit Based Ready to Serve Beverage. *Int. J. Food Stud.* **2021**, *10*, SI41–SI56. [CrossRef]
14. Schlering, C.; Zinkernagel, J.; Dietrich, H.; Frisch, M.; Schweiggert, R. Alterations in the chemical composition of spinach (*Spinacia oleracea* L.) as provoked by season and moderately limited water supply in open field cultivation. *Horticulturae* **2020**, *6*, 25. [CrossRef]
15. Kyriacou, M.C.; El-Nakhel, C.; Pannico, A.; Graziani, G.; Soteriou, G.A.; Giordano, M.; Zarrelli, A.; Ritieni, A.; De Pascale, S.; Rouphael, Y. Genotype-Specific Modulatory Effects of Select Spectral Bandwidths on the Nutritive and Phytochemical Composition of Microgreens. *Front. Plant Sci.* **2019**, *10*, 1501. [CrossRef]
16. Tomasi, N.; Pinton, R.; Dalla Costa, L.; Cortella, G.; Terzano, R.; Mimmo, T.; Scampicchio, M.; Cesco, S. New "solutions" for floating cultivation system of ready-to-eat salad: A review. *Trends Food Sci. Technol.* **2014**, *46*, 267–276. [CrossRef]
17. El-Nakhel, C.; Pannico, A.; Kyriacou, M.C.; Giordano, M.; De Pascale, S.; Rouphael, Y. Macronutrient deprivation eustress elicits differential secondary metabolites in red and green-pigmented butterhead lettuce grown in a closed soilless system. *J. Sci. Food Agric.* **2019**, *99*, 6962–6972. [CrossRef]
18. Courbet, G.; D'Oria, A.; Lornac, A.; Diquélou, S.; Pluchon, S.; Arkoun, M.; Koprivova, A.; Kopriva, S.; Etienne, P.; Ourry, A. Specificity and Plasticity of the Functional Ionome of *Brassica napus* and *Triticum aestivum* Subjected to Macronutrient Deprivation. *Front. Plant Sci.* **2021**, *12*, 1–15. [CrossRef]
19. Sardare, M.D.; Admane, S. A Review on Plant without Soil—Hydroponics. *Int. J. Res. Eng. Technol.* **2013**, *2*, 299–304. [CrossRef]
20. Hajnos, M.; Hajnos, M. Buffer Capacity of Soils BT. In *Encyclopedia of Agrophysics*; Gliński, J., Horabik, J., Lipiec, J., Eds.; Springer: Dordrecht, The Netherlands, 2011; pp. 94–95. ISBN 978-90-481-3585-1.
21. Yadav, L.P.; Koley, T.K.; Tripathi, A.; Singh, S. Antioxidant Potentiality and Mineral Content of Summer Season Leafy Greens: Comparison at Mature and Microgreen Stages Using Chemometric. *Agric. Res.* **2019**, *8*, 165–175. [CrossRef]
22. Ghoora, M.D.; Babu, D.R.; Srividya, N. Nutrient composition, oxalate content and nutritional ranking of ten culinary microgreens. *J. Food Compos. Anal.* **2020**, *91*, 103495. [CrossRef]
23. Lester, G.; Hallman, G.; Pérez, J.A. γ-Irradiation Dose: Effects on Baby-Leaf Spinach Ascorbic Acid, Carotenoids, Folate, r -Tocopherol, and Phylloquinone Concentrations. *J. Agric. Food Chem.* **2010**, *58*, 4901–4906. [CrossRef]
24. Elia, A.; Santamaria, P.; Serio, F. Nitrogen nutrition, yield and quality of spinach. *J. Sci. Food Agric.* **1998**, *76*, 341–346. [CrossRef]

25. Tamme, T.; Reinik, M.; Roasto, M. Chapter 21-Nitrates and Nitrites in Vegetables: Occurrence and Health Risks. In *Bioactive Foods in Promoting Health Fruits and Vegetables*; Watson, R.R., Preedy, V., Eds.; Academic Press: San Diego, CA, USA, 2010; pp. 307–321. ISBN 978-0-12-374628-3.
26. Erfani, F.; Hassandokht, M.R.; Jabbari, A.; Barzegar, M. Effect of cultivar on chemical composition of some Iranian spinach. *Pak. J. Biol. Sci.* **2007**, *10*, 602–606.
27. Bergquist, S.Å.M.; Gertsson, U.E.; Olsson, M.E. Influence of growth stage and postharvest storage on ascorbic acid and carotenoid content and visual quality of baby spinach (*Spinacia oleracea* L.). *J. Sci. Food Agric.* **2006**, *86*, 346–355. [CrossRef]
28. Kyriacou, M.C.; El-Nakhel, C.; Graziani, G.; Pannico, A.; Soteriou, G.A.; Giordano, M.; Ritieni, A.; De Pascale, S.; Rouphael, Y. Functional quality in novel food sources: Genotypic variation in the nutritive and phytochemical composition of thirteen microgreens species. *Food Chem.* **2019**, *277*, 107–118. [CrossRef]
29. Kampfenkel, K.; Montagu, M.; Inzé, D. Extraction and determination of ascorbate and dehydroascorbate from plant tissue. *Anal. Biochem.* **1995**, *225*, 165–167. [CrossRef] [PubMed]
30. Lichtenthaler, H.K.; Buschmann, C. Chlorophylls and Carotenoids: Measurement and Characterization by UV-VIS Spectroscopy. *Curr. Protoc. Food Anal. Chem.* **2001**, *1*, F4.3.1–F4.3.8. [CrossRef]
31. Fellegrini, N.; Ke, R.; Yang, M.; Rice-Evans, C.B.T.-M. Screening of dietary carotenoids and carotenoid-rich fruit extracts for antioxidant activities applying 2,2′-azinobis(3-ethylenebenzothiazoline-6-sulfonic acid radical cation decolorization assay. In *Oxidants and Antioxidants Part A*; Academic Press: Cambridge, MA, USA, 1999; Volume 299, pp. 379–389. ISBN 0076-6879.
32. Huang, H.; Jiang, X.; Xiao, Z.; Yu, L.; Pham, Q.; Sun, J.; Chen, P.; Yokoyama, W.; Yu, L.L.; Luo, Y.S.; et al. Red Cabbage Microgreens Lower Circulating Low-Density Lipoprotein (LDL), Liver Cholesterol, and Inflammatory Cytokines in Mice Fed a High-Fat Diet. *J. Agric. Food Chem.* **2016**, *64*, 9161–9171. [CrossRef] [PubMed]
33. El-Nakhel, C.; Pannico, A.; Graziani, G.; Kyriacou, M.C.; Gaspari, A.; Ritieni, A.; De Pascale, S.; Rouphael, Y. Nutrient Supplementation Configures the Bioactive Profile and Production Characteristics of Three *Brassica* L. Microgreens Species Grown in Peat-Based Media. *Agronomy* **2021**, *11*, 346. [CrossRef]
34. Fallovo, C.; Rouphael, Y.; Rea, E.; Battistelli, A.; Colla, G. Nutrient solution concentration and growing season affect yield and quality of *Lactuca sativa* L. var. *acephala* in floating raft culture. *J. Sci. Food Agric.* **2009**, *89*, 1682–1689. [CrossRef]
35. Murphy, C.J.; Pill, W.G. Cultural practices to speed the growth of microgreen arugula (roquette; *Eruca vesicaria* subsp. *sativa*). *J. Hortic. Sci. Biotechnol.* **2010**, *85*, 171–176. [CrossRef]
36. Wieth, A.R.; Pinheiro, W.D.; Duarte, T.D.S. Purple cabbage microgreens grown in different substrates and nutritive solution concentrations. *Rev. Caatinga* **2019**, *32*, 976–985. [CrossRef]
37. Pannico, A.; Graziani, G.; El-Nakhel, C.; Giordano, M.; Ritieni, A.; Kyriacou, M.C.; Rouphael, Y. Nutritional stress suppresses nitrate content and positively impacts ascorbic acid concentration and phenolic acids profile of lettuce microgreens. *Italus Hortus* **2020**, *27*, 41–52. [CrossRef]
38. Chen, X.G.; Gastaldi, C.; Siddiqi, M.Y.; Glass, A.D.M. Growth of a lettuce crop at low ambient nutrient concentrations: A strategy designed to limit the potential for eutrophication. *J. Plant Nutr.* **1997**, *20*, 1403–1417. [CrossRef]
39. Van Iersel, M. Fertilizer Concentration Affects Growth and Nutrient Composition of Subirrigated Pansies. *HortScience* **1999**, *34*, 660–663. [CrossRef]
40. Blom-Zandstra, M.; Lampe, J.A.N.E.M. The Role of Nitrate in the Osmoregulation of Lettuce (*Lactuca sativa* L.) Grown at Different Light Intensities. *J. Exp. Bot.* **1985**, *36*, 1043–1052. [CrossRef]
41. Ramakrishna, A.; Ravishankar, G.A. Influence of abiotic stress signals on secondary metabolites in plants. *Plant Signal. Behav.* **2011**, *6*, 1720–1731. [CrossRef]
42. Borah, K.D.; Bhuyan, J. Magnesium porphyrins with relevance to chlorophylls. *Dalt. Trans.* **2017**, *46*, 6497–6509. [CrossRef] [PubMed]
43. Bohn, T.; Walczyk, T.; Leisibach, S.; Hurrell, R. Chlorophyll-bound Magnesium in Commonly Consumed Vegetables and Fruits: Relevance to Magnesium Nutrition. *J. Food Sci.* **2004**, *69*, 347–350. [CrossRef]
44. Niroula, A.; Khatri, S.; Timilsina, R.; Khadka, D. Profile of chlorophylls and carotenoids of wheat (*Triticum aestivum* L.) and barley (*Hordeum vulgare* L.) microgreens. *J. Food Sci. Technol.* **2019**, *56*, 2758–2763. [CrossRef]
45. Pannico, A.; El-nakhel, C.; Kyriakou, M.C.; Giordano, M.; Stazi, S.R.; De Pascale, S.; Rouphael, Y. Combating Micronutrient Deficiency and Enhancing Food Functional Quality through Selenium Fortification of Select Lettuce Genotypes Grown in a Closed Soilless System. *Front. Plant Sci.* **2019**, *10*, 1495. [CrossRef] [PubMed]
46. Becker, C.; Urli, B.; Juki, M.; Kläring, H. Nitrogen Limited Red and Green Leaf Lettuce Accumulate Flavonoid Glycosides, Caffeic Acid Derivatives and Sucrose while Losing. *PLoS ONE* **2015**, *10*, e0142867. [CrossRef] [PubMed]
47. Petropoulos, S.A.; Fernandes, Â.; Dias, M.I.; Pereira, C.; Calhelha, R.; Gioia, F.D.; Tzortzakis, N.; Ivanov, M.; Sokovic, M.; Barros, L.; et al. Wild and cultivated *Centaurea raphanina* subsp. *mixta*: A valuable source of bioactive compounds. *Antioxidants* **2020**, *9*, 314. [CrossRef] [PubMed]
48. Liberal, Â.; Fernandes, Â.; Polyzos, N.; Petropoulos, S.A.; Dias, M.I.; Pinela, J.; Petrović, J.; Soković, M.; Ferreira, I.C.F.R.; Barros, L. Bioactive Properties and Phenolic Compound Profiles of Turnip-Rooted, Plain-Leafed and Curly-Leafed Parsley Cultivars. *Molecules* **2020**, *25*, 5606. [CrossRef] [PubMed]

49. Pereira, C.; Dias, M.I.; Petropoulos, S.A.; Plexida, S.; Chrysargyris, A.; Tzortzakis, N.; Calhelha, R.C.; Ivanov, M.; Stojković, D.; Soković, M.; et al. The effects of biostimulants, biofertilizers and water-stress on nutritional value and chemical composition of two spinach genotypes (*Spinacia oleracea* L.). *Molecules* **2019**, *24*, 4494. [CrossRef]
50. Petropoulos, S.A.; Fernandes, Â.; Dias, M.I.; Pereira, C.; Calhelha, R.C.; Chrysargyris, A.; Tzortzakis, N.; Ivanov, M.; Sokovic, M.D.; Barros, L.; et al. Chemical composition and plant growth of *Centaurea raphanina* subsp. *mixta* plants cultivated under saline conditions. *Molecules* **2020**, *25*, 2204. [CrossRef] [PubMed]
51. Senizza, B.; Zhang, L.; Miras-Moreno, B.; Righetti, L.; Zengin, G.; Ak, G.; Bruni, R.; Lucini, L.; Sifola, M.I.; El-Nakhel, C.; et al. The strength of the nutrient solution modulates the functional profile of hydroponically grown lettuce in a genotype-dependent manner. *Foods* **2020**, *9*, 1156. [CrossRef]

horticulturae

Article

Protein Hydrolysate Combined with Hydroponics Divergently Modifies Growth and Shuffles Pigments and Free Amino Acids of Carrot and Dill Microgreens

Christophe El-Nakhel [1,†], Michele Ciriello [1,†], Luigi Formisano [1], Antonio Pannico [1], Maria Giordano [1], Beniamino Riccardo Gentile [1], Giovanna Marta Fusco [2], Marios C. Kyriacou [3], Petronia Carillo [2,*] and Youssef Rouphael [1]

[1] Department of Agricultural Sciences, University of Naples Federico II, 80055 Portici, Italy; christophe.elnakhel@unina.it (C.E.-N.); michele.ciriello@unina.it (M.C.); luigi.formisano3@unina.it (L.F.); antonio.pannico@unina.it (A.P.); maria.giordano@unina.it (M.G.); beniaminoriccardo.gentile@unina.it (B.R.G.); youssef.rouphael@unina.it (Y.R.)
[2] Department of Environmental, Biological and Pharmaceutical Sciences and Technologies, University of Campania "Luigi Vanvitelli", 81100 Caserta, Italy; giovannamarta.fusco@unicampania.it
[3] Department of Vegetable Crops, Agricultural Research Institute, Nicosia 1516, Cyprus; m.kyriacou@ari.gov.cy
* Correspondence: petronia.carillo@unicampania.it
† These authors have contributed equally to this work.

Abstract: Microgreens are the new sophisticated commodity in horticulture that boost the human diet with bioactive metabolites and garnish it with colors and tastes. Microgreens thrive well when cultivated in soilless systems, of which closed-loop soilless systems combined with biostimulant application can provide a sustainable, innovative method of growing microgreens. *Daucus carota* L. and *Anethum graveolens* L. microgreens were grown in greenhouse conditions implementing a floating raft system combined with a protein hydrolysate of leguminous origin as root application (0.3 mL L^{-1} nutrient solution). Growth, colorimetric parameters, macronutrients, chlorophylls, carotenoids, carbohydrates, free amino acids, and soluble proteins were assessed. The use of a protein hydrolysate in the nutrient solution engendered an increase in anthocyanins (+461.7%) and total phenols (+12.4%) in carrot, while in dill, the fresh yield (+13.5%) and ascorbic acid (+17.2%) increased. In both species, soluble proteins and total free amino acids increased by 20.6% and 18.5%, respectively. The floating raft system proved to be promising for microgreens and can ease the application of biostimulants through root application. Future research should also investigate the yield and nutritional parameter responses of other species of microgreens with the aim of large-scale sustainable production.

Keywords: floating raft system; biostimulant; root application; anthocyanins; phenols; reduced sugars; carbohydrates; minerals; pigments

Citation: El-Nakhel, C.; Ciriello, M.; Formisano, L.; Pannico, A.; Giordano, M.; Gentile, B.R.; Fusco, G.M.; Kyriacou, M.C.; Carillo, P.; Rouphael, Y. Protein Hydrolysate Combined with Hydroponics Divergently Modifies Growth and Shuffles Pigments and Free Amino Acids of Carrot and Dill Microgreens. *Horticulturae* **2021**, *7*, 279. https://doi.org/10.3390/horticulturae7090279

Academic Editor: Nikolaos Katsoulas

Received: 31 July 2021
Accepted: 31 August 2021
Published: 3 September 2021

1. Introduction

In recent years, the sophisticated gastronomy market and the chain of horticultural supply have been conquered by a densely rich commodity, the so-called microgreens [1–3]. These immature greens boost the human diet with bioactive health-promoting metabolites and minerals [3,4] and generate a multitude of alluring colors and tastes [2,4]. A range of genotypes are adopted for microgreen cultivation, from commercial to local varieties and covering vast botanical families including Brassicaceae, Lamiaceae, and Apiaceae [2,3]. Microgreens can be grown in loose soil and soilless media [5]. Peat and peat-based materials are among the most adopted, followed by synthetic fibrous substrates (rockwool and polyethylene terephthalate) and natural fiber media (coconut, burlap, jute, cotton, hemp, etc.) [1,5–8]. Growth conditions proved to significantly modulate the qualitative profile of microgreens [3], of which the substrate material should be given attention [6,8], in addition to the nutrient addition strategies [3].

Innovative methods such as closed-loop hydroponics are economically sustainable for producing leafy vegetables, such as the floating raft system, where the nutrient solution composition can be managed accurately [9]. The authors of the present paper along with Cristofano and coworkers [10] also emphasized the importance of such system in accelerating the growth cycle, raising the nutrient and water use efficiency, reducing the labor cost, and avoiding suboptimal soil reactions. In addition, it can be implemented in urban agriculture projects under economic and social development, and in the reduction in the environmental impact [10].

Floating systems are appropriate for the use of plant biostimulants that act by improving the nutrient use efficiency through increasing the availability of the confined nutrients [10] and boosting plant growth, especially the biostimulants that fall in the category of biotic elicitation methods [11]. As stated by Rouphael and Colla [12], protein hydrolysates (PHs), as biostimulants, contain signaling peptides and free amino acids that can enhance seedling growth and vegetable quality. They are up taken through root absorption and can be converted to the needed plant compounds; in addition, they are highly available for plants, unlike in substrate conditions where microbial competition occurs [10]. Moreover, PHs increase root growth due to hormone-like activities, stimulate nitrogen and carbon metabolism, and modulate the antioxidant systems [13]; PHs change the root architecture from length to lateral root branching, and the root biomass, thus incrementing the root system surface area [14].

Although microgreens are grown hydroponically or semi-hydroponically, especially in peat and peat-based mixes [7], very few works have dealt with floating raft systems or the nutrient film technique (NFT). For instance, Wang and Kniel [15] adopted an NFT system for growing kale and mustard microgreens on hydroponic pads. Bulgari et al. [16] grew rocket, Swiss chard, and basil, and Puccinelli et al. [17] grew basil, both in a hydroponic floating system, adopting polystyrene cell trays filled with vermiculite. In addition, Bulgari et al. [8] grew basil and rocket microgreens in small tanks filled with three different substrates. Thus far, to our knowledge, no study has adopted a floating raft system without an additional substrate for the growth of microgreens, and no study has tested, thus far, the effect of biostimulants, particularly PHs, on the growth and quality of microgreens.

Based on the above-mentioned issues, the current study aimed to verify the possibility of growing microgreens in a floating raft system without any additional substrate, and the potential of adding a PH in the nutrient solution from day one. Therefore, in this study, biometric and colorimetric parameters of microgreens were assessed, in addition to the mineral content, pigments, and primary and secondary metabolites. Such data could be of relevant importance to microgreen growers and to scientists, in order to understand the growth of microgreens in direct contact with a nutrient solution and with biostimulant applications throughout the growth.

2. Materials and Methods

2.1. Growth Conditions, Plant Material, and Experimental Design

In order to evaluate production, bioactive compounds, minerals, carbohydrates, and the free amino acid content, two species of Apiaceae were grown in a floating system as microgreens: carrot (*Daucus carota* L.) and dill (*Anethum graveolens* L.), both purchased from Pagano Costantino & F.III S.R.L (Scafati, Salerno, Italy). The weight of 100 seeds was evaluated in triplicate (89.00 mg and 129.25 mg for carrot and dill, respectively).

The experiment was carried out in autumn 2020 in a glass greenhouse in the Department of Agriculture (DIA), University of Naples Federico II (Portici, Italy; 40°49′ N, 14°15′ E, 72 m above sea level). Both species were primed in water for one day and sown on 20 October 2020, with a density of 6 seeds cm^{-2}. Seeds were manually sown on a perforated plastic tray (total area: 588 cm^2) placed inside a plastic tank (28.5 × 22 × 6 cm) containing 1.1 L of nutrient solution (NS) (Figure 1). A quarter-strength modified Hoagland NS was prepared with osmotic water (electrical conductivity (EC) of 0.03 dS m^{-1} and pH of 6.2) as follows: 2.0 mM NO$_3{}^-$, 0.25 mM S, 0.20 mM P, 0.62 mM K, 0.75 mM Ca, 0.17 mM

Mg, 0.25 mM NH_4^+, 20 µM Fe, 9 µM Mn, 0.3 µM Cu, 1.6 µM Zn, 20 µM B, and 0.3 µM Mo, with an EC of 0.4 ± 0.05 dS m^{-1} and pH of 5.8 ± 0.2. During the growth of microgreens, fresh NS was added every other day to the tanks in order to maintain the original volume of 1.1 L. The tanks were arranged in a randomized factorial scheme (2 × 2), which involved a biostimulant application, an untreated control, and two species of microgreens (dill and carrot). Each treatment was replicated three times, with each tray consisting of a single replicate (experimental unit). Dill and carrot microgreens were harvested 22 and 25 days after sowing (DAS), respectively, when the first true leaf appeared.

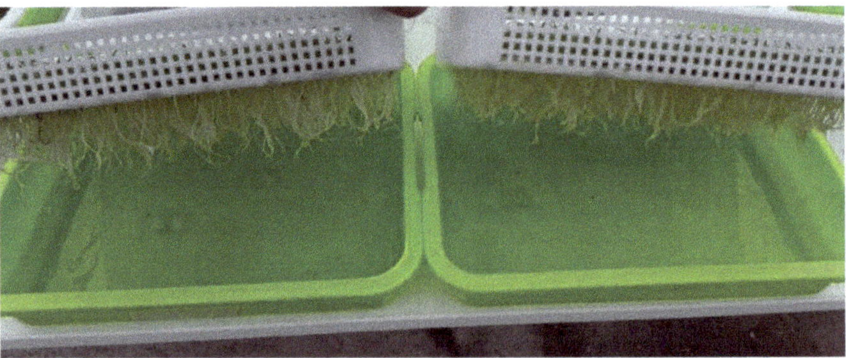

Figure 1. Hydroponic system (floating raft system) implemented in this study, showing the perforated plastic trays and the plastic tanks, and the growth of the roots directly in the nutrient solutions (not to scale).

2.2. Biostimulant Application

In this trial, a plant-derived biostimulant with a 5% organic nitrogen content obtained by enzymatic hydrolysis of leguminous biomass (Trainer®®; Hello Nature Italy SRL, Rivoli Veronese, Verona, Italy) was used. Enzymatic hydrolysis was used to release the amino acids and peptides from proteins. The final product contained mostly peptides and amino acids and, to a lesser extent, soluble carbohydrates, mineral elements, and phenolic compounds. Trainer®® has a density of 1.21 kg L^{-1}, a dry matter of 46%, and a pH of 4.0. It contains 310 g kg^{-1} of free amino acids and soluble peptides. As reported by the manufacturer, the biostimulant is composed of amino acids (Ala, Arg, Asp, Cys, Glu, Gly, His, Ile, Leu, Lys, Met, Phe, Pro, Ser, Thr, Trp, Tyr, and Val) and soluble peptides, free from plant hormones [10]. The Trainer®® content of soluble sugars is 90 g kg^{-1} f.w., and its elemental composition is as follows (g kg^{-1} f.w): N (50.0), P (0.9), K (41.1), Ca (10.9), Mg (0.5), Fe (0.024), Zn (0.010), Mn (0.001), B (0.005), and Cu (0.001). N–NO3 and N–NH4 contents are 3.13 and 6.00 mg g^{-1} f.w., respectively. The biostimulant was added to the NS at the dose of 0.3 mL L^{-1} from DAS 1.

2.3. Harvest, Biometric Parameters, and Colorimetric Indices Determination

At the stage of the first true leaf, microgreens of carrot and dill were harvested using sterilized scissors at the tray level. Fresh production was expressed as g m^{-2}. Part of the fresh microgreens was placed in a ventilated oven at 60 °C for 72 h to determine the dry weight (g m^{-2}) and the percentage of dry matter (DM). The dried material was then ground using a Wiley mill (MF10.1 Wiley laboratory mill, IKA®®, StaufenimBreisgau, Baden-Württemberg, Germany) and used for mineral determination. The remaining fresh material was placed immediately in liquid nitrogen and then stored at −80 °C, where a part was freeze dried for further analyses. At the same time, the roots were harvested and placed in a ventilated oven at 60 °C for 72 h to determine the dry weight (dw).

Before harvesting, the canopy color of the microgreens was determined with a Minolta CR-300 colorimeter (Minolta Camera Co. Ltd., Osaka, Japan) at five different points on

each tray. Before colorimetric measurements, the colorimeter was calibrated with the Minolta white standard. The measurements were obtained using the CIELAB (Commission Internationale de l'Éclairage) color space parameters, where:

L: brightness (0 = black to 100 = white), a*: greenness (green (−60) to red (+60)), and b*: yellowness (blue (−60) to yellow (+60)). Chroma (C), which represents the color intensity (chromaticity), was calculated using the following formula: $(a^2 + b^2)1/2$. The hue angle describes the qualitative color attribute in the relative amounts of redness and yellowness, Hue angle $\tan^{-1}$ (b*/a*).

2.4. Mineral Content Determination

Cations (K, Ca, Mg, Na) and anions (NO_3, SO_4, PO_4, Cl) were determined following the protocol described by Formisano et al. [18]. Briefly, 0.25 g of dried and ground microgreen material was extracted in ultrapure water, placed in a water bath at 80 °C, and shaken for 10 min. Then, after centrifuging the extracts (6000 rpm for 10 min), the supernatant was collected, filtered, and processed using an ICS3000 ion chromatograph (Thermo Scientific™ Dionex™, Sunnyvale, CA, USA) coupled to an electrical conductivity detector. Cation separation was performed with methanesulfonic acid (25 mM) using an IonPac CS12A analytical column, an IonPac CG12A precolumn, and a CERS5000 self-regenerating electrolyte suppressor (Thermo Scientific™ Dionex™, Sunnyvale, CA, USA). The separation of anions was performed using potassium hydroxide (5–30 mM) at a flow rate of 1.5 mL min^{-1} using an IonPac ATC-HC trap, an IonPac AG11-HC guard column, an IonPac AG11-HC IC column, and a DRS600 self-regenerating dynamic suppressor (Thermo Scientific™ Dionex™, Sunnyvale, CA, USA). The integration and quantification of minerals and organic acids were performed by comparing the peak areas of the samples with those of the standards. Each treatment was analyzed in triplicate, and the concentrations of anions and cations were expressed as mg g^{-1} dw.

2.5. Total Phenols Determination

According to Folin–Ciocalteu's method [19], to determine total phenols, 0.25 g of freeze-dried plant material was homogenized with 10 mL of 60% methanol and centrifuged for 15 min. An amount of 125 μL of the supernatant was added to 125 μL of Folin–Ciocalteu reagent (phosphotungstic acid + phosphomolybdic acid) in 0.5 mL of distilled water. The absorbance of the resulting mixture was read at 760 nm by UV–Vis spectrophotometry. Total phenols were expressed as mg gallic acid equivalents 100 g^{-1} dw. Each treatment was analyzed in triplicate.

2.6. Pigment and Total Ascorbic Acid Determination

As described by Wellburn [20], for the determination of photosynthetic pigments (chlorophyll a and b, and carotenoids), 0.2 g of fresh microgreen sample was extracted in ammonia acetone and centrifuged for 10 min (2000 rpm). Quantification of chlorophyll a and b and carotenoids was determined by reading the absorbances of the extracts at 647, 664, and 470 nm, respectively, using an ONDA V-10 Plus UV–Vis spectrophotometer (Giorgio Bormac srl, Carpi, Italy). Then, total chlorophyll a and b values were used to calculate total chlorophyll and the chlorophyll a/b ratio.

For the determination of total vitamin C, according to the method described by Kampfenkel et al. [21], 0.4 g of fresh sample was extracted in 2 mL of 6% TCA (trichloroacetic acid) and incubated for 15 min at −20 °C. The extract was then centrifuged for 10 min (4000 rpm), and the absorbance was read at 525 nm.

Total chlorophyll and carotenoids were expressed as mg g^{-1} fw, while total vitamin C was expressed in mg ascorbic acid 100 g^{-1} fw. The chlorophyll/carotenoid ratio was calculated as the total chlorophyll/carotenoids. Each treatment was analyzed in triplicate.

2.7. Anthocyanins, Soluble Carbohydrates, Soluble Proteins, and Amino Acids

Freeze-dried and ground samples of microgreens were used to determine anthocyanins, soluble carbohydrates, proteins, and amino acids.

Total anthocyanin analysis was performed according to the protocol described by Rouphael et al. [22] by extracting 0.2 g of sample in 180 μL of ethanol (40% *v/v*). The extract was then incubated on ice for 20 min and centrifuged (14,000 rpm; 10 min). The pellet was re-extracted by the same procedure, and the two extracts were combined and placed in a polypropylene microplate with 75 μL of 25 mM potassium chloride (pH 1.0) or 75 μL of 400 mM sodium acetate (pH 4.5). Using a Synergy HT spectrophotometer (BioTEK Instruments, Bad Friedrichshall, Stuttgart Germany), absorbance was read at 520 and 700 nm. The anthocyanin content was expressed as μg cyanidin-3-glucoside g^{-1}dw.

The extraction of soluble carbohydrates and starch was performed according to the protocol described by Carillo et al. [23]. Briefly, soluble sugars were quantified from ethanol extraction of microgreen samples, while starch content was determined from the pellet of ethanolic extract after hydrolysis to glucose, by an enzymatic assay coupled with pyridine nucleotide reduction. The increase in absorbance at 340 nm was recorded using an FLX-Xenius spectrophotometer (SAFAS, Munich City, Munich, Germany). Soluble sugars and starch were expressed as mg g^{-1} dw.

Soluble proteins were determined according to the protocol described by Ciriello et al. 2021 [9]. An aliquot of 0.2 g of sample was extracted in 1 mL of 200 mM Tris (hydroxymethyl) aminomethane hydrochloride (Tris-HCl) (pH 7.5) containing 500 mM magnesium chloride ($MgCl_2$). The extract was centrifuged, the supernatant was taken, and the protein content was measured using protein assay dye reagent concentrate (Bio-Rad, Milan, Italy). The primary amino acids of microgreens were determined in ethanolic extracts (60% *v/v*) by high-performance liquid chromatography (HPLC) after precolumn derivatization with *o*-phthaldialdehyde (OPA) according to the method described by Woodrow et al. [24]. Soluble proteins and amino acids were expressed as mg g^{-1} dw and μmol g^{-1} dw, respectively.

2.8. Statistical Analysis

Experimental data were subjected to bifactorial analysis of variance (two-way ANOVA) (species of microgreens (S) × biostimulant (B)) using IBM SPSS Statistics version 20.0 (SPSS Inc., Chicago, IL, USA). The main effects of S and B were compared according to Student's *t*-test. For the S × B interaction, significant statistical differences were determined using the Duncan multiple range test at the level of $p < 0.05$. Principal component analysis (PCA) was performed using Minitab 18.1 statistical software (Minitab LLC, State College, PA, USA) according to Ciarmiello et al. [25].

3. Results

3.1. Yield and Color Parameters

The yield of carrot and dill microgreens was dictated by the species × biostimulant interaction, where carrot was not significantly influenced by the addition of the protein hydrolysates in the nutrient solutions, while the dill yield significantly increased by 13.5% (Figure 2A). Carrot and dill microgreens in a floating system were characterized by a yield of 768.6 and 814.7 g fw m^{-2} on average, respectively. A similar trend was noted for the microgreen dry biomass; only the dill dry weight significantly increased (13.5%), whereas carrot was not influenced by the PH presence (Figure 2B). As for DM%, an interaction of both factors was clear, since only the carrot DM% significantly increased with PH application (by 11.28%), while dill was steady at 11.95% (Figure 2C).

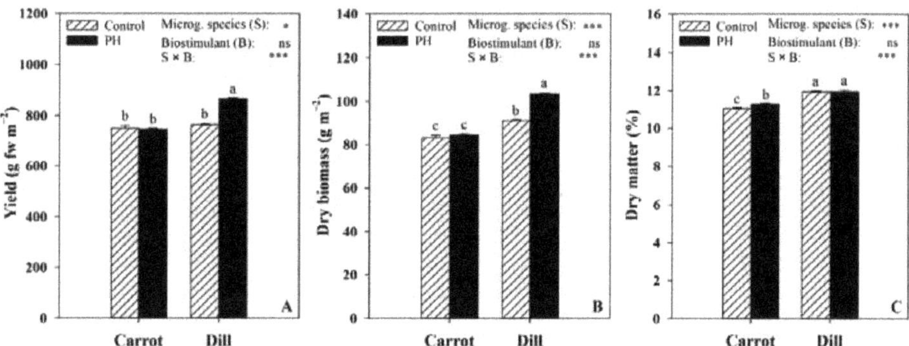

Figure 2. Fresh yield (**A**), dry biomass (**B**), and dry matter (**C**) of carrot and dill microgreens, as influenced by protein hydrolysate addition in the nutrient solution. Different letters above bars indicate significant mean differences according to Duncan's multiple range tests ($p = 0.05$). Vertical bars indicate ± SE (standard error) of means. ns, *, and *** indicate non-significant, or significant at $p \leq 0.05$ and 0.001, respectively.

All the color parameters of the carrot and dill microgreen canopies were influenced by the interaction of species × biostimulant. The brightness (L*) of the carrot canopy was not affected by the addition of the PH in the NS, whereas dill exhibited a decrease in L* (Table 1). The opposite trend was noted for the greenness parameter, where only carrot showed a darker green (+13.2%). On the other hand, carrot b* and chroma significantly increased with PH application, concomitantly with a decrease in the hue angle, whereas the opposite trend was noticed for the dill microgreen canopy (Table 1).

Table 1. Canopy colorimetric indices of carrot and dill microgreens as dictated by the application of a protein hydrolysate in the nutrient solution.

Treatments		L*	a*	b*	Chroma	Hue Angle
Microgreen species						
Carrot		33.90 ± 0.18	−10.75 ± 0.32	33.98 ± 1.72	35.64 ± 1.73	107.6 ± 0.40
Dill		37.88 ± 0.60	−8.39 ± 0.06	27.44 ± 0.52	28.70 ± 0.52	106.9 ± 0.26
Biostimulant						
Control		36.37 ± 1.18	−9.29 ± 0.37	29.38 ± 0.41	30.82 ± 0.50	107.4 ± 0.48
PH		35.41 ± 0.70	−9.84 ± 0.70	32.04 ± 2.58	33.52 ± 2.67	107.1 ± 0.19
Microgreen species	Biostimulant					
Carrot	Control	33.79 ± 0.13 c	−10.08 ± 0.23 b	30.18 ± 0.44 b	31.82 ± 0.49 b	108.5 ± 0.16 a
	PH	34.01 ± 0.36 c	−11.41 ± 0.15 c	37.78 ± 0.45 a	39.47 ± 0.43 a	106.8 ± 0.26 c
Dill	Control	38.94 ± 0.55 a	−8.49 ± 0.03 a	28.57 ± 0.08 c	29.82 ± 0.09 c	106.4 ± 0.21 c
	PH	36.82 ± 0.59 b	−8.28 ± 0.05 a	26.30 ± 0.27 d	27.58 ± 0.27 d	107.5 ± 0.09 b
Source of variance				(*p*-value)		
Microgreen species		***	***	**	**	ns
Biostimulant		ns	ns	ns	ns	ns
Microgreen species × Biostimulant		*	***	***	***	***

Non-significant (ns). *, **, and *** indicate significant at $p \leq 0.05$, 0.01, and 0.001, respectively. All data are expressed as mean ± SE (standard error), n = 3. Microgreen species and biostimulant means were compared by Student's *t*-test. Microgreen species and biostimulant interaction was compared by Duncan's multiple range test ($p = 0.05$). Different letters within each column indicate significant differences ($p = 0.05$). PH: protein hydrolysate.

3.2. Microgreen Pigments and Total Ascorbic Acid

All microgreen pigments and the related parameters were only dominated by the main effect of the PH application. Total chlorophylls and carotenoids incremented with the presence of the PH in the NS, by 43.9% and 16.8%, respectively (Table 2). In addition, the chlorophylls/carotenoids ratio also increased by 23.5% with the same treatment, whereas the chlorophyll a/b ratio decreased with the presence of the PH (−11.1%).

Table 2. Pigments (total chlorophylls and carotenoids), chlorophyll a/b ratio, chlorophylls/carotenoids ratio, anthocyanins, total ascorbic acid, and total phenols of carrot and dill microgreens as dictated by the application of a protein hydrolysate in the nutrient solution.

Treatments		Total Chlorophylls (mg g^{-1} fw)	Carotenoids (mg g^{-1} fw)	Chlorophyll a/b Ratio	Chlorophylls/Carotenoids Ratio	Anthocyanins (mg 100 g^{-1} fw)	Total Ascorbic Acid (mg AA 100 g^{-1} fw)	Total Phenols (mg gallic a. eq. 100 g^{-1} dw)
Microgreen species								
Carrot		1.146 ± 0.101	0.316 ± 0.011	1.94 ± 0.044	3.60 ± 0.20	10.35 ± 3.26	70.35 ± 3.81	4.11 ± 0.13
Dill		1.032 ± 0.083	0.316 ± 0.013	1.99 ± 0.059	3.24 ± 0.14	26.92 ± 2.31	137.8 ± 5.24	3.69 ± 0.19
Biostimulant								
Control		0.893 ± 0.036	0.291 ± 0.008	2.08 ± 0.022	3.06 ± 0.06	17.40 ± 6.41	102.3 ± 11.28	3.99 ± 0.08
PH		1.285 ± 0.042	0.340 ± 0.001	1.85 ± 0.010	3.78 ± 0.13	19.87 ± 1.27	105.9 ± 19.21	3.81 ± 0.25
Microgreen species	Biostimulant							
Carrot	Control	0.927 ± 0.041	0.292 ± 0.007	2.03 ± 0.015	3.18 ± 0.06	3.13 ± 0.65 d	77.68 ± 4.31 c	3.87 ± 0.14 b
	PH	1.365 ± 0.040	0.340 ± 0.002	1.84 ± 0.014	4.02 ± 0.12	17.58 ± 0.71 c	63.01 ± 0.57 d	4.35 ± 0.07 a
Dill	Control	0.859 ± 0.059	0.291 ± 0.017	2.12 ± 0.017	2.95 ± 0.04	31.67 ± 1.29 a	126.9 ± 3.48 b	4.10 ± 0.01 ab
	PH	1.205 ± 0.030	0.340 ± 0.001	1.86 ± 0.014	3.54 ± 0.10	22.16 ± 1.53 b	148.8 ± 2.34 a	3.27 ± 0.11 c
Source of variance					(p-value)			
Microgreen species		ns	ns	ns	ns	**	***	ns
Biostimulant		***	***	***	***	ns	ns	ns
Microgreen species × Biostimulant		ns	ns	ns	ns	***	***	***

Non-significant (ns). ** and *** indicate significant at $p \leq 0.01$ and 0.001, respectively. All data are expressed as mean ± SE (standard error), n = 3. Microgreen species and biostimulant means were compared by Student's t-test. Microgreen species and biostimulant interaction was compared by Duncan's multiple range test (p = 0.05). Different letters within each column indicate significant differences (p = 0.05). AA: ascorbic acid; gallic a. eq.: gallic acid equivalent; fw: fresh weight; dw: dry weight; PH: protein hydrolysate.

The total ascorbic acid content of the cultivated microgreens was influenced by the interaction of species × biostimulant. PH application in the NS caused a decrease of 18.9% in carrot microgreens compared to an increase of 17.3% in dill microgreens. On average, dill was characterized by a higher content of total ascorbic acid (2-fold) in comparison to carrot microgreens (70.35 mg AA 100 g^{-1} fw) (Table 2).

3.3. Microgreen Anthocyanins and Total Phenols

As listed in Table 2, the anthocyanin content incremented markedly by around 5.6-fold in carrot microgreens when the PH was applied, while it decreased by 30.0% in dill microgreens. On average, dill microgreens were 2.6-fold richer in anthocyanins in comparison to carrot. The variation in anthocyanins is in line with the total phenols in both microgreen species, where it increased by 12.4% in carrot and decreased by 20.2% in dill microgreens when the biostimulant was applied (Table 2).

3.4. Nitrate and Mineral Contents

Table 3 reports the nitrate and macronutrient contents of the cultivated microgreen species. Only nitrate, calcium, and magnesium were influenced by the interaction of the species and the biostimulant application. The carrot microgreen nitrate content was not influenced by the biostimulant application in the NS, whereas the dill nitrate content increased by 65.3% when the PH was added to the NS. Nonetheless, dill, on average, was characterized by a lower nitrate content (252.4 mg kg^{-1} fw) in comparison to carrot microgreens (602.7 mg kg^{-1} fw). As for calcium, it increased in both species in the presence of the PH, but in different percentages, with carrot registering 9.6% and dill 20.5%, whereas magnesium was modulated diversely in both species with the biostimulant application but not significantly different from the control treatment. On the other hand, phosphorous and sulfur were only dictated by the main effect of the biostimulant, where the former decreased when the PH was added, and the latter increased. Moreover, potassium and sodium contents were only dictated by the species main effect, where dill was rich in both minerals. In general, both microgreen species were high in K, followed by Ca and, ultimately, Mg, S, and P.

Table 3. Nitrate and mineral contents (P, K, Ca, Mg, S, and Na) of carrot and dill microgreens as dictated by the application of a protein hydrolysate in the nutrient solution.

Treatments		Nitrate (mg kg^{-1} fw)	P (mg g^{-1} dw)	K (mg g^{-1} dw)	Ca (mg g^{-1} dw)	Mg (mg g^{-1} dw)	S (mg g^{-1} dw)	Na (mg g^{-1} dw)
Microgreen species								
Carrot		602.7 ± 7.21	2.51 ± 0.10	10.18 ± 0.32	7.77 ± 0.16	2.58 ± 0.14	2.91 ± 0.25	0.50 ± 0.02
Dill		252.4 ± 28.12	2.70 ± 0.08	14.87 ± 0.36	7.33 ± 0.33	3.80 ± 0.21	2.28 ± 0.19	1.10 ± 0.03
Biostimulant								
Control		402.3 ± 94.99	2.80 ± 0.05	13.10 ± 1.11	7.03 ± 0.19	3.21 ± 0.43	2.12 ± 0.12	0.82 ± 0.13
PH		452.7 ± 61.97	2.41 ± 0.06	11.96 ± 1.03	8.06 ± 0.10	3.17 ± 0.18	3.07 ± 0.18	0.78 ± 0.14
Microgreen species	Biostimulant							
Carrot	Control	614.4 ± 10.85 a	2.72 ± 0.02	10.66 ± 0.45	7.41 ± 0.02 b	2.31 ± 0.03 c	2.37 ± 0.08	0.53 ± 0.02
Carrot	PH	591.0 ± 2.35 a	2.29 ± 0.02	9.71 ± 0.29	8.12 ± 0.08 a	2.85 ± 0.14 bc	3.45 ± 0.10	0.46 ± 0.01
Dill	Control	190.3 ± 4.58 c	2.87 ± 0.07	15.53 ± 0.25	6.65 ± 0.16 c	4.11 ± 0.30 a	1.88 ± 0.11	1.11 ± 0.04
Dill	PH	314.5 ± 8.69 b	2.53 ± 0.04	14.22 ± 0.39	8.01 ± 0.21 a	3.48 ± 0.19 ab	2.69 ± 0.11	1.09 ± 0.04
Source of variance					(p-value)			
Microgreen species		***	ns	***	ns	***	ns	***
Biostimulant		ns	***	ns	***	ns	**	ns
Microgreen species × Biostimulant		***	ns	ns	*	*	ns	ns

Non-significant (ns). *, **, and *** indicate significant at $p \leq 0.05$, 0.01, and 0.001, respectively. All data are expressed as mean ± SE (standard error), n = 3. Microgreen species and biostimulant means were compared by Student's *t*-test. Microgreen species and biostimulant interaction was compared by Duncan's multiple range test (p = 0.05). Different letters within each column indicate significant differences (p = 0.05). fw: fresh weight; dw: dry weight; PH: protein hydrolysate.

3.5. Starch and Reduced Sugar Contents

As listed in Table 4, no interaction between both factors was registered for the sugar content. All the analyzed sugars were dictated by the main effect of the species, and no changes were noted when the biostimulant was applied. Dill microgreens were characterized by a significant higher starch (54.16 mg g^{-1} dw), fructose (36.79 mg g^{-1} dw), and sucrose (3.58 mg g^{-1} dw) content, whereas carrot was characterized by a significant higher content of glucose (118.8 mg g^{-1} dw).

Table 4. Starch and reduced sugars (glucose, fructose, and sucrose) of carrot and dill microgreens as dictated by the application of a protein hydrolysate in the nutrient solution.

Treatments		Starch (mg g^{-1} dw)	Glucose (mg g^{-1} dw)	Fructose (mg g^{-1} dw)	Sucrose (mg g^{-1} dw)
Microgreen species					
Carrot		31.97 ± 2.05	118.8 ± 1.76	21.96 ± 0.28	1.88 ± 0.08
Dill		54.16 ± 2.63	100.4 ± 1.35	36.79 ± 0.83	3.58 ± 0.13
Biostimulant					
Control		47.23 ± 5.23	106.6 ± 3.87	30.42 ± 3.60	2.58 ± 0.39
PH		38.90 ± 5.10	112.6 ± 4.48	28.33 ± 3.06	2.89 ± 0.39
Microgreen species	Biostimulant				
Carrot	Control	36.28 ± 0.95	115.2 ± 0.45	22.40 ± 0.34	1.75 ± 0.12
	PH	27.66 ± 1.27	122.4 ± 1.46	21.52 ± 0.28	2.02 ± 0.02
Dill	Control	58.18 ± 4.02	98.04 ± 1.14	38.44 ± 0.55	3.41 ± 0.22
	PH	50.15 ± 1.50	102.8 ± 1.44	35.13 ± 0.63	3.76 ± 0.10
Source of variance			(p-value)		
Microgreen species		***	***	***	***
Biostimulant		ns	ns	ns	ns
Microgreen species × Biostimulant		ns	ns	ns	ns

Non-significant (ns). *** indicates significant at $p \leq 0.001$. All data are expressed as mean ± SE (standard error), n = 3. Microgreen species and biostimulant means were compared by Student's *t*-test. Microgreen species and biostimulant interaction was compared by Duncan's multiple range test (p = 0.05). dw: dry weight; PH: protein hydrolysate.

3.6. Soluble Protein and Amino Acid Contents

The application of the PH in the NS had different effects on the diverse amino acids tested in the dill and carrot microgreens (Table 5). The arginine, asparagine, glutamine, glutamic acid, phenylalanine, and valine contents were influenced by the interaction of species × biostimulant. Some amino acids increased in one species when the PH was applied, while they did not change significantly in the other species, or they increased intensively in one of the species. For instance, asn increased in both species when the PH was applied (by 58.9% in carrot and 54.4% in dill), whereas gln increased only in dill microgreens (by 32.1%). Moreover, the essential amino acids and the branched-chain amino acids increased only in carrot microgreens with PH application, whereas they remained statistically unchanged in dill. On the other hand, soluble proteins and total amino acids were both influenced by the main effect of the biostimulant, both increasing by 20.6% and 18.5%, respectively, when the PH was added to the NS. In addition, carrot microgreens were characterized by a higher total amino acid content on average when compared to dill. Other amino acids were also significantly influenced by the biostimulant, such as isoleucine, lysine, proline, and threonine. Finally, aspartic acid, glycine, histidine, and leucine were only dictated by the species effect, being significantly higher in carrot microgreens.

Table 5. Soluble proteins (mg g^{-1} dw) and amino acids (µmol g^{-1} dw) of carrot and dill microgreens as dictated by the application of a protein hydrolysate in the nutrient solution.

Compounds	Microgreen Species			Biostimulant			Microgreen Species × Biostimulant				ANOVA
							Carrot		Dill		
	Carrot	Dill	*t*-Test	Control	PH	*t*-Test	Control	PH	Control	PH	
Soluble proteins	43.05 ± 2.35	43.24 ± 1.42	ns	39.11 ± 0.89	47.18 ± 0.51	***	38.01 ± 1.48	48.10 ± 0.24	40.21 ± 0.71	46.26 ± 0.65	ns
Ala	4.28 ± 0.16	3.50 ± 0.33	ns	3.81 ± 0.30	3.96 ± 0.32	ns	4.27 ± 0.10	4.28 ± 0.34	3.35 ± 0.47	3.65 ± 0.55	ns
Arg	1.80 ± 0.12	1.67 ± 0.07	ns	1.70 ± 0.07	1.77 ± 0.13	ns	1.59 ± 0.08 b	2.01 ± 0.14 a	1.82 ± 0.05 ab	1.53 ± 0.06 b	**
Asn	38.63 ± 3.97	23.74 ± 2.32	**	24.25 ± 2.59	38.11 ± 4.17	*	29.84 ± 1.18 b	47.41 ± 0.29 a	18.66 ± 0.91 c	28.81 ± 0.60 b	***
Asp	5.84 ± 0.48	3.66 ± 0.37	**	5.52 ± 0.64	3.97 ± 0.44	ns	6.78 ± 0.44	4.89 ± 0.27	4.27 ± 0.50	3.05 ± 0.23	ns
GABA	14.34 ± 0.38	15.24 ± 0.47	ns	14.40 ± 0.43	15.19 ± 0.44	ns	13.83 ± 0.61	14.86 ± 0.28	14.96 ± 0.46	15.52 ± 0.89	ns
Gln	27.95 ± 0.55	31.10 ± 2.23	ns	27.83 ± 0.60	31.22 ± 2.18	ns	28.86 ± 0.36 b	27.04 ± 0.74 b	26.80 ± 0.78 b	35.40 ± 2.39 a	*
Glu	4.56 ± 0.33	1.26 ± 0.12	***	3.18 ± 0.91	2.64 ± 0.60	ns	5.16 ± 0.40 a	3.96 ± 0.16 b	1.20 ± 0.19 c	1.32 ± 0.19 c	*
Gly	1.99 ± 0.11	1.27 ± 0.11	***	1.66 ± 0.16	1.60 ± 0.23	ns	1.88 ± 0.20	2.09 ± 0.11	1.43 ± 0.17	1.11 ± 0.07	ns
His	1.49 ± 0.06	0.87 ± 0.03	***	1.17 ± 0.13	1.20 ± 0.17	ns	1.44 ± 0.03	1.54 ± 0.13	0.89 ± 0.05	0.85 ± 0.04	ns
Ile	2.18 ± 0.19	1.78 ± 0.15	ns	1.62 ± 0.09	2.34 ± 0.13	***	1.81 ± 0.07	2.55 ± 0.20	1.43 ± 0.03	2.12 ± 0.03	ns
Leu	1.66 ± 0.07	1.30 ± 0.10	*	1.33 ± 0.08	1.63 ± 0.12	ns	1.50 ± 0.02	1.82 ± 0.03	1.15 ± 0.00	1.44 ± 0.18	ns
Lys	0.529 ± 0.037	0.597 ± 0.045	ns	0.499 ± 0.021	0.627 ± 0.043	*	0.453 ± 0.003	0.605 ± 0.035	0.544 ± 0.012	0.649 ± 0.086	ns
MEA	2.62 ± 0.13	2.47 ± 0.26	ns	2.55 ± 0.23	2.54 ± 0.19	ns	2.69 ± 0.19	2.55 ± 0.21	2.40 ± 0.45	2.53 ± 0.36	ns
Met	0.254 ± 0.010	0.241 ± 0.016	ns	0.262 ± 0.007	0.233 ± 0.016	ns	0.261 ± 0.015	0.248 ± 0.016	0.262 ± 0.003	0.219 ± 0.030	ns
Orn	1.10 ± 0.05	0.85 ± 0.15	ns	1.12 ± 0.11	0.83 ± 0.11	ns	1.13 ± 0.06	1.06 ± 0.08	1.10 ± 0.23	0.60 ± 0.05	ns
Phe	0.89 ± 0.08	0.52 ± 0.04	**	0.60 ± 0.06	0.81 ± 0.11	ns	0.72 ± 0.05 b	1.06 ± 0.02 a	0.47 ± 0.01 c	0.57 ± 0.06 c	*
Pro	3.33 ± 0.27	3.79 ± 0.33	ns	3.06 ± 0.16	4.05 ± 0.28	*	2.88 ± 0.11	3.79 ± 0.39	3.25 ± 0.29	4.32 ± 0.41	ns
Ser	4.54 ± 0.32	3.43 ± 0.38	*	4.05 ± 0.52	3.93 ± 0.31	ns	5.10 ± 0.30	3.99 ± 0.33	3.00 ± 0.39	3.86 ± 0.61	ns
Thr	0.836 ± 0.08	0.647 ± 0.08	ns	0.590 ± 0.05	0.893 ± 0.06	**	0.699 ± 0.02	0.974 ± 0.11	0.481 ± 0.02	0.813 ± 0.01	ns
Trp	0.557 ± 0.04	0.571 ± 0.04	ns	0.528 ± 0.02	0.599 ± 0.04	ns	0.499 ± 0.03	0.614 ± 0.06	0.557 ± 0.02	0.584 ± 0.08	ns
Tyr	2.04 ± 0.11	1.81 ± 0.21	ns	1.76 ± 0.15	2.09 ± 0.17	ns	1.96 ± 0.17	2.11 ± 0.17	1.55 ± 0.21	2.07 ± 0.34	ns
Val	3.72 ± 0.34	2.99 ± 0.15	ns	3.10 ± 0.08	3.61 ± 0.40	ns	2.96 ± 0.08 bc	4.48 ± 0.03 a	3.23 ± 0.08 b	2.75 ± 0.23 c	***
Essential AA	13.92 ± 0.91	11.18 ± 0.21	*	11.39 ± 0.31	13.71 ± 1.00	*	11.94 ± 0.36 b	15.90 ± 0.33 a	10.84 ± 0.22 c	11.52 ± 0.23 bc	**
BCAAs	7.56 ± 0.59	6.06 ± 0.13	*	6.04 ± 0.13	7.58 ± 0.58	*	6.27 ± 0.16 b	8.85 ± 0.23 a	5.82 ± 0.11 b	6.31 ± 0.11 b	***
Total AA	125.1 ± 4.14	103.3 ± 4.82	**	104.6 ± 5.50	123.9 ± 4.54	*	116.3 ± 2.70	133.9 ± 0.72	92.82 ± 2.44	113.8 ± 0.77	ns

Non-significant (ns). *, **, and *** indicate significant at $p \leq$ 0.05, 0.01, and 0.001, respectively. All data are expressed as mean ± SE (standard error), n = 3. Microgreen species and biostimulant means were compared by Student's *t*-test. Microgreen species and biostimulant interaction was compared by Duncan's multiple range test (p = 0.05). Different letters within each column indicate significant differences (p = 0.05). dw: dry weight; PH: protein hydrolysate.

3.7. Principal Component Analysis (PCA)

A principal component analysis was performed on all analyzed dill and carrot microgreens data in relation to the biostimulant vs. control treatment, and the loading plot and scores are reported in Figure 3. The variables in the first three principal components (PCs) were highly correlated, with eigenvalues greater than 1, thus explaining 100% of the total variance, with PC1, PC2, and PC3 accounting for 64.8%, 20.5%, and 14.7%, respectively. PC1 was positively correlated with glucose, nitrate, asn, total amino acids, essential amino acids, b*, BCAAs, and glu, while it was negatively correlated with starch, a*, K, Na, fructose, TAA, and L*. Moreover, PC2 was positively correlated with carotenoids, total chlorophylls, Ca, yield, and gln, while it was negatively correlated with P and TP. The dill and carrot microgreen cultivars under different treatments were well separated and uniformly clustered with respect to PC1 and PC2. In fact, both the species and treatment factors examined in this study were relevant in the PCA clustering along PC1 and PC2, respectively. In particular, carrot microgreen treatments were distributed on the positive side of PC1, in the upper and lower right quadrants, while dill microgreen treatments were distributed on the negative side of PC1, in the upper and lower left quadrants. Moreover, the control treatment was distributed on the negative side of PC2, while the PH treatments were on the positive side of PC2 (Figure 3). Interestingly, the dill microgreens under PH treatment showed the highest yield and dry biomass, whereas the carrot microgreens were correlated with total amino acids, asn, BCAAs, and essential amino acids (Figure 3).

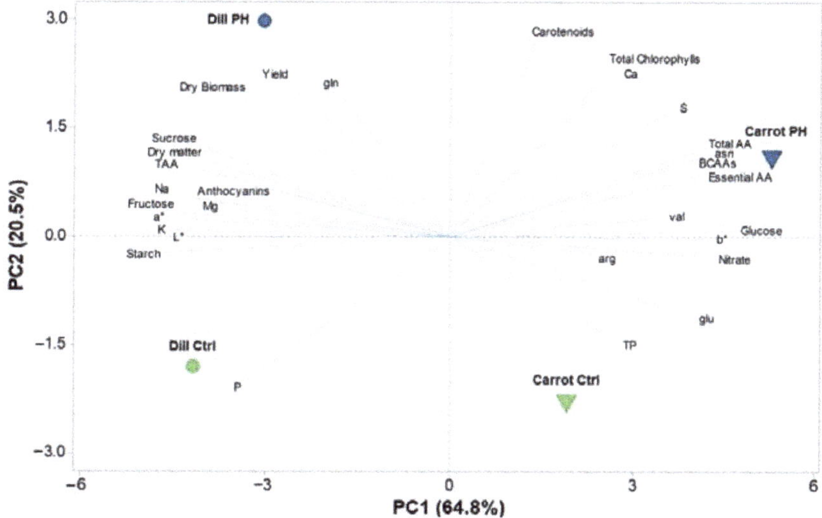

Figure 3. Principal component loading plot and scores of principal component analysis (PCA) of biometric traits, leaf colorimetry, minerals, carbohydrates, amino acids, and bioactive compounds of carrot and dill microgreens, as influenced by protein hydrolysate addition in the nutrient solution.

4. Discussion

The aim of our work was to assess the feasibility of growing microgreens in a soilless system without a substrate (floating system) by discriminating the effect of a protein hydrolysate-based biostimulant added to the nutrient solution.

In this study, regardless of the biostimulant application, the fresh production of carrot (748.5 g fw m^{-2}) and dill (814.7 g fw m^{-2}) microgreens was inconsistent with the results of the most common and used species planted in coco peat and peat-based substrates [26,27]. Ghoora et al. [26] evaluated the production performance of ten microgreen species characterized by a wide range of production (1.12–4.93 kg fw m^{-2}), indicating that the genetic aspect is a strong discriminator for fresh production. In the above study, the minor pro-

ductive species was fennel (*Foeniculum vulgare* L.), belonging to the Apiaceae family. Our research confirms the strong family-dependent productive response for microgreens. Coriander that was assessed in diverse soilless substrates exhibited a similar fresh weight in capillary mat conditions (800 g fw m^{-2}) [6], when compared to carrot and dill from the same botanical family. In the same study by Kyriacou et al. [6], all the tested species manifested around 2.2-fold more fresh weight when a peat-based substrate was adopted compared to the other soilless substrates (capillary mat, coconut fiber, Agave fiber, and cellulose sponge). Furthermore, it should be noted that the production values reported by Ghoora et al. [26] and Kyriacou et al. [27] refer to microgreens harvested at the two true leaf stage, as opposed to our study in which the harvest was carried out at the one true leaf stage. However, the lower yields obtained cannot be solely attributed to the above conditions but also to the different growth conditions under which the microgreens were grown. In fact, all studies reported in the literature evaluated the productivity of microgreens under controlled conditions (growth chambers). In contrast, the greenhouse system used in our study cannot provide stable conditions of temperature, humidity, light intensity, and quality that certainly affect productivity [1,28–30]. The variable and inconstant environmental conditions that occur in the greenhouse could be the key to explaining the high percentage of dry matter compared to the results reported by Kyriacou et al. [27] on 13 different species of microgreens, and by El-Nakhel et al. [31] on parsley (*Petroselinum crispum* (Mill.) Fuss.). Although the achievement of a high fresh yield is a crucial factor for growing microgreens, the absence of a substrate (as in our work) is also an advantage since the substrates ordinarily used are expensive and nonrenewable [8].

Under suboptimal growth conditions, the imperative to maximize production has driven the horticultural industry towards the use of biostimulants [32]. However, there have not been studies in the literature that evaluated the effectiveness of biostimulants in the production and quality performance of microgreens. Our results show that the use of a protein hydrolysate-based biostimulant (Trainer®®) in the nutrient solution increased both the fresh yield and the shoot dry weight in dill (Figure 2A,B). This increase in yield can probably be attributed to the large number of positive effects caused by the bioactive molecules in Trainer®® and not to simple nitrogen and carbon supplementation [13,14,33]. Specifically, this improvement could be related to the presence of peptides eliciting hormone-like signals capable of modulating plant growth and development [34]. The different responses of carrot production, which did not benefit from the application of the biostimulant, highlight that, even in microgreens, the response to a biostimulant is strongly species-dependent [35]. However, it should be considered that species' sensitivity to biostimulant activity is also a function of the mode and application dose [35].

The use of the biostimulant led to a change in colorimetric parameters in both tested species (Table 1), changing the perception of color (chroma), a key aspect for the acceptability and marketability of microgreens [36]. The increase in chroma in carrot and its decrease in dill due to biostimulant application could be related to the change in the anthocyanin content, which showed the same trend as chroma (Table 3). In addition to their role in plant coloration, anthocyanins bring benefits to human health by reducing the risk of chronic diseases, as documented in several clinical studies [37,38].

Interestingly, the biostimulant application in carrot increased total anthocyanins and phenols (Table 2) but did not lead to any increase in yield. It seems that the biostimulant specifically activated the secondary metabolism in carrot, improving its nutraceutical characteristics. In contrast, in dill, the use of Trainer®® promoted the biosynthesis of ascorbic acid, an essential micronutrient for the human body with a strong antioxidant power, and an enhancer of the most effective absorption of nonheme iron [39,40]. The biostimulant application in dill resulted in a higher accumulation of ascorbic acid (148.8 mg 100 g^{-1} fw) than that recorded by Xiao et al. [41] in 25 species of microgreens. Independent of the biostimulatory effect, the ascorbic acid content of dill and carrot was 62% and 1092% higher than their mature edible counterparts, respectively [42]. These results confirm that microgreens provide a source of ascorbic acid at the same level as citrus fruits, potentially

falling under foods with "high vitamin C" according to Annex II of the 2006 EU Regulation 1924/2006 [26,43]. Another key qualitative parameter that defines the healthiness of food is the nitrate content. The recognized detrimental effects on human health combined with the awareness that approximately 80% of total nitrate intake is due to fresh vegetable consumption further emphasize the relevance of this antinutritional compound content [44,45]. As with the mature counterparts, the nitrate levels of microgreens vary greatly among different species, as observed by Kyriacou et al. [27] and Bulgari et al. [8].

Regardless of the application of a biostimulant, carrot (602.7 mg kg^{-1} fw) and dill (252.4 mg kg^{-1} fw) accumulated much lower nitrates than those reported in the literature. Although data on nitrate content are currently still limited [46], especially for species belonging to the Apiaceae family, such low values could be attributable to the different growing system (floating system) that does not involve the use of a substrate, as well as the uncontrolled growth conditions compared to the ones used in comparable experiments [8,27,47]. For dill, the application of the biostimulant increased the nitrate content, but this value (314.5 mg kg^{-1} fw) was approximately half of the lowest value (687.4 mg kg^{-1} fw) recorded by Bulgari et al. [8], emphasizing that food security for this species of microgreen is not undermined at all. The increase in the nitrate content as a result of the application of the biostimulant that was observed only in dill could contribute to a better understanding of the relative increase in yield, which can be related to a probable remodulation of root growth that would have improved mineral uptake and thus nutrient acquisition [48,49].

Indeed, the biostimulant improved the uptake of nitrate, and this was evident in dill in which its content was highly increased, while in carrot, the nitrate levels were already high in the control and remained stable after PH treatment. However, the reductive assimilation of nitrate to ammonia and its incorporation into amino acids were also improved by the biostimulant given the strong increase in asn and proteins in the treated plants. The fact that the content of asparagine in carrot after treatment reached more than a third of the total free amino acid content could be a symptom of excess accumulation of ammonium. When this ion concentration becomes very high in cytosolic compartments because it is not promptly incorporated in amino acids, it has the ability to cross membranes in the neutral form, previously reacting with OH^- ($NH_4^+ + OH^- \rightarrow NH_3 + H_2O$), and then reconverting to NH_4^+ after a reaction with protons (H^+). This determines a strong dissipation of the membrane potential and proton motive force, impairing membrane transport and function [24,50]. Therefore, the synthesis of asn, the amide with the highest N-to-C ratio, may play a role in ammonia detoxification when the uptake of nitrate and its conversion to ammonium exceed the cellular needs, but because it is energetically expensive, this may affect the capacity of the biostimulant to improve growth in carrot.

The biostimulant root application improved the accumulation of calcium in both species, a mineral critical to human skeletal health, since insufficient intake increases the risk of osteoporosis in older age [51]. In agreement with the report of de la Fuente et al. [52], the most abundant element found on average in the two microgreens was potassium, followed by calcium, magnesium, phosphorus, and sulfur, minerals essential in the human diet for their recognized homeostatic and metabolic functions [53]. Although the influence of genotype on mineral concentrations in microgreens is more than established, the lower potassium values found in our experiment are more than established compared to what was recorded by Kyriacou et al. [27], which could help explain the differences in production, considering the critical physiological role played by potassium [54].

Based on these results, lower potassium concentrations would allow microgreen species grown under these specific conditions to be labeled with the nutritional claim of "reduced K" [43] and therefore recommended for patients with impaired kidney function [55]. Furthermore, the low sodium content recorded, especially in carrot (0.50 mg g^{-1}dw), is another critical nutritional aspect, as low-sodium foods reduce the incidence of hypertension and stroke [56]. In most microgreens, the content of starch, as well as soluble sugars (i.e., sucrose, fructose, and glucose), is generally low [8], especially compared to

the typical values of counterpart mature vegetables [57]. In any case, the differences in the carbohydrate profile of the two species of microgreens tested in our work confirm what was previously observed by Paradiso et al. [58], who compared the carbohydrate content of six different genotypes of microgreens. Taking into account the importance of sugars, especially in post-harvest (shelf life), as essential compounds for the maintenance of cellular metabolism, the higher levels of glucose, fructose, and sucrose compared to the average results obtained by Xiao et al. [59] could confer a better shelf life.

5. Conclusions

The remodeling of the nutritional architecture of plant-based foods is a valuable resource in which new categories of functional foods (microgreens) take the lead. In recent decades, the interest in microgreens has increased due to their outstanding nutritional properties. The results achieved prove the feasibility of producing microgreens in soilless systems without any substrate, reducing the waste of nonrenewable resources and the overall cost. The two tested species (dill and carrot) belonging to the Apiaceae family stood out positively for their low nitrate content (average 427.5 mg kg^{-1} fresh weight). The use of a protein hydrolysate (Trainer$^{®®}$) in the nutrient solution led to an increase in anthocyanins (+461.7%) and total phenols (+12.4%) in carrot, while in dill, the fresh yield (+13.5%) and ascorbic acid (+17.2%) increased. In both species, Trainer$^{®®}$ increased soluble proteins and total free amino acids by 20.6% and 18.5%, respectively. In light of the encouraging results achieved using our cultivation system, future research should also investigate the yield and nutritional parameter responses of other microgreen species with the aim of large-scale sustainable production, in addition to depicting the adequate application dose for each species.

Author Contributions: Conceptualization, C.E.-N. and Y.R.; methodology, C.E.-N., M.C., M.G. and L.F.; software, A.P.; validation, C.E.-N., M.C. and L.F.; formal analysis, C.E.-N., M.C., B.R.G. and L.F.; investigation, C.E.-N.; resources, G.M.F. and P.C.; data curation, C.E.-N., G.M.F. and P.C.; writing—original draft preparation, C.E.-N., M.C., L.F. and P.C.; writing—review and editing, C.E.-N., M.C.K., P.C. and Y.R.; visualization, all authors; supervision, Y.R.; project administration, C.E.-N. and Y.R.; funding acquisition, P.C. and Y.R. All authors have read and agreed to the published version of the manuscript.

Funding: This research received no external funding.

Institutional Review Board Statement: Not applicable.

Informed Consent Statement: Not applicable.

Data Availability Statement: The datasets generated for this study are available on request to the corresponding author.

Acknowledgments: The authors would like to thank Luigi Giuseppe Duri and Francesco Cristofano for their technical support during the greenhouse experiment.

Conflicts of Interest: The authors declare no conflict of interest.

References

1. Kyriacou, M.C.; Rouphael, Y.; Di Gioia, F.; Kyratzis, A.; Serio, F.; Renna, M.; De Pascale, S.; Santamaria, P. Micro-scale vegetable production and the rise of microgreens. *Trends Food Sci. Technol.* **2016**, *57*, 103–115. [CrossRef]
2. Caracciolo, F.; El-Nakhel, C.; Raimondo, M.; Kyriacou, M.C.; Cembalo, L.; de Pascale, S.; Rouphael, Y. Sensory attributes and consumer acceptability of 12 microgreens species. *Agronomy* **2020**, *10*, 1043. [CrossRef]
3. Kyriacou, M.C.; El-Nakhel, C.; Pannico, A.; Graziani, G.; Zarrelli, A.; Soteriou, G.A.; Kyratzis, A.; Antoniou, C.; Pizzolongo, F.; Romano, R. Ontogenetic Variation in the Mineral, Phytochemical and Yield Attributes of Brassicaceous Microgreens. *Foods* **2021**, *10*, 1032. [CrossRef] [PubMed]
4. Kyriacou, M.C.; El-Nakhel, C.; Soteriou, G.A.; Graziani, G.; Kyratzis, A.; Antoniou, C.; Ritieni, A.; De Pascale, S.; Rouphael, Y. Preharvest Nutrient Deprivation Reconfigures Nitrate, Mineral, and Phytochemical Content of Microgreens. *Foods* **2021**, *10*, 1333. [CrossRef]
5. Enssle, N. Microgreens: Market Analysis, Growing Methods and Models. 2020. Available online: https://csusm-dspace.calstate.edu/handle/10211.3/217154 (accessed on 4 August 2020).

6.   Kyriacou, M.C.; El-Nakhel, C.; Pannico, A.; Graziani, G.; Soteriou, G.A.; Giordano, M.; Palladino, M.; Ritieni, A.; De Pascale, S.; Rouphael, Y. Phenolic constitution, phytochemical and macronutrient content in three species of microgreens as modulated by natural fiber and synthetic substrates. *Antioxidants* **2020**, *9*, 252. [CrossRef]

7.   Di Gioia, F.; De Bellis, P.; Mininni, C.; Santamaria, P.; Serio, F. Physicochemical, agronomical and microbiological evaluation of alternative growing media for the production of rapini (*Brassica rapa* L.) microgreens. *J. Sci. Food Agric.* **2017**, *97*, 1212–1219. [CrossRef]

8.   Bulgari, R.; Negri, M.; Santoro, P.; Ferrante, A. Quality Evaluation of Indoor-Grown Microgreens Cultivated on Three Different Substrates. *Horticulturae* **2021**, *7*, 96. [CrossRef]

9.   Ciriello, M.; Formisano, L.; Pannico, A.; El-Nakhel, C.; Fascella, G.; Duri, L.G.; Cristofano, F.; Gentile, B.R.; Giordano, M.; Rouphael, Y. Nutrient Solution Deprivation as a Tool to Improve Hydroponics Sustainability: Yield, Physiological, and Qualitative Response of Lettuce. *Agronomy* **2021**, *11*, 1469. [CrossRef]

10. Cristofano, F.; El-Nakhel, C.; Pannico, A.; Giordano, M.; Colla, G.; Rouphael, Y. Foliar and Root Applications of Vegetal-Derived Protein Hydrolysates Differentially Enhance the Yield and Qualitative Attributes of Two Lettuce Cultivars Grown in Floating System. *Agronomy* **2021**, *11*, 1194. [CrossRef]

11. Galieni, A.; Falcinelli, B.; Stagnari, F.; Datti, A.; Benincasa, P. Sprouts and microgreens: Trends, opportunities, and horizons for novel research. *Agronomy* **2020**, *10*, 1424. [CrossRef]

12. Rouphael, Y.; Colla, G. Biostimulants in Agriculture. *Front. Plant Sci.* **2020**, *11*. [CrossRef]

13. Colla, G.; Hoagland, L.; Ruzzi, M.; Cardarelli, M.; Bonini, P.; Canaguier, R.; Rouphael, Y. Biostimulant action of protein hydrolysates: Unraveling their effects on plant physiology and microbiome. *Front. Plant Sci.* **2017**, *8*, 2202. [CrossRef]

14. Ceccarelli, A.V.; Miras-Moreno, B.; Buffagni, V.; Senizza, B.; Pii, Y.; Cardarelli, M.; Rouphael, Y.; Colla, G.; Lucini, L. Foliar application of different vegetal-derived protein hydrolysates distinctively modulates tomato root development and metabolism. *Plants* **2021**, *10*, 326. [CrossRef] [PubMed]

15. Wang, Q.; Kniel, K.E. Survival and Transfer of Murine Norovirus within a Hydroponic System during Kale and Mustard Microgreen Harvesting. *Appl. Environ. Microbiol.* **2016**, *82*, 705–713. [CrossRef] [PubMed]

16. Bulgari, R.; Baldi, A.; Ferrante, A.; Lenzi, A. Yield and quality of basil, Swiss chard, and rocket microgreens grown in a hydroponic system. *N. Z. J. Crop Hortic. Sci.* **2017**, *45*, 119–129. [CrossRef]

17. Puccinelli, M.; Malorgio, F.; Rosellini, I.; Pezzarossa, B. Production of selenium-biofortified microgreens from selenium-enriched seeds of basil. *J. Sci. Food Agric.* **2019**, *99*, 5601–5605. [CrossRef] [PubMed]

18. Formisano, L.; Ciriello, M.; El-Nakhel, C.; De Pascale, S.; Rouphael, Y. Dataset on the Effects of Anti-Insect Nets of Different Porosity on Mineral and Organic Acids Profile of *Cucurbita pepo* L. Fruits and Leaves. *Data* **2021**, *6*, 50. [CrossRef]

19. Singleton, V.L.; Orthofer, R.; Lamuela-Raventós, R.M. Analysis of total phenols and other oxidation substrates and antioxidants by means of folin-ciocalteu reagent. *Methods Enzymol.* **1999**, *299*, 152–178. [CrossRef]

20. Wellburn, A.R. The Spectral Determination of Chlorophylls a and b, as well as Total Carotenoids, Using Various Solvents with Spectrophotometers of Different Resolution. *J. Plant Physiol.* **1994**, *144*, 307–313. [CrossRef]

21. Kampfenkel, K.; Van Montagu, M.; Inzé, D. Extraction and determination of ascorbate and dehydroascorbate from plant tissue. *Anal. Biochem.* **1995**, *225*, 165–167. [CrossRef]

22. Rouphael, Y.; Corrado, G.; Colla, G.; De Pascale, S.; Dell'Aversana, E.; D'Amelia, L.I.; Fusco, G.M.; Carillo, P. Biostimulation as a Means for Optimizing Fruit Phytochemical Content and Functional Quality of Tomato Landraces of the San Marzano Area. *Foods* **2021**, *10*, 926. [CrossRef]

23. Carillo, P.; Kyriacou, M.C.; El-Nakhel, C.; Pannico, A.; dell'Aversana, E.; D'Amelia, L.; Colla, G.; Caruso, G.; De Pascale, S.; Rouphael, Y. Sensory and functional quality characterization of protected designation of origin 'Piennolo del Vesuvio' cherry tomato landraces from Campania-Italy. *Food Chem.* **2019**, *292*, 166–175. [CrossRef]

24. Woodrow, P.; Ciarmiello, L.F.; Annunziata, M.G.; Pacifico, S.; Iannuzzi, F.; Mirto, A.; D'Amelia, L.; Dell'Aversana, E.; Piccolella, S.; Fuggi, A.; et al. Durum wheat seedling responses to simultaneous high light and salinity involve a fine reconfiguration of amino acids and carbohydrate metabolism. *Physiol. Plant.* **2017**, *159*, 290–312. [CrossRef] [PubMed]

25. Ciarmiello, L.F.; Piccirillo, P.; Carillo, P.; De Luca, A.; Woodrow, P. Determination of the genetic relatedness of fig (*Ficus carica* L.) accessions using RAPD fingerprint and their agro-morphological characterization. *South Afr. J. Bot.* **2015**, *97*, 40–47. [CrossRef]

26. Ghoora, M.D.; Babu, D.R.; Srividya, N. Nutrient composition, oxalate content and nutritional ranking of ten culinary microgreens. *J. Food Compos. Anal.* **2020**, *91*. [CrossRef]

27. Kyriacou, M.C.; El-Nakhel, C.; Graziani, G.; Pannico, A.; Soteriou, G.A.; Giordano, M.; Ritieni, A.; De Pascale, S.; Rouphael, Y. Functional quality in novel food sources: Genotypic variation in the nutritive and phytochemical composition of thirteen microgreens species. *Food Chem.* **2019**, *277*, 107–118. [CrossRef]

28. Murphy, C.J.; Llort, K.F.; Pill, W.G. Factors affecting the growth of microgreen table beet. *Int. J. Veg. Sci.* **2010**, *16*, 253–266. [CrossRef]

29. Turner, E.R.; Luo, Y.; Buchanan, R.L. Microgreen nutrition, food safety, and shelf life: A review. *J. Food Sci.* **2020**, *85*, 870–882. [CrossRef] [PubMed]

30. Murphy, C.J.; Pill, W.G. Cultural practices to speed the growth of microgreen arugula (roquette; *Eruca vesicaria* subsp. sativa). *J. Hortic. Sci. Biotechnol.* **2010**, *85*, 171–176. [CrossRef]

31. El-Nakhel, C.; Pannico, A.; Graziani, G.; Giordano, M.; Kyriacou, M.C.; Ritieni, A.; De Pascale, S.; Rouphael, Y. Mineral and Antioxidant Attributes of Petroselinum crispum at Different Stages of Ontogeny: Microgreens vs. Baby Greens. *Agronomy* **2021**, *11*, 857. [CrossRef]
32. Colla, G.; Rouphael, Y. Biostimulants in horticulture. *Sci. Hortic.* **2015**, *196*, 1–2. [CrossRef]
33. Lucini, L.; Miras-Moreno, B.; Rouphael, Y.; Cardarelli, M.; Colla, G. Combining Molecular Weight Fractionation and Metabolomics to Elucidate the Bioactivity of Vegetal Protein Hydrolysates in Tomato Plants. *Front. Plant Sci.* **2020**, *11*. [CrossRef] [PubMed]
34. Moreno-Hernández, J.M.; Benítez-García, I.; Mazorra-Manzano, M.A.; Ramírez-Suárez, J.C.; Sánchez, E. Strategies for production, characterization and application of protein-based biostimulants in agriculture: A review. *Chil. J. Agric. Res.* **2020**, *80*, 274–289. [CrossRef]
35. Colla, G.; Nardi, S.; Cardarelli, M.; Ertani, A.; Lucini, L.; Canaguier, R.; Rouphael, Y. Protein hydrolysates as biostimulants in horticulture. *Sci. Hortic.* **2015**, *196*, 28–38. [CrossRef]
36. Barrett, D.M.; Beaulieu, J.C.; Shewfelt, R. Color, flavor, texture, and nutritional quality of fresh-cut fruits and vegetables: Desirable levels, instrumental and sensory measurement, and the effects of processing. *Crit. Rev. Food Sci. Nutr.* **2010**, *50*, 369–389. [CrossRef]
37. D'Cunha, N.M.; Georgousopoulou, E.N.; Dadigamuwage, L.; Kellett, J.; Panagiotakos, D.B.; Thomas, J.; McKune, A.J.; Mellor, D.D.; Naumovski, N. Effect of long-term nutraceutical and dietary supplement use on cognition in the elderly: A 10-year systematic review of randomised controlled trials. *Br. J. Nutr.* **2018**, *119*, 280–298. [CrossRef]
38. Khoo, H.E.; Azlan, A.; Tang, S.T.; Lim, S.M. Anthocyanidins and anthocyanins: Colored pigments as food, pharmaceutical ingredients, and the potential health benefits. *Food Nutr. Res.* **2017**, *61*, 1361779. [CrossRef]
39. Padayatty, S.J.; Katz, A.; Wang, Y.; Eck, P.; Kwon, O.; Lee, J.-H.; Chen, S.; Corpe, C.; Dutta, A.; Dutta, S.K. Vitamin C as an antioxidant: Evaluation of its role in disease prevention. *J. Am. Coll. Nutr.* **2003**, *22*, 18–35. [CrossRef] [PubMed]
40. Lopez, H.W.; Leenhardt, F.; Coudray, C.; Remesy, C. Minerals and phytic acid interactions: Is it a real problem for human nutrition? *Int. J. food Sci. Technol.* **2002**, *37*, 727–739. [CrossRef]
41. Xiao, Z.; Lester, G.E.; Luo, Y.; Wang, Q. Assessment of Vitamin and Carotenoid Concentrations of Emerging Food Products: Edible Microgreens. *J. Agric. Food Chem.* **2012**, *60*, 7644–7651. [CrossRef] [PubMed]
42. USDA (United State Departement of Agriculture). Available online: https://fdc.nal.usda.gov/ (accessed on 30 July 2021).
43. Regulation (EC) No. 1924/2006 of the European Parliament and of the Council of 20 December 2016 on nutritional and health claims made on foods. *Off. J. Eur. Union* **2006**, *49*, 9–25.
44. Hord, N.G.; Tang, Y.; Bryan, N.S. Food sources of nitrates and nitrites: The physiologic context for potential health benefits. *Am. J. Clin. Nutr.* **2009**, *90*, 1–10. [CrossRef]
45. Colla, G.; Kim, H.J.; Kyriacou, M.C.; Rouphael, Y. Nitrate in fruits and vegetables. *Sci. Hortic.* **2018**, *237*, 221–238. [CrossRef]
46. Xiao, Z.; Codling, E.E.; Luo, Y.; Nou, X.; Lester, G.E.; Wang, Q. Microgreens of Brassicaceae: Mineral composition and content of 30 varieties. *J. Food Compos. Anal.* **2016**, *49*, 87–93. [CrossRef]
47. Kyriacou, M.C.; El-Nakhel, C.; Pannico, A.; Graziani, G.; Soteriou, G.A.; Giordano, M.; Zarrelli, A.; Ritieni, A.; De Pascale, S.; Rouphael, Y. Genotype-Specific Modulatory Effects of Select Spectral Bandwidths on the Nutritive and Phytochemical Composition of Microgreens. *Front. Plant Sci.* **2019**, *10*. [CrossRef] [PubMed]
48. Colla, G.; Rouphael, Y.; Canaguier, R.; Svecova, E.; Cardarelli, M. Biostimulant action of a plant-derived protein hydrolysate produced through enzymatic hydrolysis. *Front. Plant Sci.* **2014**, *5*, 448. [CrossRef]
49. Rouphael, Y.; Colla, G.; Giordano, M.; El-Nakhel, C.; Kyriacou, M.C.; De Pascale, S. Foliar applications of a legume-derived protein hydrolysate elicit dose-dependent increases of growth, leaf mineral composition, yield and fruit quality in two greenhouse tomato cultivars. *Sci. Hortic.* **2017**, *226*, 353–360. [CrossRef]
50. Howitt, S.M.; Udvardi, M.K. Structure, function and regulation of ammonium transporters in plants. *Biochim. Biophys. Acta Biomembr.* **2000**, *1465*, 152–170. [CrossRef]
51. Theobald, H.E. *Dietary Calcium and Health*; Wiley Online Library: Hoboken, NJ, USA, 2005; Volume 30.
52. De la Fuente, B.; López-García, G.; Máñez, V.; Alegría, A.; Barberá, R.; Cilla, A. Evaluation of the bioaccessibility of antioxidant bioactive compounds and minerals of four genotypes of Brassicaceae microgreens. *Foods* **2019**, *8*, 250. [CrossRef]
53. Gharibzahedi, S.M.T.; Jafari, S.M. The importance of minerals in human nutrition: Bioavailability, food fortification, processing effects and nanoencapsulation. *Trends Food Sci. Technol.* **2017**, *62*, 119–132. [CrossRef]
54. Tränkner, M.; Tavakol, E.; Jákli, B. Functioning of potassium and magnesium in photosynthesis, photosynthate translocation and photoprotection. *Physiol. Plant.* **2018**, *163*, 414–431. [CrossRef] [PubMed]
55. Renna, M.; Castellino, M.; Leoni, B.; Paradiso, V.M.; Santamaria, P. Microgreens production with low potassium content for patients with impaired kidney function. *Nutrients* **2018**, *10*, 675. [CrossRef] [PubMed]
56. Choi, M.; Scholl, U.I.; Yue, P.; Björklund, P.; Zhao, B.; Nelson-Williams, C.; Ji, W.; Cho, Y.; Patel, A.; Men, C.J.; et al. K+ channel mutations in adrenal aldosterone-producing adenomas and hereditary hypertension. *Science* **2011**, *331*, 768–772. [CrossRef]
57. Cavaiuolo, M.; Cocetta, G.; Bulgari, R.; Spinardi, A.; Ferrante, A. Identification of innovative potential quality markers in rocket and melon fresh-cut produce. *Food Chem.* **2015**, *188*, 225–233. [CrossRef] [PubMed]
58. Paradiso, V.M.; Castellino, M.; Renna, M.; Gattullo, C.E.; Calasso, M.; Terzano, R.; Allegretta, I.; Leoni, B.; Caponio, F.; Santamaria, P. Nutritional characterization and shelf-life of packaged microgreens. *Food Funct.* **2018**, *9*, 5629–5640. [CrossRef] [PubMed]
59. Xiao, Z.; Lester, G.E.; Park, E.; Saftner, R.A.; Luo, Y.; Wang, Q. Evaluation and correlation of sensory attributes and chemical compositions of emerging fresh produce: Microgreens. *Postharvest Biol. Technol.* **2015**, *110*, 140–148. [CrossRef]

Article

Nutritional Value of *Apiaceae* Seeds as Affected by 11 Species and 43 Cultivars

Nadezhda Golubkina [1,*], Viktor Kharchenko [1], Anastasia Moldovan [1], Vladimir Zayachkovsky [1], Viktor Stepanov [1], Viktor Pivovarov [1], Agnieszka Sekara [2], Alessio Tallarita [3] and Gianluca Caruso [3]

[1]   Federal Scientific Center of Vegetable Production, Selectsionnaya 14, VNIISSOK, Odintsovo District, 143072 Moscow, Russia; kharchenkoviktor777@gmail.com (V.K.); nastiamoldovan@mail.ru (A.M.); vladimir89854217114@mail.ru (V.Z.); vstepanov8848@mail.ru (V.S.); pivovarov@vniissok.ru (V.P.)

[2]   Department of Horticulture, Faculty of Biotechnology and Horticulture, University of Agriculture, 31-120 Krakow, Poland; agnieszka.sekara@urk.edu.pl

[3]   Department of Agricultural Sciences, University of Naples Federico II, Naples, 80055 Portici, Italy; lexvincentall@gmail.com (A.T.); gcaruso@unina.it (G.C.)

*   Correspondence: segolubkina45@gmail.com

Abstract: The fragmentary literature data on *Apiaceae* seed antioxidant potential elicited a comparative evaluation work of seed biochemical profile between 11 species and 43 cultivars grown in similar conditions: anise, lovage, fennel, coriander, caraway, parsley, celery, dill, carrot, parsnip and chervil. Among the different solvents, temperature and duration regimes applied, 70% EtOH, 80 °C and 1 h running showed the best extraction efficiency of antioxidants. Total antioxidant activity (AOA) decreased as follows: lovage > anise > parsley > celery > fennel = dill > coriander > caraway > parsnip > carrot > chervil. Lovage, anise and fennel demonstrated the highest levels of total phenolics (TP), AOA and potassium. A positive correlation was recorded between total dissolved solids (TDS) and K and between AOA and TP content (r = 0.86 and r = 0.79 respectively, at $p < 0.001$). Varietal differences in AOA and TP levels were much lower than those relevant to TDS, K and water soluble protein (WSP), while the highest differences were found for selenium (Se). Two parsley cultivars showed anomalously high Se content and four dill cultivars unusually high levels of TDS and potassium. A positive correlation arose between Se and WSP levels in parsley seeds (r = 0.85 at $p < 0.05$).

Keywords: *Apiaceae*; seeds; antioxidants; potassium; total dissolved solids; protein

Citation: Golubkina, N.; Kharchenko, V.; Moldovan, A.; Zayachkovsky, V.; Stepanov, V.; Pivovarov, V.; Sekara, A.; Tallarita, A.; Caruso, G. Nutritional Value of *Apiaceae* Seeds as Affected by 11 Species and 43 Cultivars. *Horticulturae* **2021**, *7*, 57. https://doi.org/10.3390/horticulturae7030057

Academic Editor: Sergio Ruffo Roberto

Received: 1 March 2021
Accepted: 19 March 2021
Published: 21 March 2021

Publisher's Note: MDPI stays neutral with regard to jurisdictional claims in published maps and institutional affiliations.

1. Introduction

Modern nutritiology is characterized by the significant expansion of food products and biologically active food supplements, based on substances produced at all stages of herbs' growth and development: seeds [1], sprouts [2,3], microgreens [4], all parts and tissues of mature plants, including agricultural wastes [5]. Among seeds of different crops, *Apiaceae* seeds are of special interest due to their wide utilization as spices containing high levels of essential oil [6] and their high value in traditional medicine [7].

The *Apiaceae* family includes more than 3500 species, among which celery, parsley, dill, coriander, carrot, parsnip, fennel, anise, caraway, lovage and chervil are the most common. Interestingly, all parts of the above *Apiaceae* representatives are edible and demonstrate high biological activity [1,6]. High levels of antioxidants in *Apiaceae* plants normalize digestion, express powerful antibiotic properties [7], increase immunity and have anticarcinogenic, cardioprotective and hypolipidemic effects. Celery and lovage seeds normalize spermatogenesis, while seeds of caraway, coriander, carrot and anise promote insulin production [8]. Carrot seed extracts possess cardio- and hepatoprotective effect, normalize cognitive function, decrease cholesterol level and show anti-bacterial, anti-fungal, anti-inflammatory, and analgesic effect and promote wound healing [9].

With the exception of the essential oil components of *Apiaceae* seeds, seed antioxidant system has been studied rather fragmentarily up to date. In this respect, the difficulty lies in the utilization need of seeds from numerous *Apiaceae* crops grown in similar conditions, as environment is known to cause a dominant effect on seed quality, and in particular on antioxidant status, protein and mineral content [1,10]. Previous studies of *Apiaceae* seed antioxidant status were carried out by: Christova-Bagdassarian et al. [11] on antioxidant status of fennel, anise, dill, coriander and caraway; Martins et al. [12] on phenolic content and antioxidant activity of anise and coriander; Wangensteen et al. [13] on phenolic content in coriander seeds; Marques and Farah [14] on anise phenolic profile. Unfortunately, all these works dealt only with a few representatives of *Apiaceae* family or used samples gathered from different geographical areas [12]. In the latter respect, the efficiency of seed utilization in medicine and as a functional food is determined to a large extent not only by species but also by cultivar peculiarities in accumulating antioxidants, proteins and minerals. Unfortunately, up to date no evaluation of antioxidant status differences between *Apiaceae* cultivars seed has been carried out, contrary to deep investigations of the mature plant antioxidant status [15] and the relationship between *Apiaceae* seed quality and N, P K content [16]. Taking into account the scant information available in the literature, the present investigation aimed to comparatively evaluate species and cultivar peculiarities in accumulating antioxidants and other biologically active compounds in seeds of 11 *Apiaceae* species, including 43 cultivars.

2. Materials and Methods

2.1. Experimental Protocol and Growing Conditions

A research was carried out in 2018 and 2019 at the experimental fields of Federal Scientific Center of Vegetable Production (Moscow region, Russia, 55°39.51′ N, 37°12.23′ E) for obtaining seeds from 11 *Apiaceae* species and 43 cultivars of the mentioned scientific Center selection. All the plants were grown in similar conditions, in order to minimize the effect of biotic and abiotic factors and to obtain reliable significance of species and varietal nutritional peculiarities.

Plants were grown on a sod-podzolic clay-loam soil with pH 6.8, 2.1% organic matter, 1.1 g kg^{-1} N, 0.045 g kg^{-1} P_2O_5, 0.357 g kg^{-1} K_2O. *Apiaceae* collection included perennial/biennial and annual plants. The first group was represented by the following cultivars: Udalets and Maslichny, of fennel (*Foeniculum vulgare* Mill.); Lider, of lovage (*Levisticum officinale Koch.*); Elixir, Samurai, Atlant, Gribovsky, Zakhar, Egor, Dobrynya, Judinka, of leafy, stem and root celery (*Apium graveolens* L.); Sakharnaya, Zolushka, Nezhnost, Moskvichka, Breeze, Krasotka, of parsley (*Petroselinum crispum* L.); Krugly, Zhemchug, Bely aist, of parsnip (*Pastinaca sativa* L.); and Moskovskaya zimnya, F_1, Nadezhda, Minor, Nantskaya 11, Riff F_1, Marlinka, Shantane, of carrot (*Daucus carota* subsp. sativus). The second group included the following cultivars: Gribovsky, Alligator, Rusich, Zontik, Spartak, Kibrai, Lesnogorodsky, Kulinar, Salut, of dill (*Anethum graveolens* L.); Stimul and prospect specimen, of coriander (*Coriandrum sativum* L.); Vitiaz, of anise (*Pimpinella anisum* L.); Peresvet, of caraway (*Carum carvi* L.); and prospect specimen of chervil (*Anthriscus cerefolium*–L.–Hoffm.) (21-20; 22-20; 24-20).

Seed harvesting was carried out manually at different dates depending on the species and cultivar peculiarities: on 15 to 31 August for carrot and parsnip; at the dates reported in Table S1 for the other species and cultivars. All harvested seeds were ripened in a stem dryer and kept at +10 °C in refrigerator prior to analysis.

As the parameters analyzed may greatly depend on meteorological conditions, seed position inside the umbrella [16], plant density and agrochemicals used [17], a two-year experiment (2018–2019) was carried out using the same technical management for plant cultivation, seed collection and storage [18].

Mean values of temperature (°C) and rainfall in 2018 and 2019 are presented in Table 1.

Table 1. Mean temperature and rainfall in 2018–2019.

Month	2018		2019	
	Mean Temperature (°C)	Rainfall (mm)	Mean Temperature (°C)	Rainfall (mm)
May	16.2	61	16.3	57
June	17.3	56	19.6	64
July	20.5	92	16.8	69
August	19.8	28	16.4	57

To eliminate the influence of environmental conditions on the results, equal amounts (10 g) of seeds of 2018 and 2019 harvest were weighed, mixed and used for the analysis, prior to which the seeds were dried to constant weight at 70 °C and homogenized using a Kenwood dough mixer (Model A 907 D).

2.2. Total Dissolved Solids (TDS)

TDS were determined in water extracts using TDS-3 conductometer (HM Digital, Inc., Seoul, Korea). About half a g of seed powder homogenate was ground in a mortar with 50 mL distilled water and left at room temperature for 1 h. The concentration of dissolved solids determined in water solution by conductometer was re-calculated per kg d.w. using the formula:

$$\text{TDS (mg·kg}^{-1}\text{ d.w.)} = A \times 50{:}a,$$

where A is the conductometer reading in mg L^{-1};

50: extract volume (mL);
a: seed powder weight (g).

2.3. Total Polyphenols (TP)

Total polyphenols were determined in 70% ethanol extract using the Folin–Ciocalteu colorimetric method as previously described [19]. One gram of dry seed homogenates was extracted with 20 mL of 70% ethanol at 80 °C for 1 h. The mixture was cooled down and quantitatively transferred to a volumetric flask, and the volume was adjusted to 25 mL. The mixture was filtered through filter paper, and 1 mL of the resulting solution was transferred to a 25 mL volumetric flask, to which 2.5 mL of saturated Na_2CO_3 solution and 0.25 mL of diluted (1:1) Folin–Ciocalteu reagent were added. The volume was brought to 25 mL with distilled water. One hour later the solutions were analyzed through a spectrophotometer (Unico 2804 UV, Suite E Dayton, NJ, USA), and the concentration of polyphenols was calculated according to the absorption of the reaction mixture at 730 nm. As an external standard, 0.02% gallic acid was used.

2.4. Antioxidant Activity (AOA)

The antioxidant activity of seeds was assessed using a redox titration method [19] via titration of 0.01 N $KMnO_4$ solution with ethanolic extracts of dry samples, produced as described in the Section 2.3. The reduction of $KMnO_4$ to colorless Mn^{+2} in this process reflects the quantity of antioxidants dissolvable in 70% ethanol. The values were expressed in mg gallic acid equivalents (GAE) g^{-1} d.w.

2.5. Selenium

Selenium was analyzed using the fluorimetric method previously described for tissues and biological fluids [20]. About 0.1 g of dried homogenized samples were digested via sequential heating with a mixture of 1.5 mL nitric-chloral acids (10:7) at 120 °C (1 h), 150 °C (1 h) and 180 °C (1 h). To eliminate traces of nitric acid, the samples were heated during 10 min at 150 °C with 2 drops of 30% H_2O_2. Subsequent reduction of selenate (Se^{+6}) to selenite (Se^{+4}) was achieved via heating of samples with 1 mL solution of 6 N HCl at 120 °C during 10 min. The formation of a complex between Se^{+4} and 2,3-diaminonaphtalene

(DAN) was elicited at 53 °C (30 min) using 1 mg mL^{-1} solution of DAN in 1% HCl. After cooling, the obtained piazoselenol solution was extracted with 3 mL of hexane and the extracts were subjected to fluorescence analysis at 519 nm λ emission and 376 nm λ excitation (Fluorimeter 02-4M, Lumex marketing, St. Petersburg, Russia). Each determination was done in triplicate. The precision of the results was verified using a reference standard–lyophilized cabbage in each determination with Se concentration of 150 µg·kg^{-1} d.w.

2.6. Water Soluble Protein

Water soluble protein levels were detected spectrophotometrically using the Bradford method based on utilization of Coomassie Brilliant Blue 250 and 0.05 M Tris buffer, at pH 8 [21]. Half a g of homogenized seed powder was carefully ground in a mortar with 15 mL of freshly prepared Tris buffer and left at room temperature for phases separation (about 1 h). One hundred µl of the resulting supernatant was mixed with 0.9 mL of Tris buffer and 3 mL of Coomassie reagent and the reaction mixtures were subjected to spectrophotometer for the determination of absorption value at 595 nm. Inner standard–bovine albumin (Sigma).

2.7. Potassium (K)

Potassium content was determined by an ionomer Expert-001 (Econix, Russia), using ion selective electrode on water extracts of seeds (1 g of seed homogenate per 50 mL of distilled water).

2.8. Statistical Analysis

Data were processed by analysis of variance and mean separations were performed through the Duncan multiple range test, with reference to 0.05 probability level, using SPSS software version 21. Data expressed as percentage were subjected to angular transformation before processing.

3. Results and Discussion

3.1. Total Dissolved Solids (TDS), Potassium (K) and Water Soluble Protein Content (WSP)

Water extracts of seeds may provide an important information about seed quality, in particular potassium and soluble protein content. Highly useful integral indicator, the so called TDS, may be successfully applied for related characteristics both of seeds and plant parts [22].

Despite significant species differences in TDS, K and WSP levels, the mean values of these parameters were similar in seeds of both annual and biennial/perennial *Apiaceae* plants (Tables 2 and 3).

The highest seed TDS levels were recorded in fennel, lovage and anise and the lowest in carrot and chervil. The analysis of TDS and K variations allowed to identify several dill cultivars (Aligator, Spartak, Kibray, Salut) with anomalously high levels of the above parameters: TDS from 32.3 to 38.8 g kg^{-1} d.w. and K levels from 20.4 to 25.2 kg^{-1} d.w., whereas the celery cultivar Atlant (stem form) demonstrated unusually high levels of TDS (46.2 g kg^{-1} d.w) and potassium content (20.5 g kg^{-1} d.w). Overall, TDS and K content of Apiaceae seeds showed a significant positive correlation (r = 0.86 at $p < 0.001$; Figure 1), which suggests that K derivatives are the main components of *Apiaceae* seed water soluble compounds accounting for about 50% of the TDS. Furthermore, taking into account the simplicity of TDS determination the latter can be recommended as a fast method for implementation of potassium content comparative evaluation in seeds of agricultural crops.

Table 2. Weight, total dissolved solids (TDS), potassium and water-soluble protein (WSP) content in seeds of perennial and biennial *Apiaceae* plants.

Species	Cultivar	TDS (g kg^{-1} d.w.)	K (g kg^{-1} d.w.)	WSP (%)	Weight of 1000 Seeds (g)
Fennel	Udalets	45.45 [a]	19.28 [a]	4.1 [b]	3.8 [b]
	№ 15-07	37.20 [b]	15.93 [b]	6.4 [a]	6.1 [a]
Lovage	Leader	36.90 [b]	14.17 [b]	6.1 [a]	3.0 [c]
Leafy celery	Elixir	21.90 [d]	11.28 [c]	2.27 [c,d]	0.5 [j]
	Samurai	23.75 [d,e]	11.27 [c]	2.53 [c]	0.6 [j]
	Zakhar	21.25 [d]	11.50 [c]	2.32 [c,d]	0.4 [k]
Stem celery	Atlant	46.15 [a]	20.48 [a]	2.0 [d,e]	0.5 [j]
	Gribovsky	33.00 [b,c]	17.86 [a]	1.19 [g]	0.4 [k]
Root celery	Egor	28.85 [c]	13.04 [b,c]	1.59 [f]	0.5 [j]
	Dobrynya	27.05 [c,e]	12.69 [b,c]	1.59 [f]	0.6 [j]
	Judinka	33.10 [b,c]	19.71 [a]	1.0 [g]	0.4 [k]
Leafy parsley	Nezhnost	32.15 [b]	20.62 [a]	1.56 [f]	1.9 [d]
	Moskvichka	18.20 [f]	9.20 [d]	1.75 [e,f]	1.8 [d,e]
Curly parsley	Breeze	28.35 [c]	14.00 [b]	1.81 [e,f]	1.6 [e]
	Krasotka	29.90 [c]	20.64 [a]	1.7 [e,f]	1.3 [f]
Root parsley	Sakharnaya	32.55 [b,c]	17.73 [a]	1.7 [e,f]	1.2 [f,g]
	Zolushka	18.95 [f]	9.06 [d]	1.77 [e]	1.3 [f]
Parsnip	Krugly	21.05 [d]	14.33 [b]	1.72 [e,f]	6.5 [a]
	Bely aist	22.20 [d]	13.09 [b,c]	1.8 [e]	7.1 [a]
	Zhemchug	22.70 [d]	13.30 [b,c]	1.96 [d,e]	6.4 [a]
Carrot	Moskovskaya zimnya	19.40 [f]	5.60 [e]	1.2 [g]	1.0 [g]
	F1 Nadezhda	15.15 [g]	8.16 [d]	1.18 [g]	1.2 [f,g]
	Minor	16.25	8.95 [d]	1.2 [g]	1.6 [e]
	Nantskaya-11	14.50 [g,h]	9.56 [d]	1.28 [g]	1.3 [f]
	F1 Riff	12.45 [h]	6.91 [d]	1.25 [g]	2.1 [d]
	Marlinka	12.40 [h]	6.93 [d]	1.2 [g]	1.6 [e]
	Shantane	18.75 [f]	9.27 [d]	1.2 [g]	1.3 [f]
	CV (%)	30.2	32.0	40.0	71.4

CV: coefficient of variation. Within each column, values with the same letters do not differ significantly according to Duncan test at $p < 0.05$.

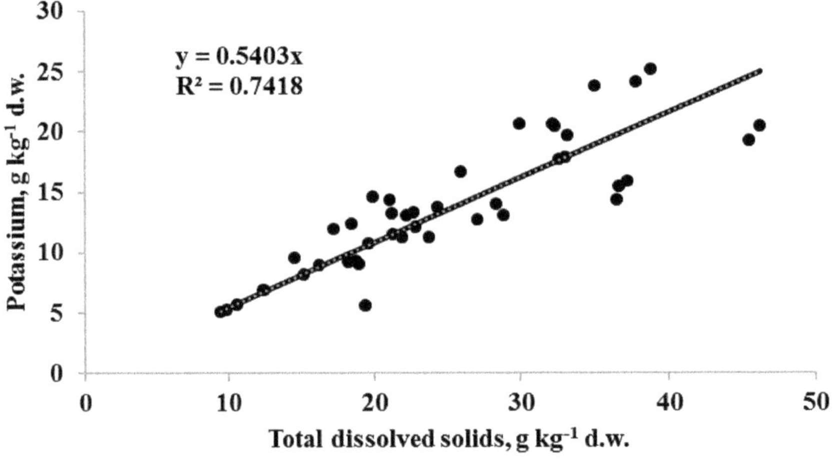

Figure 1. Correlation between potassium and total dissolved solids (TDS) content in *Apiaceae* seeds. (r = 0.86 at $p < 0.001$; $n = 43$).

Table 3. Weight, total dissolved solids (TDS), potassium and water-soluble protein (WSP) content in seeds of annual *Apiaceae* plants.

Species	Cultivar	TDS (g kg^{-1} d.w.)	K (g kg^{-1} d.w.)	WSP (%)	Weight of 1000 Seeds (g)
Anise	Vityaz	36.65 [a]	15.47 [d,e]	6.3 [a]	2.4 [b]
Dill	Alligator	37.75 [a]	24.17 [a]	1.1 [e]	1.4 [e,f]
	Spartak	38.80 [a]	25.20 [a]	1.06 [e]	1.6 [d,e]
	Zontik	21.15 [b,c]	13.23 [e]	1.43 [c,d]	1.2 [f,g]
	Kibray	32.25 [a]	20.44 [b,c]	1.08 [e]	1.3 [e,f]
	Lesnogorodsky	24.30 [b]	13.77 [e]	1.23 [d]	1.8 [c,d]
	Salut	35.00 [a]	23.83 [a,b]	1.43 [c,d]	1.1 [g]
	Culinar	22.80 [b,c]	12.10 [e]	1.14 [e]	2.1 [b]
	Gribovsky	19.60 [c,d]	10.77 [e]	1.06 [e]	2.0 [b,c]
	Rusich	18.45 [d]	12.39 [e]	1.48 [c]	1.5 [e]
Coriander	Stimul	17.25 [d]	11.98 [e]	5.3 [a]	5.2 [a]
	№ 07-19	25.90 [b]	16.68 [c,d]	5.8 [a]	5.4 [a]
Caraway	Peresvet	19.85 [c,d]	14.59 [d,e]	4.2 [b]	2.2 [b]
Chervil	№21-20	9.85 [e]	5.30 [f]	1.66 [c]	1.8 [c,d]
	№22-20	10.60 [e]	5.70 [f]	1.49 [c]	1.7 [c,d]
	№24-20	9.45 [e]	5.09 [f]	2.30 [b]	2.1 [c]
	CV %	31.6	31.3	66.7	40.9

CV: coefficient of variation. Within each column, values with the same letters do not differ significantly according to Duncan test at $p < 0.05$.

The highest differences in seed TDS and K levels were recorded between dill, celery and parsley cultivars, while seeds of 8 celery cultivars demonstrated the highest variations in protein content (Tables 2 and 3). The latter phenomenon is connected with morphological differences between leafy, stalk and root celery forms, with the highest seed WSP levels in leafy forms (Tables 2 and 3). Notably, carrot, parsley and parsnip seeds showed the lowest variability in WSP content between cultivars.

The content of water soluble proteins (WSP) in seeds is considered an important aspect of N, C and S supply, as these compounds are known to demonstrate enhanced biological activity and bioavailability [23]. Their concentration is an important parameter of seed germination [24], and their content increase in growing plants reflects the stress degree. Tables 2 and 3 data indicate that water soluble protein levels in anise, lovage, fennel and coriander thrice exceeded the mean level recorded in all *Apiaceae* seeds studied, whereas the WSP median reached 1.46 % value both in seeds of annual and biennial/perennial plants.

Both annual and perennial/biennial *Apiaceae* plants were characterized by high variations in seed weight (Tables 2 and 3). Among the agricultural crops investigated, the highest 1000 seeds weight was recorded in coriander, fennel and parsnip (5.2 to 7.1 g), and the lowest in celery (0.4 to 0.6 mg). These peculiarities are not related to WSP, K content and TDS values, whereas WSP concentration did not show a significant correlation with the above mentioned parameters.

3.2. Antioxidants

Antioxidant status of agricultural crops constantly attracts research attention in connection with their importance for the pharmacological value of the different plant parts [25]. According to the literature reports, *Apiaceae* seeds are rich not only in essential oils but also in antioxidants such as phenolics. Though most of the results have been obtained from studies on *Apiaceae* leaves, some investigations carried out on seeds have identified caffeic acid in dill, fennel and celery, cumaric acid in coriander and celery, resveratrol in parsley and dill and rutin in anise, caraway, fennel and dill (Table 4). A research of Faudale et al. [26] revealed great fluctuations both of the total phenolics content (TP) in fennel leaves and seeds and of polyphenol composition. The data presented in Table 3

indicate that TP content in *Apiaceae* seeds showed a wide concentration range from 5–7 to 116 mg GAE g^{-1} d.w., while a significantly narrower range was recorded in the present work (4.5–16.1 mg GAE g^{-1} d.w.) (Tables 4 and 5). The phenomenon may be connected with many factors: geographical and cultivar peculiarities, stress factors, different methods of extraction and calculation (per g of seeds or extracts d.w.). Indeed, Tomsone and Kruma [27] indicated that the method of TP extraction may result in values differing from each other by 100% in lovage seeds (Table 4). Similar results were obtained by Uddin et al. [28] in celery seeds, where TP content reached 63.46 mg GAE g^{-1} d.w. in MeOH and only 36.6 mg GAE g^{-1} d.w. in EtOH extract.

Table 4. Literature data relevant to polyphenol composition and content in seeds of *Apiaceae* plants.

Species	Polyphenol Composition	Seed TP **	Extraction Conditions	References
Lovage	No data available	5.68–10.43	95% EtOH	[27]
Fennel *	Caffeic acid and quercetin derivatives, rosmarinic acid		80% EtOH sonication	[26]
Anise		46.17		
Caraway	Rutin, tannin	25.96	80% MeOH	[11,29]
Fennel		115.96		
Coriander		17.04		
	p-hydroxybenzoic acid, cumarin; p-cumaric acid	15.55	MeOH	[30]
Caraway		3.99	50% EtOH + 0.1% formic acid	[31]
Dill		0.94		
Celery	Gallic, Caffeic, Trans-ferulic, o-cumaric acids	-	-	[32]
	caffeic acid, p-coumaric acid, ferulic acid; apigenin, luteolin, and kaempferol.	63.46–36.60	MeOH	[28]
Parsley	Resveratrol, pyrogallol, salicylic acids, benzoic acid, naringin	91.29	MeOH	[33]
	No data	67.25	MeOH	[34]
Dill	cafeic acid, epicatechin, resveratrol, rutin, quercetin, kaempherol	26.41	EtOH, 60 min	[35]
Carrot	No data available	7.08	MeOH	[36]
Chervil		No data available		
Parsnip				

* leaves data; ** TP, total phenolics in mg GAE g^{-1} d.w.

Table 5. Seed total phenolics (TP), Se content and total antioxidant activity (AOA) of perennial and biennial *Apiaceae* plants.

Species	Cultivar	AOA mg GAE g^{-1} d.w.	TP mg GAE g^{-1} d.w.	Se µg kg^{-1} d.w.
Lovage	Leader	61.2 [a]	16.1 [a]	26 [g]
Leafy parsley	Nezhnost	54.1 [a,b]	10.4 [e]	23 [g]
	Mokvichka	53.1 [a,b]	12.5 [c,d]	142 [a]
	Breeze	55.3 [a,b]	10.3 [e]	137 [a]
Curley parsley	Krasotka	48.2 [b]	9.4 [e]	23 [g]
Root parsley	Sakharnaya	51.2 [a,b]	10.0 [e]	22 [g]
	Zolushka	46.5 [b,c]	10.3 [e]	89 [b]

Table 5. *Cont.*

Species	Cultivar	AOA mg GAE g^{-1} d.w.	TP mg GAE g^{-1} d.w.	Se μg kg^{-1} d.w.
Leafy celery	Elixir	37.5 [d,e]	12.2 [c,d]	26 [g]
	Samurai	37.5 [d,e]	10.4 [e]	28 [g]
Stem celery	Zakhar	36.2 [d,e]	12.9 [b,c]	25 [g]
	Atlant	41.5 [c,d]	13.0 [b,c]	38 [d]
Root celery	Gribovsky	40.0 [c,d]	13.0 [b,c]	10 [j]
	Egor	42.5 [c,d]	15.1 [a,b]	30 [g]
	Dobrynya	40.9 [c,d]	13.8 [a,b]	25 [g]
	Judinka	40.0 [c,d]	11.8 [c,d,e]	13 [j]
Fennel	Udalets	32.4 [e]	12.2 [c,d]	10 [j]
	№ 15-07	36.7 [d,e]	14.3 [b]	49 [c,e]
Parsnip	Krugly	17.9 [f,g]	9.8 [e]	37 [e]
	Bely aist	19.0 [f]	11.3 [d,e]	51 [c]
	Zhemchug	17.9 [f,g]	9.5 [e]	31 [f,g]
Carrot	Moskovskaya zimnya	15.6 [g,h]	9.2 [e,g]	56 [c]
	F1 Nadezhda	13.4 [h,j]	7.6 [f,g]	40 [d,e]
	Minor	15.8 [g,h]	10.3 [e]	34 [f]
	Nantskaya-11	14.1 [h]	7.5 [f]	33 [f]
	F1 Riff	12.5 [j]	6.5 [f]	54 [c]
	Marlinka	13.0 [j]	7.8 [f,g]	38 [d]
	Shantane	12.5 [j]	7.1 [f]	42 [d,e]
	CV %	40.8	18.3	50.4

CV: coefficient of variation. Within each column, values with the same letters do not differ significantly according to Duncan test at $p < 0.05$.

In the latter respect, selection of appropriate solvent and temperature regime for the extraction of antioxidants seems to be decisive. We achieved appropriate comparisons with variations of temperature, solvent and duration of extraction on seeds of three *Apiaceae* species: lovage, showing the highest AOA and essential oil content, chervil with the lowest level of AOA and low level of essential oil content after seeds drying, and coriander occupying an intermediate position. Data presented in Figure 2 indicate that changes in AOA and TP displayed similar trends. In fact, comparison of AOA and TP values in seeds obtained in different extraction conditions demonstrated the least efficiency of EtOH both at ambient and high temperature. In all cases, dilution of alcohol with water significantly improved the efficiency of extraction, while the best results were obtained with 70% EtOH and heating of the extracts at 80 °C for one hour. Furthermore, increase of the extraction temperature not only increased the efficiency of the extraction, but also changed positively the proportion of TP referred to total AOA level. The latter phenomenon may be connected with the possible loss of essential oil, which also contributes the AOA level of extracts.

Taking into account the obtained results, utilization of 70% EtOH at high temperature for a relatively short time was the best treatment for evaluating *Apiaceae* seed antioxidant status. In the latter conditions (Tables 5 and 6), the AOA values of *Apiaceae* seeds ranged from 12.5 to 61.2 mg GAE g^{-1} d.w. for perennial/biennial plants and from 7.6 to 58.0 mg GAE g^{-1} d.w. for annual representatives. Species differences in AOA values reflected the following sequence: lovage, anise, parsley > celery, dill, fennel > parsnip, coriander, carrot, caraway > chervil and were not associated with the belonging of plants to annual or perennial/biennial groups. High antioxidant activity of anise seeds was in accordance with that reported earlier by Martins et al. [12]. On the other hand, a positive correlation between AOA and TP (Figure 3) makes the seed TP/AOA ratio a reliable indicator of plant antioxidant status. In this respect, parsnip, chervil, carrot and caraway are characterized by the highest TP input to AOA, while parsley, lovage, dill and anise by the lowest, and it may be partially connected with species differences in essential oil content and its composition.

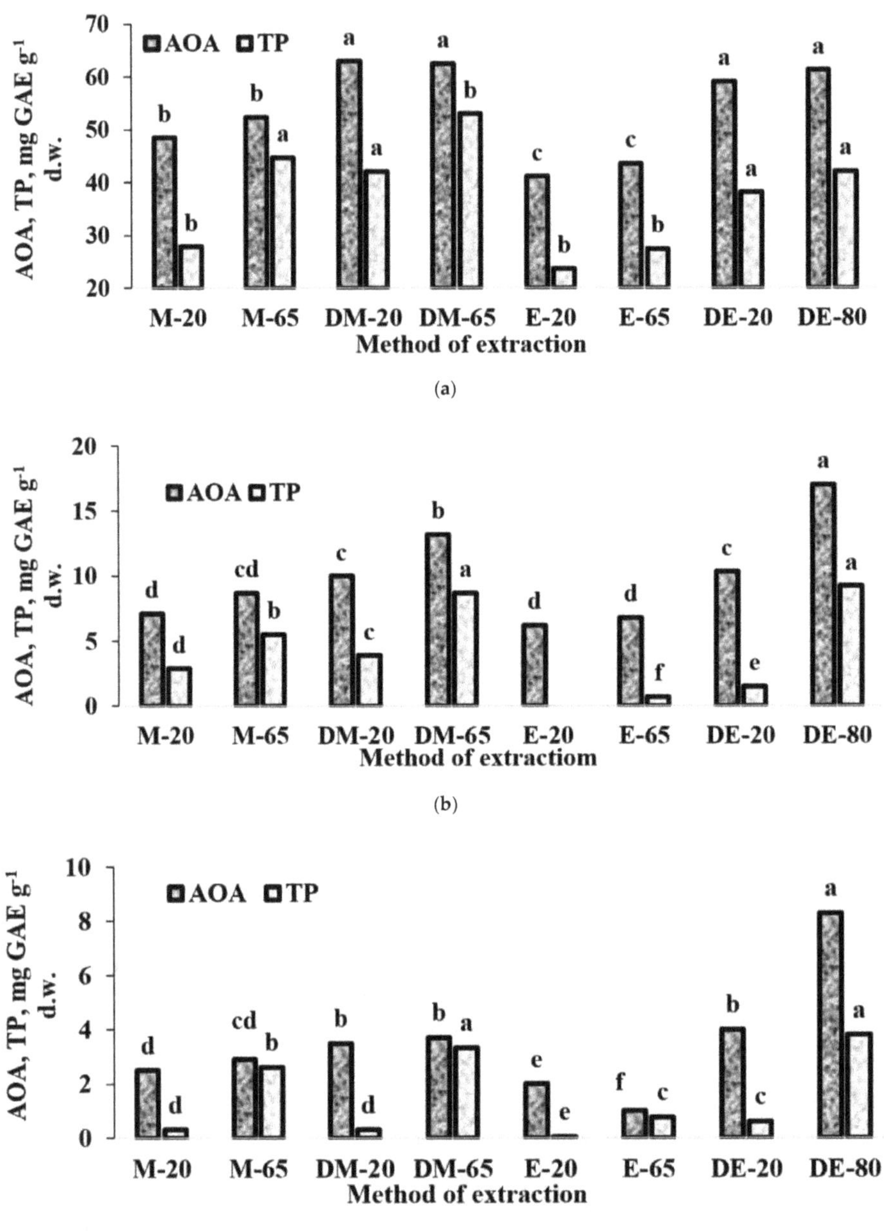

Figure 2. Effect of extraction conditions on total antioxidant activity (AOA) and total phenolics (TP) content in lovage (**a**), coriander (**b**) and chervil (**c**) seeds: (M-20)-methanol, 20 °C, 18 h; (M-65)-methanol. 65 °C, 1 h; (DM-20)—70% Methanol, 20 °C, 18 h; (DM-65)—70% Methanol, 65 °C, 1 h; (E-20)-ethanol, 20 °C, 18 h; (E-65)-ethanol 65 °C 1 h; (DE-20)—70% ethanol, 20 °C, 18 h; (DE-80)—70% ethanol, 80 °C, 1 h.

Table 6. Seed total phenolics (TP), Se content and total antioxidant activity (AOA) of annual *Apiaceae* plants.

Species	Cultivar	AOA mg GAE g^{-1} d.w.	TP mg GAE g^{-1} d.w.	Se µg kg^{-1} d.w.
Anise	Vityaz	58.0 [a]	15.6 [a]	41 [c]
Dill	Alligator	26.7 [c]	6.8 [e]	35 [e]
	Spartak	29.4 [c]	8.3 [c,d]	32 [e,f]
	Zontik	36.4 [b]	10.7 [b]	45 [c]
	Kibray	37.4 [b]	10.1 [b,c]	35 [e]
	Lesnogorodsky	37.9 [b]	8.5 [c,d]	44 [c,d]
	Salut	38.3 [b]	8.1 [d,e]	45 [c]
	Culinar	38.3 [b]	11.8 [b]	38 [d,e]
	Gribovsky	38.5 [b]	10.9 [b]	40 [c,d]
	Rusich	39.0 [b]	10.1 [b,c]	44 [c,d]
Coriander	Stimul	18.3 [d]	8.2 [c,d]	27 [f,g]
	№ 07-19	17.0 [d]	7.8 [e]	28 [f,g]
Caraway	Peresvet	12.9 [e]	8.6 [c,d]	25 [g]
Chervil	№21-20	8.3 [f]	4.6 [f]	110 [a]
	№22-20	9.8 [f]	4.9 [f]	85 [b]
	№24-20	7.6 [f]	4.5 [f]	82 [b]
	CV (%)	38.4	22.2	31.2

CV: coefficient of variation. Within each column, values with the same letters do not differ significantly according to Duncan test at $p < 0.05$.

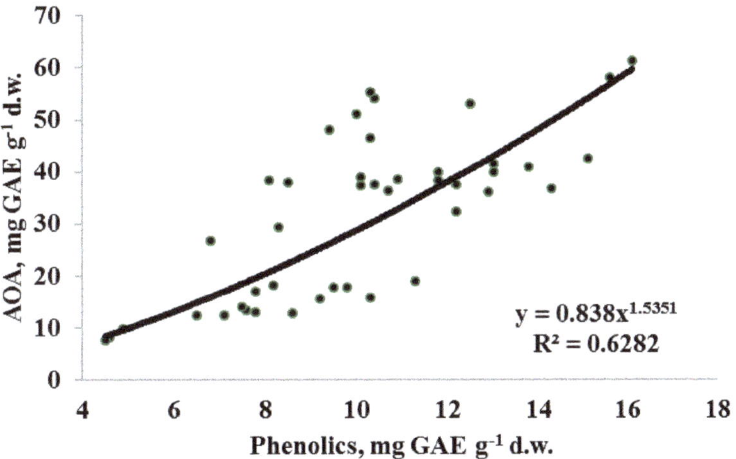

Figure 3. Correlation between total antioxidant activity (AOA) and total phenolics (TP) content in seeds of *Apiaceae* plants (r = 0.79 at $p < 0.001$; $n = 43$).

Among natural antioxidants, selenium plays a special role, as it is highly valuable in human organism due to its strong antioxidant activity and ability to protect against viral, cardiovascular diseases, cancer and covid infection [37]. However, this trace element is not essential for plants, though it demonstrates a protective effect against biotic and abiotic stresses [38]. *Apiaceae* plants belong to a group of plants highly sensitive to toxic Se concentrations (the so-called 'Se non accumulators'). Indeed, Se levels in *Apiaceae* seeds are low (Tables 5 and 6), which is in agreement with the above statement. On the other hand, a comparison of variation coefficients for AOA, TP, TDS, K, WSP and Se indicates lower CV values of AOA and TP compared to TDS, K and WSP, contrary to Se data with variation coefficient ranging from 11 to 70 %. Indeed, differences among cultivars in the studied

parameters were highly expressed for Se in parsley and celery, while phenolics content in most cases was rather stable. Furthermore, the largest differences between cultivars for TP did not exceed 15 % (dill seeds; Table 6). The highest levels of Se accumulation were recorded in Breeze and Moskvichka parsley cultivars (Table 4), which arises high prospects of these cultivars seed utilization as important functional food supplements characterized by high Se (137 to 142 µg kg^{-1} d.w.) and high AOA (53 to 55 mg GAE g^{-1} d.w.). In this respect, a well-known synergism between Se and other natural antioxidants in human organism may be considered as an additional benefit.

According to literature reports, plants are the main dietary Se source for humans, able to convert inorganic forms of the element to highly bioavailable organic ones, and in particular to Se-containing amino acids (selenomethionine, SeMet and selenocysteine, SeCys), and their methylated forms, possessing high anti-carcinogen activity [39]. As far as *Apiaceae* seeds are concerned, no correlation between Se and WSP content was revealed, which may be partly connected with both low Se levels and significant part of seed storage protein present in insoluble form. Nevertheless, despite the small sample analyzed, a significant correlation between Se and water soluble proteins was recorded in parsley cultivars (r = 0.85 at $p < 0.05$; $n = 6$) characterized by the highest Se CV values (Tables 5 and 6).

In other cases, Se accumulation in *Apiaceae* seeds demonstrated no significant correlation with the total antioxidant activity (AOA) and phenolics content, which makes it suppose the need of special Se biofortification of plants for producing functional food with high Se and antioxidants content.

4. Conclusions

The results obtained in the present research showed the high nutritional value of *Apiaceae* seeds, with anise, parsley, lovage and celery seeds containing the highest levels of antioxidants. The comparisons were carried out with the aim of identifying the differences of seed biochemical characteristics between species and cultivars, and may be useful to orient the *Apiaceae* seed utilization as dietary supplements and natural food conservatives.

Supplementary Materials: The following are available online at https://www.mdpi.com/2311-7524/7/3/57/s1, Table S1: Dates of Apiaceae seed harvesting.

Author Contributions: Conceptualization, V.K., V.P. and G.C.; data curation, V.Z.; formal analysis, V.S. and A.T.; investigation, N.G., V.K., A.M. and V.Z.; methodology, N.G., V.S., V.P. and A.T.; supervision, A.S., V.S. and V.P.; validation, G.C.; draft manuscript writing, N.G., A.M. and A.T.; manuscript revision and final editing, A.S., V.K. and G.C. All authors have read and agreed to the published version of the manuscript.

Funding: This research did not receive any grants from public, commercial or non-for-profit agencies.

Conflicts of Interest: The authors declare that they have no conflicts of interest.

References

1. Aćimović, M.; Kostadinović, M.L.; Popović, S.J.; Dojčinović, N.S. Apiaceae seeds as functional food. *J. Agr. Sci.* **2015**, *60*, 237–246. [CrossRef]
2. Ebert, A.W. Sprouts, microgreens, and edible flowers: The potential for high value specialty produce in Asia. In Proceedings of the SEAVEG 2012 Regional Symposium, Chiang Mai, Thailand, 24–26 January 2012; pp. 216–227.
3. Idowu, A.T.; Olatunde, O.O.; Adekoya, A.E.; Idowu, S. Germination: An alternative source to promote phytonutrients in edible seeds. *FQS* **2020**, *4*, 129–133. [CrossRef]
4. Koley, T.K. Microgreens from Vegetables: More Nutrition for Better Health. In *Training Manual on "Advances in Genetic Enhancement of Underutilized Vegetable Crops"*; Indian Institute of Vegetable Research: Varanasi, India, 18–27 October 2016; pp. 194–197.
5. Sagar, N.A.; Pareek, S.; Sharma, S.; Yahia, E.M.; Lobo, M.G. Fruit and Vegetable Waste: Bioactive Compounds, Their Extraction, and Possible Utilization Comprehensive Reviews. *J. Food Sci. Food Saf.* **2018**, *17*, 512–531. [CrossRef]
6. Ahmad, B.S.; Talou, T.; Saad, Z.; Hijazi, A.; Merah, O. The Apiaceae: Ethnomedicinal family as source for industrial uses. *Ind. Crops Prod.* **2017**, *109*, 661–671. [CrossRef]
7. Lee, J.B.; Yamagishi, C.; Hayashi, K.; Hayashi, T. Antiviral and immune stimulating effects of lignin-carbohydrate-protein complexes from *Pimpinella anisum. Biosci. Biotechnol. Biochem.* **2011**, *75*, 459–465. [CrossRef] [PubMed]

8. Rajeshwari, C.U.; Vinay Kumar, A.V.; Andallu, B. *Protective Role of Aniseeds (Pimpinella anisum L.) in Type 2 Diabetes: Under Supervision of Dr (Mrs.) B Andallu*; LAP Lambert Academic Publishing: Saarbrücken, Germany, 2010.
9. Pavlyuk, I.; Stadnytska, N.; Jasicka-Misiak, I.; Górka, B.; Wieczorek, P.P.; Novikov, V. A Study of the Chemical Composition and Biological Activity of Extracts from Wild Carrot (*Daucus carota* L.) Seeds Waste. *Res. J. Pharm. Biol. Chem. Sci.* **2015**, *6*, 603–611.
10. Fenner, M. The effects of the parent environment on seed germinability. *Seed Sci. Res.* **1991**, *1*, 75–84. [CrossRef]
11. Christova-Bagdassarian, V.; Bagdassarian, K.S.; Atanassova, M. Phenolic Profile, Antioxidant and Antibacterial Activities from the Apiaceae Family Dry Seeds. *Mintage J. Pharm. Med. Sci.* **2013**, *2*, 26–31.
12. Martins, N.; Barros, L.; Santos-Buelga, C.; Ferreira, I.C.F.R. Antioxidant potential of two Apiaceae plant extracts: A comparative study focused on the phenolic composition. *Ind. Crops Prod.* **2016**, *79*, 188–194. [CrossRef]
13. Wangensteen, H.; Samuelsen, A.B.; Malterud, K.E. Antioxidant activity in extracts from coriander. *Food Chem.* **2004**, *88*, 293–297. [CrossRef]
14. Marques, V.; Farah, A. Chlorogenic acids and related compounds in medicinal plants and infusions. *Food Chem.* **2009**, *113*, 1370–1376. [CrossRef]
15. Sebnem, S.I.; Ayten, S. Antioxidant Potential of Different Dill (*Anethum Graveolens* L.) Leaf Extracts. *Int. J. Food Prop.* **2011**, *14*, 894–902. [CrossRef]
16. Kołodziejek, J. Effect of seed position and soil nutrients on seed mass, germination and seedling growth in *Peucedanum oreoselinum* (Apiaceae). *Sci. Rep.* **2017**, *7*, 1959. [CrossRef]
17. Martínez-Ballesta, M.C.; Egea-Gilabert, C.; Conesa, E.; Ochoa, J.; Vicente, M.J.; Franco, J.A.; Bañon, S.; Martínez, J.J.; Fernández, J.A. The Importance of Ion Homeostasis and Nutrient Status in Seed Development and Germination. *Agronomy* **2020**, *10*, 504. [CrossRef]
18. Shevchenko, J.P.; Kharchenko, V.A.; Shevchenko, G.S.; Soldatenko, A.V. *Green and Spicy-Flavoring Crops*; Federal Scientific Center of Vegetable Production: Moscow, Russia, 2019; ISBN 978-5-901695-80-7.
19. Golubkina, N.A.; Kekina, H.G.; Molchanova, A.V.; Antoshkina, M.S.; Nadezhkin, S.M.; Soldatenko, A.V. *Plants Antioxidants and Methods of Their Determination*; Infra-M: Moscow, Russia, 2020. [CrossRef]
20. Alfthan, G.V. A micromethod for the determination of selenium in tissues and biological fluids by single-test-tube fluorimetry. *Anal. Chim. Acta* **1984**, *165*, 187–194. [CrossRef]
21. Bradford, M.M. A rapid sensitive method for the quantification of microgram quantities of protein utilizing the principle of protein-dye binding. *Anal. Biochem.* **1976**, *72*, 248–254. [CrossRef]
22. Kharchenko, V.A.; Moldovan, A.I.; Golubkina, N.A.; Koshevarov, A.A.; Caruso, G. Antioxidant status of celery (*Apium graveolens* L.). *Veg. Crops Russia* **2020**, *2*, 82–86. (In Russian) [CrossRef]
23. Zhang, D.; Lü, H.; Chu, S.; Zhang, H.; Zhang, H.; Yang, Y.; Li, H.; Yu, D. The genetic architecture of water-soluble protein content and its genetic relationship to total protein content in soybean. *Sci. Rep.* **2017**, *7*, 5053. [CrossRef]
24. Zhao, M.; Zhang, H.; Yan, H.; Qiu, L.; Baskin, C.C. Mobilization and role of starch, protein and fat reserves during seed germination of six wild grassland species. *Front. Plant Sci.* **2018**, *9*. [CrossRef]
25. Christova-Bagdassarian, V.L.; Bagdassarian, K.S.; Atanassova, M.S. Phenolic Compounds and Antioxidant Capacity in Bulgarian Plans (dry seeds). *Int. J. Adv. Res.* **2013**, *1*, 186–197.
26. Faudale, M.; Vilasomat, F.; Bastida, J.; Poli, F.; Codina, C. Antioxidant Activity and Phenolic Composition of Wild, Edible, and Medicinal Fennel from Different Mediterranean Countries. *J. Agric. Food Chem.* **2008**, *56*, 1912–1920. [CrossRef] [PubMed]
27. Tomsone, L.; Kruma, Z. Comparison of different extraction methods for isolating phenolic compounds from lovage (*Levisticum officinale* L.) seeds. In Proceedings of the Conference on Innovations in Science, Education and Business, Kaliningrad, Russia, 25–27 September 2013; pp. 94–197. (In Russian).
28. Uddin, Z.; Shad, A.A.; Bakht, J.; Ullah, I.; Jan, S. In vitro antimicrobial, antioxidant activity and phytochemical screening of *Apium graveolens*. *Pak. J. Pharm. Sci.* **2015**, *28*, 1699–1720. [PubMed]
29. Odeh, A.; Allaf, A.W. Determination of polyphenol component fractions and integral antioxidant capacity of Syrian aniseed and fennel seed extracts using GC–MS, HPLC analysis, and photochemiluminescence assay. *Chem. Pap.* **2017**, *71*, 1731–1737. [CrossRef]
30. Sriti, J.; Aidi Wannes, W.; Talou, T.; Ben Jemia, M.; Elyes Kchouk, M.; Marzouk, B. Antioxidant properties and polyphenol contents of different parts of coriander (*Coriandrum sativum* L.) fruit. *La Riv. Ital. Delle Sostanze Grasse* **2012**, *LXXXIX*, 253–262.
31. Vallverdú-Queralt, A.; Regueiro, J.; Alvarenga, J.F.R.; Martinez-Huelamo, M.; Leal, L.N.; Lamuela-Raventos, R.M. Characterization of the phenolic and antioxidant profiles of selected culinary herbs and spices: Caraway, turmeric, dill, marjoram and nutmeg. *Food Sci. Technol. Campinas.* **2015**, *35*, 189–195. [CrossRef]
32. Atalar, M.N.; Karadağ, M.; Koyuncu, M.; Alma, M.H. Determination of Phenolic Components of *Apium graveolens* (Celery) Seeds. In Proceedings of the Aromatic Plants and Cosmetics Symposium, Iğdir, Turkey, 3–6 October 2019.
33. Tadros, L.K.; El-Rafey, H.H.; Elfadaly, H.A.; Taher, M.A.; Elhafny, A. Phenolic Profile, Essential oil Composition, Purification of Kaempferol 3-arabinofuranoside and Antimicrobial Activity of Parsley Cultivated in Dakhalia Governorate. *J. Agric. Chem. Biotechnol. Mansoura Univ.* **2017**, *8*, 183–189. [CrossRef]
34. Mert, A.; Timur, M. Essential Oil and Fatty Acid Composition and Antioxidant Capacity and Total Phenolic Content of Parsley Seeds (*Petroselinum crispum*) Grown in Hatay Region. *Ind. J. Pharm. Educ. Res.* **2017**, *51*, s437–s440. [CrossRef]

35. Paven, C.S.J.; Radu, D.; Alexa, E.; Pintilie, L.S.; Rivis, A. Anethum Graveolens- an important source of antiozidant compounds for food industry. In Proceedings of the 18th International Multidisciplinary Scientific GeoConference SGEM, Albena, Bulgaria, 2–8 July 2018.
36. Ksouri, A.; Dob, T.; Belkebir, A.; Krimat, S.; Chelghoum, C. Chemical composition and antioxidant activity of the essential oil and the methanol extract of Algerian wild carrot *Daucus carota* L. ssp. carota. (L.). *J. Mater. Environ. Sci.* **2015**, *6*, 784–791.
37. Zhang, L.; Liu, Y. Potential interventions for novel coronavirus in China: A systematic review. *J. Med. Vir.* **2020**, *92*, 479–490. [CrossRef] [PubMed]
38. Schiavon, M.; Nardi, S.; dalla Vecchia, F.; Ertani, A. Selenium biofortification in the 21st century: Status and challenges for healthy human nutrition. *Plant Soil* **2020**, *453*, 245–270. [CrossRef] [PubMed]
39. Gupta, M.; Gupta, S. An overview of selenium content metabolism and toxicity in plant. *Front. Plant Sci.* **2017**, *7*, 2074. [CrossRef] [PubMed]

horticulturae

Article

Modulating Vapor Pressure Deficit in the Plant Micro-Environment May Enhance the Bioactive Value of Lettuce

Chiara Amitrano, Youssef Rouphael, Stefania De Pascale * and Veronica De Micco

Department of Agricultural Sciences, University of Naples Federico II, Via Università 100, 80055 Portici (Naples), Italy; chiara.amitrano@unina.it (C.A.); youssef.rouphael@unina.it (Y.R.); demicco@unina.it (V.D.M.)
* Correspondence: depascal@unina.it

Abstract: Growing demand for horticultural products of accentuated sensory, nutritional, and functional quality traits has been driven by the turn observed in affluent societies toward a healthy and sustainable lifestyle relying principally on plant-based food. Growing plants under protected cultivation facilitates more precise and efficient modulation of the plant microenvironment, which is essential for improving vegetable quality. Among the environmental parameters that have been researched for optimization over the past, air relative humidity has always been in the background and it is still unclear if and how it can be modulated to improve plants' quality. In this respect, two differentially pigmented (green and red) Salanova® cultivars (*Lactuca sativa* L. var. *capitata*) were grown under two different Vapor Pressure Deficits (VPDs; 0.69 and 1.76 kPa) in a controlled environment chamber in order to appraise possible changes in mineral and phytochemical composition and in antioxidant capacity. Growth and morpho-physiological parameters were also analyzed to better understand lettuce development and acclimation mechanisms under these two VPD regimes. Results showed that even though Salanova plants grown at low VPD (0.69 kPa) increased their biomass, area, number of leaves and enhanced Fv/Fm ratio, plants at high VPD increased the levels of phytochemicals, especially in the red cultivar. Based on these results, we have discussed the role of high VPD facilitated by controlled environment agriculture as a mild stress aimed to enhance the quality of leafy greens.

Keywords: air humidity (RH); *Lactuca sativa* L. var. *capitata*; controlled environment agriculture (CEA); bioactive compounds; leaf gas exchange; minerals profile; genetic material

Citation: Amitrano, C.; Rouphael, Y.; De Pascale, S.; De Micco, V. Modulating Vapor Pressure Deficit in the Plant Micro-Environment May Enhance the Bioactive Value of Lettuce. *Horticulturae* **2021**, *7*, 32. https://doi.org/10.3390/horticulturae7020032

Academic Editor: Othmane Merah

Received: 29 January 2021
Accepted: 18 February 2021
Published: 22 February 2021

Publisher's Note: MDPI stays neutral with regard to jurisdictional claims in published maps and institutional affiliations.

1. Introduction

Air humidity (RH), and more specifically the Vapor Pressure Deficit (VPD), is one of the most important microclimate factors affecting plant transpiration rate in Controlled Environment Agriculture (CEA). Consequently, VPD affects all physiological and biochemical processes associated with the transpiration, such as: water balance, cooling, gas-exchange, and ion translocation, thus affecting plant growth and productivity [1,2]. It is well established that plants grown under a reduced VPD (high RH) enhance carbon gain by opening their stomata, usually improving at the same time, dry matter production [3]. Moreover, plants enhance growth under high RH levels, as long as the transpiration rate is still enough to support the uptake and distribution of essential macronutrients (Ca^{2+}, Mg^{2+}, K^+) and phytohormones (auxin, cytokinin) [4]. Furthermore, in lettuce, high air humidity, especially during night, appears to prevent Ca^{2+} deficiency, a common physiological disorder known as tipburn, which negatively affects the nutritional quality and marketability of the product [5]. Under high VPD levels (low RH), plants try to avoid dehydration and water loss by closing their stomata, which negatively affect photosynthetic efficiency, thus determining a major reduction in plant growth and yield [6,7]. Nevertheless, high VPD in indoor cultivation has proven to enhance vegetable quality, for example increasing ascorbate, lycopene, β-carotene, rutin, and caffeic acid concentrations in greenhouse tomato, often connected to high irradiance during sunny hours when greenhouses are subjected to high VPD [6,8].

In greenhouse cherry tomato cv. Naomi, Rosales et al. [9] found an increment in ascorbic acid synthesis in plants grown under high VPD levels (2–3 kPa), probably due to the occurrence of oxidative stress [10,11]. This is consistent with other "controlled" stress like drought or salinity that, if moderately applied to plants, can increase product quality [12,13]. For instance, Favati et al. [14] found improved quality of tomato fruit subjected to deficit irrigation, in particular due to the enhancement of ascorbic acid and β-carotene. Moreover, controlled drought stress increased the levels of carotenoids in edible organs of pepper and carrot as well as the levels of sugars in tomato and cucumber fruits [12].

Notwithstanding the positive outcomes of recent research, little is known about the effects of VPD modulation on leafy greens nutritional and functional quality. Indeed, humidity is one of the most difficult environmental factors to control in CEA (instrumentally and economically), thus often being neglected by growers [6]. However, over the past two decades there has been a growing demand for high quality horticultural products [13,15], with consumers always looking for fresh and high nutritional food [16]. Bioactive compounds, also known as phytochemicals, are already present in leafy green vegetables and especially in lettuce, where red-leaved cultivars present very high content of vitamin C, polyphenols and antioxidant activities compared to their green counterparts [17,18]. Phytochemicals-rich-food are in great demand due to their ability to reduce the risk of cardiovascular diseases, some forms of cancer, and stimulate cognitive health against age-related problems [19]. Even though the genetic material (i.e., genotype) is the principal factor in determining how much phytochemicals a plant will accumulate during its life cycle, the influence of microclimatic factors affecting greenhouse and indoor growing modules vegetables, cannot be neglected. Several scientific papers have been published regarding genotype, and microclimate (e.g., air and root zone temperature, light quantity, and quality) effects on the quality of controlled environments vegetables [20,21], whereas the effects of VPD on leafy greens quality is still poorly explored.

In light of the foregoing, the aim of the current study was to assess how the modulation of VPD influences the nutritional and functional quality of green and red-leaved lettuce (*Lactuca sativa* L. var. *capitata*). For this purpose, a growth chamber experiment under controlled climatic conditions was conducted, growing plants under two different VPDs (0.6 kPa and 1.7 kPa), considered respectively low- and high- VPD. The development of lettuces in terms of anatomical structure of the leaf lamina, plant growth, as well as some plant physiological responses (Fv/Fm ratio and chlorophyll content) were examined. Treatments were compared in terms of leaf colorimetry coordinates, antioxidant activity, minerals profile, polyphenols, and total ascorbic acid content.

2. Materials and Methods

2.1. Experimental Design, Lettuce Genotypes, and Controlled Growing Conditions

The experiment was carried out on two butterhead Salanova® lettuce cultivars (*Lactuca sativa* L. var. *capitata*), with green and red leaves. Two-week old transplants were purchased from a local provider and grown at the Department of Agricultural Sciences (University of Naples Federico II, Italy) in two consecutive cycles, in a growth-chamber (KBP-6395F, Termaks, Bergen, Norwey) equipped with a Light-Emitting Diode (LED) panel unit (K5 Series XL750, Kind LED, Santa Rosa, CA, USA), with an emission wavelength range of 400–700 nm. The two cultivation cycles were identical in terms of agricultural practices and microclimatic conditions (light intensity, quality, photoperiod, air, and zone temperature), except for the VPD levels. More specifically, the first cycle was performed under an average VPD of 0.69 kPa and the second under a VPD of 1.76 kPa. The two VPDs were achieved keeping air temperature (T) at 24 ± 1° C and changing the RH accordingly. RH and T were controlled by the growth chamber and monitored inside the chamber by means of mini-sensors (Testo 174 H), equipped with a data-logger which collected data every 15 min.

In each cycle, 9 green and 9 red Salanova lettuces were transplanted into plastic trays (14 × 19 × 6 cm: W × L × D) on peat:perlite substrate (1:1 *v/v*). Daily rotation of the trays was performed to ensure homogenous light and humidity across the shelf surface. Plants

were grown for 23 days under a red-green-blue (RGB) light of 315 μmol m^{-2} s^{-1}, 12 h photoperiod (13.6 Daily Light Integral; DLI). All plants were fertigated to field capacity with a modified Hoagland solution (8.2 mM N-NO$_3^-$, 2.0 mM S, 2.7 mM K$^+$, 5.8 mM Ca^{2+}, 1.4 mM Mg^{2+}, 1.0 mM NH$_4^+$, 15.0 μM Fe, 9.0 μM Mn, 0.3 μM Cu, 1.6 μM Zn, 20 μM B, and 0.3 μM Mo), resulting in an electrical conductivity of 1.4 dS m^{-1} and a pH of 5.8.

2.2. Plant Growth Parameters, Biomass Production, and Leaf Colorimetry

Harvesting of all experimental units was performed 23 days after transplanting (DAT). Before harvesting, each plant was photographed from the top and digital images were used to assess plant total area (PA) through ImageJ 1.45 software (U.S. National Institutes of Health, Bethesda, MD, USA). The number of leaves (LN) was counted for all plants, which were then weighted to determine the above-ground fresh biomass (FB). For the dry biomass (DB) determination, samples of fresh leaf tissues (about 15 g per plant) were oven-dried at 70 °C for 3 days, until they reached a constant weight. On the harvesting day, leaf color was measured on the upper part of three representative leaves per plant, using a Minolta CR-300 Chroma Meter (Minolta Camera Co. Ltd., Osaka, Japan). The meter was calibrated with the standard white plate before measurements. Leaf chromaticity was performed following the *Commission Internationale de l'Eclairage* and expressed as: lightness (L*), b* (+b* yellowness) used to calculate chroma (C* = (a*2 + b*2)1/2) and Hue angle (H$^\circ$ = arctan (b*/a*)).

2.3. Anatomical Analyses of Leaves

At 23 DAT, one complete life-span leaf per plant was collected from the median part of the canopy and promptly stored in F.A.A. fixative solution (40% formaldehyde, glacial acetic acid, 50% ethanol, 5: 5: 90 by volume). Each leaf was dissected to remove the apical and basal portions, while keeping the median region of the lamina. 5 × 5 mm portions of the leaf lamina were dehydrated in an ethanol series (50, 70, and 95%) and embedded in the JB4 acrylic resin (Polysciences, Warrington, PA, USA). Thin cross sections (5 μm thick) were cut by means of a rotary microtome, stained with 0.025% toluidine blue [22] and mounted with mineral oil for microscopy. Sections were analyzed under the BX60 transmitted light microscope (Olympus, Hamburg, Germany), and digital images were collected and analyzed through the Olympus AnalySIS software (AnalySIS 3.2, Olympus). The following functional anatomical traits were quantified: upper and lower epidermis thickness (UET; LET) (μm); palisade parenchyma thickness (PT) (μm); spongy parenchyma thickness (ST) (μm); total leaf lamina thickness (LT) (μm) and percentage of intercellular spaces (IS) (%). All the thickness measurements were taken in 6 position along the lamina, avoiding veins and damaged areas. The IS was measured as percentage of area occupied by intercellular spaces over a given surface of parenchyma, in three regions of the leaf lamina, as reported in [23].

2.4. Mineral Composition in Leaf Tissue

Dried material was used for the evaluation of mineral leaf composition in terms of cations (K$^+$, Mg^{2+}, Ca^{2+} Na^{2+}), anions (NO$_3^-$, SO$_4^{2-}$, PO$_4^{3-}$) and acids (malate, tartrate, citrate, isocitrate). Dried leaves (0.25 g per replicate) were suspended in 50 mL of ultrapure water (Milli-Q, Merk Millipore, Darmstadt, Germany), frozen and then shook for 10 min in a water bath (ShakeTemp SW22, Julabo, Seelbach, Germany) at 80 °C. The mixture was then centrifuged at 6000 rpm for 10 min (R-10M, Remi Elektrotechnik, India) and the supernatant, was filtered to 0.45 μm, and stored at -20 °C until analysis. Anions and cations were separated and quantified by ion chromatography equipped with a conductivity detection (ICP 3000 Dionex, Thermo fisher Scientific Inc., MA, USA).

2.5. Extraction and Quantification of Total Ascorbic Acid, Polyphenols, Lipophylic, and Hydrophilic Antioxidant Activities

All phytochemical analyses were performed on 9 green and 9 red Salanova lettuces (one leaf per replicate). Total ascorbic acid (TAA) was assessed spectrophotometrically

based on the protocol of Kampfenkel, Montagu, and Inze [24]. The phenolic content (PH) was determined using the Folin-Cicolteau procedure [25] using gallic acid (Sigma Aldrich Inc, St Louis, MO, USA) as a standard. The hydrophilic antioxidant activity (HAA) was measured using *N,N*-dimethyl-p-phenylenediamine (DMPD) method [26], whereas the lipophilic antioxidant activity (LAA) was measured following the ABTS method [27].

2.6. Soil Plant Analysis Development Index and Chlorophyll a Fluorescence Emission

At 12 and 23 DAT (middle and final point of experiments), the Soil Plant Analysis Development (SPAD) index was measured on 9 fully expanded leaves per condition by means of a portable chlorophyll meter SPAD-502 (Konica Minolta, Japan), avoiding major veins, leaflet margins, and damaged areas. On the same dates, measurements of leaf chlorophyll "a" fluorescence emission, were performed on the same leaves to calculate the maximum quantum efficiency of PSII photochemistry (Fv/Fm) on 30′ dark-adapted leaves, with a portable fluorometer, equipped with a light sensor (ADC BioScientific Ltd., Hoddesdon, United Kingdom).

2.7. Statistics

Data were initially subjected to a two-way analysis of variance (ANOVA). Interactions between cultivar and VPD (C × VPD) were further addressed through specific one-way ANOVA and treatment means were compared using Duncan's multiple range test performed at $p \leq 0.05$ using the SPSS 20 software package (IBM, Armonk, NY, USA). Moreover, multivariate analysis was used to perform an agglomerative hierarchical cluster analysis (HCA) of the data sets. For HCA, the paired group (UPGMA) and Euclidean distances were used for clustering. Results of HCA were displayed as a tree-shaped dendrogram, where the horizontal distance between clusters represented data dissimilarity, and a heatmap, through the web tool (Clustvis; https://biit.cs.ut.ee/clustvis/).

3. Results

3.1. Plant Growth Parameters, Biomass Production, and Leaf Colorimetry

As presented in Table 1, cultivar and VPD had a significant effect on Salanova plant area (PA), leaves number (LN), fresh biomass (FB), dry biomass (DB), as main factors and in interaction. More specifically, red cultivar (R) presented higher values of all growth parameters (PA, FB, DB) enhanced by 10, 11, and 4%, with an exception made for LN. At the same time, 0.69 kPa increased all growth parameters (PA, LN, FB, FB) by 17, 12, 15, and 47%, always showing highest values in red cultivar (0.69 R), followed by 0.69 G, 1.76 R, and 1.76 G. Leaf colorimetry parameters were also influenced by C and VPD as main factors and by their interaction (C × VPD). In this case, b*, leaf brightness (L*) and Chroma were higher in G cultivar by 91, 54, and 23% and Hue in R cultivar (82%). Whereas, 0.69 kPa elicited increments in b* Chroma and Hue (41, 83 and 23%), while 1.76 kPa enhanced L* (32%). Concerning the interaction, the three colorimetry coordinates had a completely different trends among treatment, with increments in L* in 1.76 G followed by 0.69 G, 1.76 R, and 0.69 G. b* and chroma values incremented in 0.69 G followed by 1.76 G, 0.6 R, and 1.76 R; whereas Hue values increased in 0.69 R followed by 1.76 R and 0.69 G.

Table 1. Growth analyses consisting of plant area (PA), leaf number (LN), fresh biomass (FB), dry Biomass (DB), and leaf colorimetry coordinates (L*, Chroma and Hue angle) in green (G) and red (R) lettuce plants grown under the two Vapor Pressure Deficit (VPD) levels (0.69 and 1.76 kPa).

	PA (cm^2 Plant^{-1})	LN (No. Plant^{-1})	FB (g Plant^{-1})	DB (g Plant^{-1})	b *	L *	Chroma	Hue
Cultivar								
G	196 ± 1.02 b	51.9 ± 0.72 a	33.2 ± 0.50 b	3.60 ± 0.03 a	40.3 ± 1.21a	49.4 ± 0.46 a	28.1 ± 0.44 a	107 ± 4.46 b
R	214 ± 0.37 a	49.3 ± 0.81 b	36.8 ± 0.29 a	3.75 ± 0.02 a	3.62 ± 0.78 b	22.7 ± 0.47 b	21.5 ± 0.45 b	195 ± 3.34 a
VPD								
0.69 kPa	224 ± 1.01 a	53.9 ± 0.76 a	37.9 ± 0.48 a	4.80 ± 0.06 a	28.8 ± 1.16 a	34.0 ± 0.33 b	42.3 ± 0.62 a	171 ± 3.15 a
1.76 kPa	186 ± 0.33 b	47.3 ± 0.77 b	32.1 ± 0.31 b	2.55 ± 0.03 b	17.1 ± 0.88 b	38.1 ± 0.34 a	7.24 ± 0.66 b	132 ± 2.02 b
Int.								
0.69 G	211 ± 0.17 b	53.4 ± 0.24 a	35.6 ± 0.17 b	4.75 ± 0.02 a	44.5 ± 0.81 a	48.5 ± 0.42 b	46.2 ± 0.74 a	106 ± 0.44 c
0.69 R	237 ± 0.03 a	54.4 ± 0.40 a	40.4 ± 0.22 a	4.85 ± 0.03 a	9.25 ± 0.80 c	19.5 ± 0.31 d	9.84 ± 0.73 c	236 ± 0.88 a
1.76 G	180 ± 0.20 d	50.4 ± 0.25 b	30.8± 0.23 d	2.46 ± 0.01 b	36.2 ± 0.70 b	50.3 ± 0.77 a	38.3 ± 0.68 b	109 ± 0.28 c
1.76 R	192 ± 0.21 c	44.2 ± 0.49 c	33.4 ± 0.07 c	2.66 ± 0.01 b	-2.00 ± 0.16 d	25.9 ± 0.13 c	4.60 ± 0.11 d	154 ± 2.18 b
Sig.								
C	***	*	***	NS	***	***	***	***
VPD	***	***	***	**	***	***	***	***
C × VPD	***	***	***	***	*	***	*	***

All data are expressed as mean ± standard error. ***, **, *, NS refer to $p \leq$ 0.001, 0.01, 0.05 and Non-significant, respectively. Lower case letters indicate the significant differences of the interaction.

3.2. Morpho-Anatomical Analyses

As shown in Figure 1 the morpho-anatomical structure of the leaf lamina was not different among the four different combinations of cultivar and VPD. Cultivar and VPD alone showed no significant differences on Salanova lettuces morpho-anatomical parameters, with an exception made for LET where G cultivar showed an increment of 13% (Table 2). However, their interaction (C × VPD) elicited a significant difference in the upper and lower epidermis thickness (UET and LET). More specifically, UET was the highest in 0.69 G and the lowest in 1.76 G, while no significant differences were found in R cultivars between 0.69 and 1.76 kPa. Differently, LET was the highest in 0.69 R, followed by 0.69 G and was the lowest in 1.76 with no significant differences between G and R.

Table 2. Morpho-anatomical analyses consisting of upper epidermis thickness (UET), lower epidermis thickness (LET), palisade thickness (PT), spongy thickness (ST), lamina thickness (LT) and percentage of intercellular spaces (IS) in green (G) and red (R) lettuce plants grown under the two VPD levels (0.69 and 1.76 kPa).

	UET (μm)	LET (μm)	PT (μm)	ST (μm)	LT (μm)	IS (%)
Cultivar						
G	23.4 ± 0.66 a	16.7 ± 1.03 a	97.3 ± 4.69 a	148 ± 6.29 a	287 ± 8.86 a	46.2 ± 2.87 a
R	22.4 ± 1.05 a	14.8 ± 0.65 b	94.2 ± 3.71 a	149 ± 11.8 a	282 ± 11.3 a	45.3 ± 2.05 a
VPD						
0.69 kPa	22.9 ± 0.64 a	15.4 ± 0.79 a	95.5 ± 5.06 a	146 ± 7.57 a	281 ± 8.76 a	45.6 ± 1.89 a
1.76 kPa	22.8 ± 1.11 a	16.1 ± 0.74 a	95.9 ± 5.31 a	151 ± 9.15 a	287 ± 13.4 a	46.8 ± 2.50 a
Int.						
0.69 G	24.2 ± 0.63 a	16.2 ± 0.79 ab	94.5 ± 3.10 a	144 ± 3.56 a	280 ± 4.85 a	44.2 ± 1.96 a
0.69 R	22.7 ± 0.64 ab	17.2 ± 0.79 a	99.8 ± 3.93 a	152 ± 8.02 a	293 ± 11.73 a	44.9 ± 1.33 a
1.76 G	21.7 ± 0.70 b	14.6 ± 0.48 b	96.5 ± 3.19 a	147 ± 5.47 a	281 ± 8.02 a	48.1 ± 1.83 a
1.76 R	23.1 ± 0.82 ab	15.1 ± 0.52 b	92.1 ± 4.24 a	151 ± 7.56 a	282 ± 10.82 a	45.6 ± 1.35 a
Sig.						
C	NS	*	NS	NS	NS	NS
VPD	NS	NS	NS	NS	NS	NS
C × VPD	*	*	NS	NS	NS	NS

All data are expressed as mean ± standard error. ***, **, *, NS refer to $p \leq$ 0.001, 0.01, 0.05, and Non-significant, respectively. Lower case letters indicate the significant differences of the interaction.

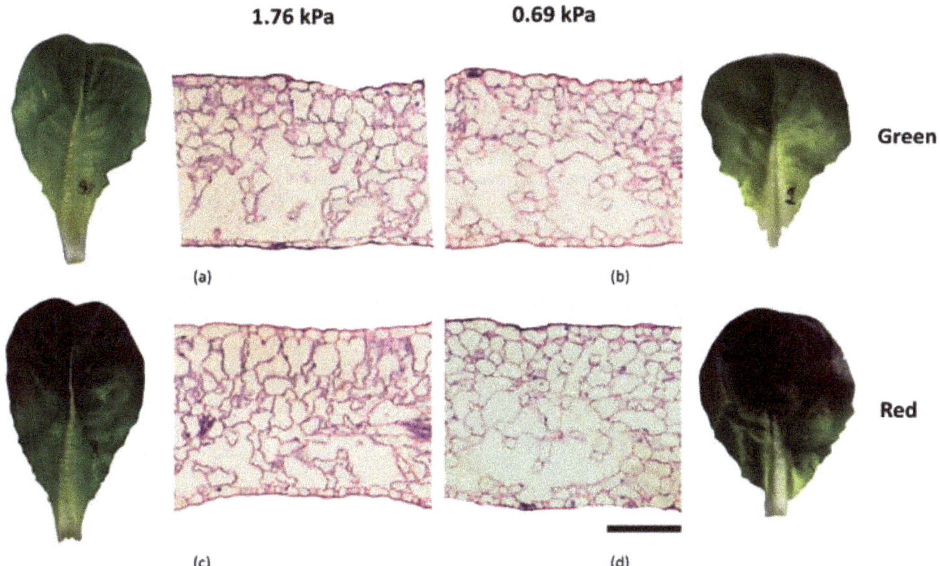

Figure 1. Light microscopy views of cross-sections of green (**a**,**b**) and red (**c**,**d**) leaf lamina of lettuces grown under the two VPD levels 1.76 (**a**,**c**) and 0.69 (**b**,**d**). Bar = 100 μm.

3.3. Mineral Composition

Results from ion chromatography are showed in Table 3. Mineral content varied among treatments. More specifically, red cultivar enhanced the content of NO_3^-, Ca^{2+} and malate by 22, 24, and 50%, whereas green cultivar enhanced the content of K^+, tartrate, and isocitrate by 20, 45, and 26%. No significant differences among cultivar were detected in the other minerals and organic acids. Differently 0.69 kPa enhanced the content of PO_4^{3-}, Ca^{2+}, malate and tartrate by 24, 19, 53, and 25%, whereas 1.76 kPa enhanced the content of NO_3^-, SO_4^{2-}, and K^+ by 9, 47, and 46%. No significant differences between 0.69 and 1.76 kPa were found in the other minerals and organic acids. Concerning the interaction (C × VPD), no significant changes were found in Na^{2+}, Malate and Citrate. Whereas, NO_3^-, SO_4^{2-}, and K^+ followed the same trend with highest values in 1.76 G and no significant differences among the other treatments (0.69 G, 1.76R, 0.69 R). Furthermore, PO_4^{3-} showed highest content under 0.69 both G and R, followed by 1.76 R and 1.76 G; Ca^{2+} content increased under 0.69 R, not showing any significant differences among other treatments; Mg^{2+} content increased under 0.69 R, followed by 1.76 G, 1.76 R, and 0.69 G; tartrate content was more elevated in 0.69 G, followed by 1.76 G, 1.76 R, and 0.69 R and isocitrate content presented highest values in G, with no significant differences between 0.69 and 1.76 kPa, followed by 1.76 R and 0.69 R.

Table 3. Minerals in leaves of green (G) and red (R) lettuce plants grown under the two VPD levels (0.69 and 1.76 kPa).

	NO_3^- (mg/kg FW)	PO_4^{3-} (g/kg DW)	SO_4^{2-} (g/kg DW)	K^+ (g/kg DW)	Ca^{2+} (g/kg DW)	Mg^{2+} (g/kg DW)	Na^{2+} (g/kg DW)	Malate (g/kg DW)	Tartrate (g/kg DW)	Citrate (g/kg DW)	Isocitrate (g/kg DW)
Cultivar											
G	4013 ± 711 b	6.92 ± 0.73 a	2.01 ± 0.25 a	58.2 ± 2.36 a	14.7 ± 0.86 b	3.46 ± 0.74 a	3.36 ± 0.40 a	56.5 ± 4.82 b	3.64 ± 0.28 a	12.2 ± 1.86 a	0.23 ± 0.03 a
R	4911 ± 625 a	7.44 ± 0.67a	1.78 ± 0.29 a	46.6 ± 5.54 b	18.3 ± 2.44 a	3.74 ± 0.64 a	4.11 ± 0.98 a	84.8 ± 8.60 a	1.98 ± 0.25 b	14.5 ± 1.31 a	0.17 ± 0.03 b
VPD											
0.69 kPa	4513 ± 746 b	8.19 ± 0.36 a	1.53 ± 0.21b	42.6 ± 1.56 b	18.3 ± 1.15 a	3.56 ± 0.76 a	3.80 ± 0.56 a	92.6 ± 8.19 a	3.21 ± 0.26 a	13.7 ± 1.41 a	0.18 ± 0.03 a
1.76 kPa	4911 ± 553 a	6. 24 ±1.39 b	2.25 ± 0.37a	62.2 ± 7.15 a	14.7 ± 1.87 b	3.64 ± 0.59 a	3.65 ± 0.66 a	43.3 ± 1.86 b	2.41 ± 0.28 b	12.9 ± 2.45 a	0.21 ± 0.03 a
Int.											
0.69 G	3246 ± 556 d	8.18 ± 0.27 a	1.26 ± 0.13 b	37.5 ± 0.48 b	15.6 ± 0.27 b	3.04 ± 0.64 c	3.09 ± 0.24 a	76.5 ± 4.28 b	4.65 ± 0.22 a	13.4 ± 0.74 a	0.23 ± 0.02 a
0.69 R	3980 ± 380 c	8.21 ± 0.19 a	1.80 ± 0.15 b	47.7 ± 2.16 b	21.1 ± 1.76 a	4.08 ± 0.25 a	4.52 ± 0.64 a	108.6 ± 7.83 a	1.77 ± 0.08 c	14.2 ± 1.34 a	0.15 ± 0.01 b
1.76 G	4080 ± 310 b	5.65 ± 0.91 b	2.75 ± 0.23 a	78.9 ± 3.77 a	13.9 ± 1.19 b	3.87 ± 0.21 ab	3.62 ± 0.32 a	31.5 ± 1.09 c	2.62 ± 0.12 b	10.9 ± 1.49 a	0.23 ± 0.01 a
1.76 R	4942 ± 490 a	6.84 ± 0.96 ab	1.75 ± 0.28 b	45.5 ± 6.77 b	15.6 ± 1.36 b	3.41 ± 0.77 bc	3.69 ± 0.68 a	55.1 ± 1.54 b	2.19 ± 0.33 bc	14.9 ± 1.93 a	0.19 ± 0.03 ab
Sig.											
C	***	NS	NS	*	*	NS	NS	**	***	NS	*
VPD	*	*	**	***	*	NS	NS	***	**	NS	NS
C × VPD	***	*	**	***	*	**	NS	NS	***	NS	*

All data are expressed as mean ± standard error. ***, **, *, NS refer to $p \leq 0.001$, 0.01, 0.05 and Non-significant, respectively. Lower case letters indicate the significant differences of the interaction.

3.4. Antioxidant Activities and Phytochemicals

Antioxidant activity and phytochemical content were influenced by C, VPD, and their interaction (Table 4). Cultivar had a significant effect on TAA, PH, and LAA, with increments in the red cultivar by 27, 12, and 40% compared to the green one; whereas VPD had a significant effect on TAA, PH, and HAA, with increments in the 1.76 kPa plants by 22, 47, and 8% compared to the low VPD condition. However, the interaction (C × VPD) was always significant. More specifically, TAA content resulted enhanced in 1.76 R followed by 1.76 G, 0.69 R, and 0.69 G. Differently, PH and LAA showed a common trend, with highest values in 1.76 R and 0.69 R; these values were significantly higher than those detected in 1.76 G which in turn showed significantly higher values than 0.69 G. HAA showed highest values once again in 1.76 R which was not significantly different from 0.69R; the latter showed intermediate values between 1.76R and 1.76G, while the lowest values were found in 0.69 G that was significantly different from all the other conditions.

Table 4. Total Ascorbic Acid (TAA), Phenols (PH), Hydrophilic antioxidant activity (HAA) and lipophilic antioxidant activity (LAA) in leaves of green (G) and red (R) lettuce plants grown under the two VPD levels (0.69 and 1.76 kPa).

	TAA mg100 g^{-1} FW	PH mg GA eq. 100 g^{-1} DW	HAA mmol AA eq. 100 g^{-1} DW	LAA mmol trolox eq. 100 g^{-1} DW
Cultivar				
G	76.6 ± 5.22 b	9.21 ± 0.24 b	14.9 ± 0.52 a	30.2 ± 1.73 b
R	97.7 ± 3.74 a	10.3 ± 0.48 a	16.6 ± 0.36 a	42.3 ± 0.70 a
VPD				
0.69 kPa	63.7 ± 3.65 b	8.08 ± 0.21 b	10.3 ± 0.57 b	36.1 ± 1.20 a
1.76 kPa	111 ± 6.86 a	11.9 ± 0.54 a	11.2 ± 0.25 a	36.5 ± 1.77 a
Int.				
0.69 G	52.5 ± 2.93 d	5.83 ± 0.14 c	9.71 ± 0.41 c	28.9 ± 1.01 c
0.69 R	75.1 ± 1.44 c	10.3 ± 0.13 a	10.8 ± 0.32 ab	43.3 ± 0.38 a
1.76 G	101 ± 4.57 b	6.77 ± 0.19 b	10.4 ± 0.21 b	31.5 ± 1.45 b
1.76 R	120 ± 4.59 a	10.3 ± 0.71 a	11.6 ± 0.09 a	41.4 ± 0.64 a
Sig.				
C	**	***	NS	***
VPD	***	*	*	*
C × VPD	*	*	*	*

All data are expressed as mean ± standard error. ***, **, *, NS refer to $p \leq 0.001, 0.01, 0.05$ and Non-significant, respectively. Lower case letters indicate the significant differences of the interaction.

3.5. Soil Plant Analysis Development Index and Chlorophyll a Fluorescence Emission

Results from SPAD and Fv/Fm are showed in Table 5, separated for data (12 and 23 DAT). At 12 and 23 DAT, C and VPD had a significant effect as main factors and in interaction on SPAD index, showing enhanced values in R cultivar (49, 47%) and under 0.69 kPa (17, 4%). Differently, at both 12 and 23 DAT, cultivar did not elicit significant differences in Fv/Fm, whereas VPD had a significant effect with enhanced values under 0.69 kPa (2, 5%). Concerning the interaction, at both 12 and 23 DAT, SPAD index showed higher values in R cultivar under 0.69 kPa, followed by 1.76 R and 1.76 G and 0.69 G, with no differences among them. Differently, Fv/Fm ratio presented significantly higher values in 0.69 kPa with no differences between cultivars, followed by 1.76 kPa, again with no differences between cultivars.

Table 5. Soil Plant Analysis Development (SPAD) index and Fv/Fm in leaves of green (G) and red (R) lettuce plants grown under the two VPD levels (0.69 and 1.76 kPa) at 12 and 23 DAT (days after transplanting).

	SPAD Index		Fv/Fm	
	12 DAT	**23 DAT**	**12 DAT**	**23 DAT**
Cultivar				
G	28.07 ± 0.74 b	27.9 ± 1.46 b	0.77 ± 0.10 a	0.80 ± 0.07a
R	41.85 ± 0.91 a	41.1 ± 1.17 a	0.77 ± 0.30 a	0.80 ± 0.12a
VPD				
0.69 kPa	36.70 ± 0.85 a	35.2 ± 1.00 a	0.78 ± 0.11 a	0.82 ± 0.12a
1.76 kPa	30.51 ± 0.69 b	33.8 ± 1.63 b	0.76 ± 0.20 b	0.78 ± 0.13b
Int.				
0.69 G	28.5 ± 0.57 c	28.5 ± 0.54 c	0.79 ± 0.06 a	0.82 ± 0.02 a
0.69 R	44.9 ± 0.57 a	41.8 ± 0.46 a	0.78 ± 0.06 a	0.82 ± 0.01 a
1.76 G	27.6 ± 0.35 c	27.3 ± 0.92 c	0.75 ± 0.13 b	0.79 ± 0.06 b
1.76 R	38.2 ± 0.69 b	33.4 ± 0.71 b	0.77 ± 0.07 ab	0.78 ± 0.07 b
Sig.				
C	***	*	NS	NS
VPD	***	***	*	***
C × VPD	***	*	*	*

All data are expressed as mean $\pm$ standard error. ***, **, *, NS refer to $p \leq 0.001, 0.01, 0.05$ and Non-significant, respectively. Lower case letters indicate the significant differences of the interaction.

3.6. Hierarchical Clustering of Functional and Nutritional Aspects of Green and Red Salanova

A heat map providing an integrated overview of the effects of cultivar and VPD on the physiological and qualitative traits of Salanova lettuce is displayed in Figure 2. In the left dendrogram, the heat map identified two main clusters, separated by the different cultivar (0.69 G and 1.76 G on one cluster and 0.69 R and 1.76 R on the other); furthermore, as visible in the upper dendrogram also variables grouped together. More specifically, our results indicated that 0.69 R separated from the other treatments because of its highest positive relation with the cations content (Na^{2+}, Ca^{2+}), Hue and malate content, and negative relation with isocitrate content and L*. Whereas, 1.76 R separated from the other treatments mainly due to its antioxidant content, especially $NO_3{}^-$, HF, TAA, and its negative relation with chroma. Differently, 0.69 G presented a higher positive variation of tartrate, isocitrate and the colorimetry parameter chroma. Finally, 1.76 G separated from the others because of its higher accumulation of $SO_4{}^{2-}$ and K^+ and negative variation of citrate.

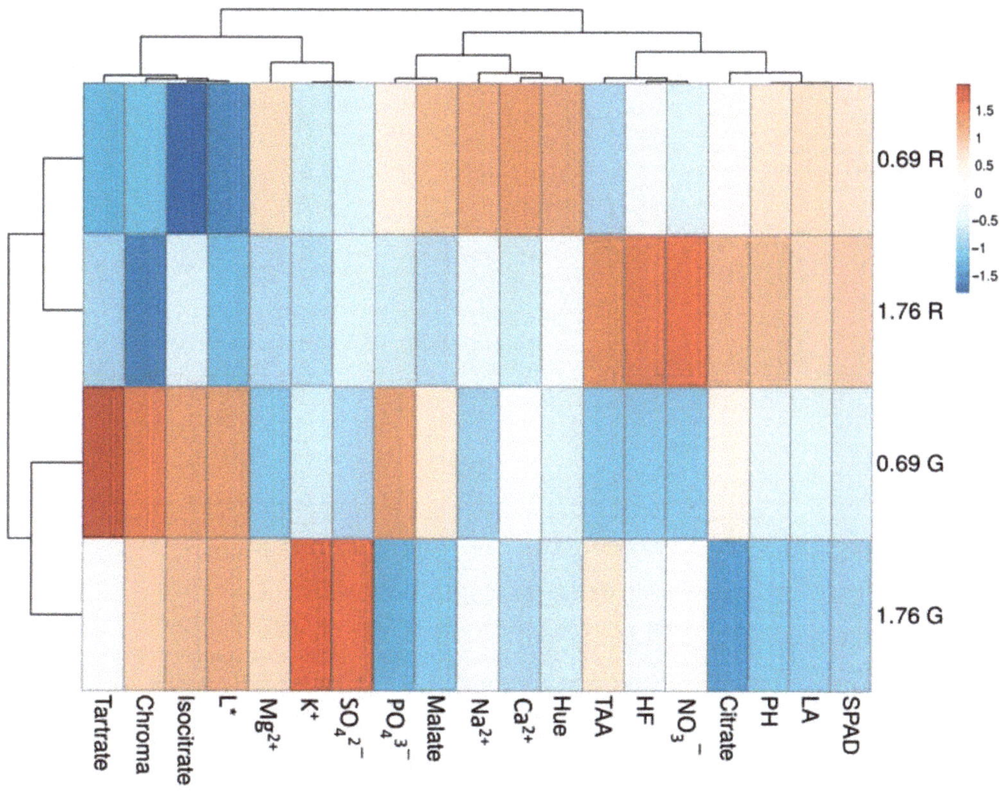

Figure 2. Heat map of qualitative and physiological aspects of green (G) and red (R) lettuce plants grown under the two VPD levels (0.69 and 1.76 kPa).

4. Discussion

Modulating the microclimate in indoor module-cultivation can positively affect crop morpho-physiological development, also leading to differences in appearance and in product quality, especially influencing the content of plant secondary metabolites [1,13,28]. In general, high VPD can limit plant growth and dry matter accumulation, reducing yield and photosynthesis, which are major constrains for crop production [29]. Our results are consistent with this general statement, always showing a lower biomass, number of leaves and canopy area in plants exposed to 1.76 kPa. Moreover, lettuce developed under a low-VPD environment (0.69 kPa), apart from increased growth and biomass, also presented a higher Fv/Fm and chlorophyll content (SPAD index) both at 12 and 23 DAS, overall suggesting a better performance of the photosynthetic apparatus. Indeed, a high content of photosynthetic pigments in plants is often associated with high Fv/Fm values [30] and even small increases in the photosynthetic rates are known to cause wide improvements in crop biomass and yield [31,32]. In low-VPD-exposed *S. lycopersicum* plants, a higher photosynthesis, mostly due to a better regulation of stomatal closure, and consequently high values of Fv/Fm, have been found in correlation with improved yield and biomass [29]. Fv/Fm values lower than 0.8, which is a threshold level for unstressed plants, are common in plants facing the onset of photodamage [33]. Indeed, the chlorophyll "a" fluorescence parameter Fv/Fm reflects the PSII (photosystem II) maximum quantum efficiency and consequently has been widely used as a screening for early stress detection in plants and for improvements in crop production in CEA [34,35]. For example, several researches found a decrease in Fv/Fm in different tomato [36] cultivars subjected to heat stress or

a combination of heat and drought stresses [37]. In the present study, plants at 1.76 kPa always presented values lower than 0.8, suggesting that plants may sense the dry air, characteristic of high VPD as a mild-stress, similarly to what happens in conditions of heat stress or drought. However, microscopy observation in lettuce samples did not evidence VPD-induced differences in lamina thickness and intercellular spaces patterns. It is known that VPD levels can bring to a different morpho-anatomical development of leaf lamina, changing the whole mesophyll structure, thus changing the resistance/conductance to water vapor and CO_2 within the leaf [38,39]. From these different morpho-anatomical characteristics depend the photosynthetic rates and the whole plant physiological behavior. Indeed, although some intra- and inter- species variation is observed, physiological responses cannot overcome plant morpho-anatomical structure [40]. In the present study, the reductions in Fv/Fm, photosynthetic pigments, yield, and biomass were not due to VPD-driven changes in morpho-anatomical structure of leaf lamina probably because the microclimate around the developing leaves under the two VPD treatments was not enough different to induce any differential cell differentiation leading to different mesophyll structure. Therefore, the observed reductions in growth and photosynthetic traits were likely linked with the oxidative stress which typically occurs under unfavorable environmental conditions and which can change crop quality [41].

Several authors have demonstrated that mild to moderate stress stresses could produce higher quality products, especially crop rich in phytochemicals, depending on: the type of stress (environmental, nutritional, etc.), the time of exposure, the intensity of application, as well as the crop species/cultivars [12]. For instance, Da Ge et al. [42], found in maize grains subjected to water stress, increments in Ca^{2+}, Mg^{2+}, Cu^{2+}, and Zn^{2+}. Additionally, El-Nakhel et al. [43] reported that mineral eustress (half strength nutrient solution) was able to boost the phenolic and carotenoids profile in butterhead Salanova in particular in the red-pigmented ones.

Minerals are essential elements in human diet, necessary as co-factors for several enzyme activities [44], and leafy greens are among the prime sources of these nutrients [45]. In our study, 1.76 kPa incremented the concentration of K^+, which is involved as a carrier ion, transporting solutes and hormones in xylem and phloem, other than be involved in enzyme activation, osmotic potential, and synthesis of protein [46]. However, Salanova lettuces at high VPD also presented a high nitrate content, especially in red cultivar. Since leafy greens are usually harvested at vegetative growth stages, and the edible parts can accumulate relatively large amounts of nitrate, these crops have been found to be the major source of nitrate uptake by humans. In the present study, however, nitrate concentration in both Salanova cultivars, were inferior to the European Commission regulation No 1258/2011 [47] which set the NO_3 content for protected-grown lettuce at 5000 mg NO_3 kg^{-1} per fresh weight [48]. The lower concentration of nitrates has been associated with yellowish leaves characterized by a decreased hue angle and increased L*, b*, and chroma [49]. In our study, although presenting a higher content of nitrates, 1.76 VPD lettuces decreased b* and L* producing less dark leaves but with vivid colors (increased chroma). The analysis of color is an important consideration for edible food, since the most common property to measure quality of any material is its appearance and consumers can easily be influenced by a fruit or vegetable color which they consider inappropriate [50]. Furthermore, different research also reported similar relationships between total N/NO_3 concentrations, chlorophyll content and chromaticity parameters (especially L*), so much to suggest the use of colorimeter reader or SPAD meter to predict the total content of chlorophyll and nitrate in a time-saving non-destructive analytical method [51]. Our results did not show such correlation with L*, however red cultivar presented an enhanced SPAD index as well as a highest nitrate content (Tables 3 and 4), compared to green one. The highest chlorophyll concentration (SPAD index) of red lettuces might seem odd and could be explained by the highest content of nitrates in this cultivar.

Moreover, red lettuces also showed a highest content of phytochemicals, compared to green cultivar, especially under 1.76 VPD (Table 5). Many studies have demonstrated that

red-pigmented-leafy-green-cultivars contained highest amounts of metabolites compared to their green counterparts [52]. Just to mention a few, El-Nakhel et al. [43] found in red Salanova lettuces higher quantities of phenolic compounds compared to green Salanova plants. Other studies also found an enhanced quantity of ascorbic acid in red-pigmented lettuce leaves [52,53]. Both phenolic compounds and ascorbic acid are potent antioxidants which confers valuable nutritional properties to vegetables [52,54].

Ascorbic acid, like other vitamins, cannot be synthesized by humans endogenously, so it represents an essential dietary component [55]; thus, ascorbic acid-rich-lettuces could represent an added value for the marketability of the products. It is interesting that increments in ascorbate, polyphenols, and antioxidant capacity were reported in lettuce grown under various types of stress. For instance, in lettuces subjected to moderate stress (heat shock, chilling, high light intensity), Oh et al. [56] found a two/three-fold increase in the total phenolic content and a significant increase in the antioxidant capacity, with no adverse effects on the general plant growth. In our study, lettuces exposed to high VPD always enhanced their phytochemical content, compared to those exposed to low VPD, probably sensing the surrounding environment as a mild stress not able to induce permanent structural changes neither cell shrinkage, but still enough to modulate chlorophyll content, Fv/Fm ratio and biomass which resulted reduced under the 1.76 kPa treatment. Levels of antioxidant molecules, such as ascorbate metabolites, phenolic compounds, and α-tocopherol, higher in high VPD, can indicate a defense against oxidative stress [41]. In this study, we examined the total amount of phenolic compounds in leaves; however, as a future perspective it would be valuable to focus on individual phenolic components to have a more comprehensive idea of plant phytochemical's synthesis in response to VPD.

The most common effect of low humidity rates on crops is to induce leaf water stress, since under this environmental condition the uptake of water from the soil is not enough to cope with the high transpiration rates [6,57]. Indeed, when subjected to high VPD, plants begin to dehydrate and start to physically translocate a larger volume of soil water through the plant system, which can also exacerbate the stress if in interaction with other adverse environmental conditions like high EC rates, bringing to the accumulation of additional salts within the plant [58]. Still, the use of these mild-stress during cultivation techniques has proven to increase tomato fruit dry matter content [59], which is an important parameter in improving yield and nutritional quality [60] also increasing sugar content, the ratio of sugar:acids [61], and the synthesis of secondary metabolites and antioxidants [9,62]. There is evidence that many antioxidants play a key role in plant adaptation to abiotic and biotic stresses [56,63]. Additionally, a significant part of antioxidants produced by plants in response to stress is secondary metabolites, including some simple and complex phenolic compounds derived primarily via the phenylpropanoid pathway [64].

As a number of these are phytochemicals with health-promoting qualities in the human diet, in the light of the above results, it would be feasible to use VPD, among other mild environmental stresses, to enhance the phytochemical content of lettuce or other common leafy vegetable. To date there are no clear indications on how to use high VPD levels, in a sort of plant "hardening off", to ameliorate the nutraceutical value of leafy greens. A next step could be to grow plants under optimal conditions and then subject them to short periods of high VPD to promptly increase their antioxidant levels, without reducing plant photosynthesis and consequently crop production.

Author Contributions: Conceptualization, C.A. and V.D.M.; methodology, C.A., Y.R. and V.D.M.; formal analysis, C.A.; investigation, C.A.; resources, Y.R., S.D.P., V.D.M.; data curation, C.A. and V.D.M.; writing—original draft preparation, C.A.; writing—review and editing, C.A., Y.R., S.D.P., V.D.M.; supervision, V.D.M.; funding acquisition, Y.R., S.D.P., V.D.M. All authors have read and agreed to the published version of the manuscript.

Funding: This research was conducted in the framework of the Ph.D. sponsored by the Italian Ministry of University and Research (PON research and innovation).

Institutional Review Board Statement: Not applicable.

Informed Consent Statement: Not applicable.

Data Availability Statement: Data available on request due to restrictions eg privacy or ethical. The data presented in this study are available on request from the corresponding author.

Acknowledgments: We wish to thank Carmen Arena and Chiara Cirillo for support in chlorophyll "a" fluorescence measurements and data analysis. We also wish to thank Antonio Pannico for technical support during the cultivation and Maria Giordano for technical support during qualitative analysis.

Conflicts of Interest: The authors declare no conflict of interest (financial or nonfinancial) for this research.

References

1. Gruda, N. Impact of Environmental Factors on Product Quality of Greenhouse Vegetables for Fresh Consumption. *Crit. Rev. Plant Sci.* **2005**, *24*, 227–247. [CrossRef]
2. Bakker, J. Physiological disorders in cucumber under high humidity conditions and low ventilation rates in greenhouses. *ISHS Acta Hortic.* **1984**, 257–264. [CrossRef]
3. Amitrano, C.; Arena, C.; Rouphael, Y.; De Pascale, S.; De Micco, V. Vapour pressure deficit: The hidden driver behind plant morphofunctional traits in controlled environments. *Ann. Appl. Biol.* **2019**, *175*, 313–325. [CrossRef]
4. Krug, H. *Gemuseproduktion: Ein Lehr-und Nachschlagewerk fur Studium und Praxis*; Verlag Paul Parey: Singhofen, Germany, 1986.
5. Collier, G.F. Tipburn of Littuce. In *Horticultural Reviews*; Wiley Blackwell: Hoboken, NJ, USA, 1982; Volume 4, pp. 49–65.
6. Leonardi, C.; Guichard, S.; Bertin, N. High vapour pressure deficit influences growth, transpiration and quality of tomato fruits. *Sci. Hortic.* **2000**, *84*, 285–296. [CrossRef]
7. Sinclair, T.R.; Devi, J.; Shekoofa, A.; Choudhary, S.; Sadok, W.; Vadez, V.; Riar, M.; Rufty, T. Limited-transpiration response to high vapor pressure deficit in crop species. *Plant Sci.* **2017**, *260*, 109–118. [CrossRef] [PubMed]
8. Dorai, M.; Papadopoulos, A.; Gosselin, A. Influence of electric conductivity management on greenhouse tomato yield and fruit quality. *Agronomie* **2001**, *21*, 367–383. [CrossRef]
9. Rosales, M.A.; Cervilla, L.M.; Sanchez-Rodriguez, E.; Rubio-Wilhelmi, M.M.; Blasco, B.; Rios, J.J.; Soriano, T.; Castilla, N.; Romero, L.; Ruiz, J.M. The effect of environmental conditions on nutritional quality of cherry tomato fruits: Evaluation of two experimental Mediterranean greenhouses. *J. Sci. Food Agric.* **2011**, *91*, 152–162. [CrossRef] [PubMed]
10. Davey, M.W.; Montagu, M.V.; Inze, D.; Sanmartin, M.; Kanellis, A.; Smirnoff, N.; Benzie, I.J.J.; Strain, J.J.; Favell, D.; Fletcher, J. Plant L-ascorbic acid: Chemistry, function, metabolism, bioavailability and effects of processing. *J. Sci. Food Agric.* **2000**, *80*, 825–860. [CrossRef]
11. Rosales, M.A.; Ruiz, J.M.; Hernández, J.; Soriano, T.; Castilla, N.; Romero, L. Antioxidant content and ascorbate metabolism in cherry tomato exocarp in relation to temperature and solar radiation. *J. Sci. Food Agric.* **2006**, *86*, 1545–1551. [CrossRef]
12. Wang, Y.; Frei, M. Stressed food—The impact of abiotic environmental stresses on crop quality. *Agric. Ecosyst. Environ.* **2011**, *141*, 271–286. [CrossRef]
13. Bisbis, M.B.; Gruda, N.; Blanke, M. Potential impacts of climate change on vegetable production and product quality—A review. *J. Clean. Prod.* **2018**, *170*, 1602–1620. [CrossRef]
14. Favati, F.; Lovelli, S.; Galgano, F.; Miccolis, V.; Di Tommaso, T.; Candido, V. Processing tomato quality as affected by irrigation scheduling. *Sci. Hortic.* **2009**, *122*, 562–571. [CrossRef]
15. Kyriacou, M.C.; Rouphael, Y. Towards a new definition of quality for fresh fruits and vegetables. *Sci. Hortic.* **2018**, *234*, 463–469. [CrossRef]
16. Amitrano, C.; Arena, C.; De Pascale, S.; De Micco, V. Light and Low Relative Humidity Increase Antioxidants Content in Mung Bean (*Vigna radiata* L.) Sprouts. *Plants* **2020**, *9*, 1093. [CrossRef] [PubMed]
17. Llorach, R.; Martinez-Sanchez, A.; Tomas-Barberan, F.A.; Gil, M.I.; Ferreres, F. Characterisation of polyphenols and antioxidant properties of five lettuce varieties and escarole. *Food Chem.* **2008**, *108*, 1028–1038. [CrossRef] [PubMed]
18. El-Nakhel, C.; Pannico, A.; Graziani, G.; Kyriacou, M.C.; Giordano, M.; Ritieni, A.; De Pascale, S.; Rouphael, Y. Variation in Macronutrient Content, Phytochemical Constitution and In Vitro Antioxidant Capacity of Green and Red Butterhead Lettuce Dictated by Different Developmental Stages of Harvest Maturity. *Antioxidants* **2020**, *9*, 300. [CrossRef]
19. Slavin, J.L.; Lloyd, B. Health benefits of fruits and vegetables. *Adv. Nutr.* **2012**, *3*, 506–516. [CrossRef]
20. Gil, M.I. Preharvest factors and fresh-cut quality of leafy vegetables. *Acta Hortic.* **2016**, 57–64. [CrossRef]
21. Rouphael, Y.; Kyriacou, M.C.; Petropoulos, S.A.; De Pascale, S.; Colla, G. Improving vegetable quality in controlled environments. *Sci. Hortic.* **2018**, *234*, 275–289. [CrossRef]
22. Feder, N.; O'Brien, T. Plant microtechnique: Some principles and new methods. *Am. J. Bot.* **1968**, *55*, 123–142. [CrossRef]
23. De Micco, V.; Amitrano, C.; Stinca, A.; Izzo, L.G.; Zalloni, E.; Balzano, A.; Barile, R.; Conti, P.; Arena, C. Dust accumulation due to anthropogenic impact induces anatomical and photochemical changes in leaves of Centranthus ruber growing on the slope of the Vesuvius volcano. *Plant Biol.* **2019**, *22*, 93–102. [CrossRef]

24. Kampfenkel, K.; Van Montagu, M.; Inzé, D. Effects of iron excess on Nicotiana plumbaginifolia plants (implications to oxidative stress). *Plant Physiol.* **1995**, *107*, 725–735. [CrossRef] [PubMed]
25. Singleton, V.L.; Orthofer, R.; Lamuela-Raventós, R.M. Analysis of total phenols and other oxidation substrates and antioxidants by means of folin-ciocalteu reagent. *Methods Enzymol.* **1999**, *299*, 152–178.
26. Fogliano, V.; Verde, V.; Randazzo, G.; Ritieni, A. Method for measuring antioxidant activity and its application to monitoring the antioxidant capacity of wines. *J. Agric. Food Chem.* **1999**, *47*, 1035–1040. [CrossRef]
27. Re, R.; Pellegrini, N.; Proteggente, A.; Pannala, A.; Yang, M.; Rice-Evans, C. Antioxidant activity applying an improved ABTS radical cation decolorization assay. *Free Radic. Biol. Med.* **1999**, *26*, 1231–1237. [CrossRef]
28. Gruda, N. Do soilless culture systems have an influence on product quality of vegetables? *J. Appl. Bot. Food Qual.* **2009**, *82*, 141–147.
29. Zhang, D.; Du, Q.; Zhang, Z.; Jiao, X.; Song, X.; Li, J. Vapour pressure deficit control in relation to water transport and water productivity in greenhouse tomato production during summer. *Sci. Rep.* **2017**, *7*, 43461. [CrossRef]
30. Carillo, P.; Cirillo, C.; De Micco, V.; Arena, C.; De Pascale, S.; Rouphael, Y. Morpho-anatomical, physiological and biochemical adaptive responses to saline water of Bougainvillea spectabilis Willd. trained to different canopy shapes. *Agric. Water Manag.* **2019**, *212*, 12–22. [CrossRef]
31. Parry, M.A.; Reynolds, M.; Salvucci, M.E.; Raines, C.; Andralojc, P.J.; Zhu, X.-G.; Price, G.D.; Condon, A.G.; Furbank, R.T. Raising yield potential of wheat. II. Increasing photosynthetic capacity and efficiency. *J. Exp. Bot.* **2011**, *62*, 453–467. [CrossRef]
32. Zhu, X.-G.; Long, S.P.; Ort, D.R. Improving photosynthetic efficiency for greater yield. *Annu. Rev. Plant Biol.* **2010**, *61*, 235–261. [CrossRef]
33. Ogaya, R.; Peñuelas, J.; Asensio, D.; Llusià, J. Chlorophyll fluorescence responses to temperature and water availability in two co-dominant Mediterranean shrub and tree species in a long-term field experiment simulating climate change. *Environ. Exp. Bot.* **2011**, *71*, 123–127. [CrossRef]
34. Sharma, D.K.; Andersen, S.B.; Ottosen, C.O.; Rosenqvist, E. Wheat cultivars selected for high Fv/Fm under heat stress maintain high photosynthesis, total chlorophyll, stomatal conductance, transpiration and dry matter. *Physiol. Plant.* **2015**, *153*, 284–298. [CrossRef]
35. Baker, N.R.; Rosenqvist, E. Applications of chlorophyll fluorescence can improve crop production strategies: An examination of future possibilities. *J. Exp. Bot.* **2004**, *55*, 1607–1621. [CrossRef] [PubMed]
36. Poudyal, D.; Rosenqvist, E.; Ottosen, C.-O. Phenotyping from lab to field—Tomato lines screened for heat stress using Fv/Fm maintain high fruit yield during thermal stress in the field. *Funct. Plant Biol.* **2019**, *46*, 44–55. [CrossRef]
37. Nankishore, A.; Farrell, A.D. The response of contrasting tomato genotypes to combined heat and drought stress. *J. Plant Physiol.* **2016**, *202*, 75–82. [CrossRef] [PubMed]
38. Du, Q.; Liu, T.; Jiao, X.; Song, X.; Zhang, J.; Li, J. Leaf anatomical adaptations have central roles in photosynthetic acclimation to humidity. *J. Exp. Bot.* **2019**, *70*, 4949–4962. [CrossRef]
39. Tosens, T.; Niinemets, U.; Vislap, V.; Eichelmann, H.; Diez, P.C. Developmental changes in mesophyll diffusion conductance and photosynthetic capacity under different light and water availabilities in Populus tremula: How structure constrains function. *Plant Cell Env.* **2012**, *35*, 839–856. [CrossRef]
40. Murphy, M.R.C.; Jordan, G.J.; Brodribb, T.J. Acclimation to humidity modifies the link between leaf size and the density of veins and stomata. *Plant Cell Environ.* **2014**, *37*, 124–131. [CrossRef]
41. Lihavainen, J.; Keinänen, M.; Keski-Saari, S.; Kontunen-Soppela, S.; Sõber, A.; Oksanen, E. Artificially decreased vapour pressure deficit in field conditions modifies foliar metabolite profiles in birch and aspen. *J. Exp. Bot.* **2016**, *67*, 4367–4378. [CrossRef] [PubMed]
42. Da Ge, T.; Sui, F.G.; Nie, S.; Sun, N.B.; Xiao, H.; Tong, C.L. Differential responses of yield and selected nutritional compositions to drought stress in summer maize grains. *J. Plant Nutr.* **2010**, *33*, 1811–1818. [CrossRef]
43. El-Nakhel, C.; Giordano, M.; Pannico, A.; Carillo, P.; Fusco, G.M.; De Pascale, S.; Rouphael, Y. Cultivar-specific performance and qualitative descriptors for butterhead Salanova lettuce produced in closed soilless cultivation as a candidate salad crop for human life support in space. *Life* **2019**, *9*, 61. [CrossRef]
44. Soetan, K.; Olaiya, C.; Oyewole, O. The importance of mineral elements for humans, domestic animals and plants—A review. *Afr. J. Food Sci.* **2010**, *4*, 200–222.
45. Gupta, K.; Wagle, D. Nutritional and antinutritional factors of green leafy vegetables. *J. Agric. Food Chem.* **1988**, *36*, 472–474. [CrossRef]
46. Peuke, A.D.; Jeschke, W.D.; Hartung, W. Flows of elements, ions and abscisic acid in Ricinus communis and site of nitrate reduction under potassium limitation. *J. Exp. Bot.* **2002**, *53*, 241–250. [CrossRef] [PubMed]
47. Statement on possible public health risks for infants and young children from the presence of nitrates in leafy vegetables. *EFSA J.* **2010**, *8*, 1935. [CrossRef]
48. Colla, G.; Kim, H.-J.; Kyriacou, M.C.; Rouphael, Y. Nitrate in fruits and vegetables. *Sci. Hortic.* **2018**, *237*, 221–238. [CrossRef]
49. Navarro, A.; Elia, A.; Conversa, G.; Campi, P.; Mastrorilli, M. Potted mycorrhizal carnation plants and saline stress: Growth, quality and nutritional plant responses. *Sci. Hortic.* **2012**, *140*, 131–139. [CrossRef]
50. Jha, S.N. Colour Measurements and Modeling. In *Nondestructive Evaluation of Food Quality*; Springer: Berlin/Heidelberg, Germany, 2010; pp. 17–40.

51. Tuncay, O. Relationships between nitrate, chlorophyll and chromaticity values in rocket salad and parsley. *Afr. J. Biotechnol.* **2011**, *10*, 17152–17159.
52. Kim, D.E.; Shang, X.; Assefa, A.D.; Keum, Y.S.; Saini, R.K. Metabolite profiling of green, green/red, and red lettuce cultivars: Variation in health beneficial compounds and antioxidant potential. *Food Res. Int.* **2018**, *105*, 361–370. [CrossRef]
53. Baslam, M.; Morales, F.; Garmendia, I.; Goicoechea, N. Nutritional quality of outer and inner leaves of green and red pigmented lettuces (*Lactuca sativa* L.) consumed as salads. *Sci. Hortic.* **2013**, *151*, 103–111. [CrossRef]
54. El-Nakhel, C.; Petropoulos, S.A.; Pannico, A.; Kyriacou, M.C.; Giordano, M.; Colla, G.; Troise, A.D.; Vitaglione, P.; De Pascale, S.; Rouphael, Y. The bioactive profile of lettuce produced in a closed soilless system as configured by combinatorial effects of genotype and macrocation supply composition. *Food Chem.* **2020**, *309*, 125713. [CrossRef] [PubMed]
55. Li, Y.; Schellhorn, H.E. New developments and novel therapeutic perspectives for vitamin C. *J. Nutr.* **2007**, *137*, 2171–2184. [CrossRef]
56. Oh, M.M.; Carey, E.E.; Rajashekar, C.B. Environmental stresses induce health-promoting phytochemicals in lettuce. *Plant Physiol. Biochem.* **2009**, *47*, 578–583. [CrossRef]
57. Grange, R.; Hand, D. A review of the effects of atmospheric humidity on the growth of horticultural crops. *J. Hortic. Sci.* **1987**, *62*, 125–134. [CrossRef]
58. Tack, J.; Singh, R.K.; Nalley, L.L.; Viraktamath, B.C.; Krishnamurthy, S.L.; Lyman, N.; Jagadish, K.S. High vapor pressure deficit drives salt-stress-induced rice yield losses in India. *Glob. Chang. Biol.* **2015**, *21*, 1668–1678. [CrossRef] [PubMed]
59. Ho, L.; Grange, R.; Picken, A. An analysis of the accumulation of water and dry matter in tomato fruit. *Plant Cell Environ.* **1987**, *10*, 157–162.
60. Boyer, J.S. Plant productivity and environment. *Science* **1982**, *218*, 443–448. [CrossRef]
61. Bertin, N.; Guichard, S.; Leonardi, C.; Longuenesse, J.; Langlois, D.; Navez, B. Seasonal evolution of the quality of fresh glasshouse tomatoes under Mediterranean conditions, as affected by air vapour pressure deficit and plant fruit load. *Ann. Bot.* **2000**, *85*, 741–750. [CrossRef]
62. Rosales, M.A.; Ríos, J.J.; Cervilla, L.M.; Rubio-Wilhelmi, M.M.; Blasco, B.; Ruiz, J.M.; Romero, L. Environmental conditions in relation to stress in cherry tomato fruits in two experimental Mediterranean greenhouses. *J. Sci. Food Agric.* **2009**, *89*, 735–742. [CrossRef]
63. Burritt, D.J.; Mackenzie, S. Antioxidant metabolism during acclimation of Begonia × erythrophylla to high light levels. *Ann. Bot.* **2003**, *91*, 783–794. [CrossRef]
64. Dixon, R.A.; Paiva, N.L. Stress-Induced Phenylpropanoid Metabolism. *Plant Cell* **1995**, *7*, 1085. [CrossRef] [PubMed]

Article

Estimation of Glucosinolates and Anthocyanins in Kale Leaves Grown in a Plant Factory Using Spectral Reflectance

Milon Chowdhury [1,2], Viet-Duc Ngo [3], Md Nafiul Islam [1], Mohammod Ali [1], Sumaiya Islam [2], Kamal Rasool [1], Sang-Un Park [4] and Sun-Ok Chung [1,2,*]

1   Department of Agricultural Machinery Engineering, Graduate school, Chungnam National University, Daejeon 34134, Korea; chowdhurym90@cnu.ac.kr (M.C.); nafiulislam@cnu.ac.kr (M.N.I.); sdali77@o.cnu.ac.kr (M.A.); kamalrasool@o.cnu.ac.kr (K.R.)
2   Department of Smart Agricultural Systems, Graduate School, Chungnam National University, Daejeon 34134, Korea; dina0075@o.cnu.ac.kr
3   Division of Computational Mechatronics, Institute for Computational Science, Ton Duc Thang University, Ho Chi Minh City 700000, Vietnam; ngovietduc@tdtu.edu.vn
4   Department of Crop Science, Graduate school, Chungnam National University, Daejeon 34134, Korea; supark@cnu.ac.kr
*   Correspondence: sochung@cnu.ac.kr; Tel.: +82-42-821-6712; Fax: +82-42-823-6246

Abstract: The spectral reflectance technique for the quantification of the functional components was applied in different studies for different crops, but related research on kale leaves is limited. This study was conducted to estimate the glucosinolate and anthocyanin components of kale leaves cultivated in a plant factory based on diffuse reflectance spectroscopy through regression methods. Kale was grown in a plant factory under different treatments. After specific periods of transplantation, leaf samples were collected, and reflectance spectra were measured immediately from nine different points on each leaf. The same leaf samples were freeze-dried and stored for analysis of the functional components. Regression procedures, such as principal component regression (PCR), partial least squares regression (PLSR), and stepwise multiple linear regression (SMLR), were applied to relate the functional components with the spectral data. In the laboratory analysis, progoitrin and glucobrassicin, as well as cyanidin and malvidin, were found to be dominating components in glucosinolates and anthocyanins, respectively. From the overall analysis, the SMLR model showed better performance, and the identified wavelengths for estimating the glucosinolates and anthocyanins were in the early near-infrared (NIR) region. Specifically, reflectance at 742, 761, 787, 796, 805, 833, 855, 932, 947, and 1000 nm showed a strong correlation.

Keywords: protected horticulture; crop sensor; functional components; reflectance spectroscopy

Citation: Chowdhury, M.; Ngo, V.-D.; Islam, M.N.; Ali, M.; Islam, S.; Rasool, K.; Park, S.-U.; Chung, S.-O. Estimation of Glucosinolates and Anthocyanins in Kale Leaves Grown in a Plant Factory Using Spectral Reflectance. *Horticulturae* **2021**, *7*, 56. https://doi.org/10.3390/horticulturae7030056

Academic Editors: Christophe El-Nakhel and Elazar Fallik

Received: 1 January 2021
Accepted: 18 March 2021
Published: 21 March 2021

1. Introduction

Kale (*Brassica oleracea var. alboglabra*) is one of the major sources of phytonutrient components (e.g., glucosinolates, anthocyanins, carotenoids, amino acids, and sugars), from which glucosinolates and anthocyanins are well known for containing cancer-chemo preventive compounds. In basic terms, glucosinolates and anthocyanins are the combination of secondary metabolites, enriched with nitrogen and sulfur-containing glycosides, available in species of the Brassicaceae families [1]. In-vitro and in-vivo studies reported that glucosinolates and their breakdown components inhibited many cancer development steps, such as phase I and II modulation of detoxification enzymes [2]. Consumption of anthocyanins reduces the risk of cardiovascular problems, diabetes, and cancer due to their anti-inflammatory and antioxidant activities [3,4].

Quantification of functional components in vegetables and fruits is important to nutritionists and researchers but also essential to farmers for producing nutrient-rich crops. Usually, functional components are analyzed in the laboratory using different

reagents, tools, and equipment, which is time-consuming and labor-intensive work and sometimes hazardous to humans. High-performance liquid chromatography (HPLC) analysis, commonly used in laboratories, is one of the popular methods for determining the level of functional components. This method requires specialized technicians, time, and usually involves high costs. The stability of results and the health of humans and the environment can be affected by the reagents used during the extraction steps and the HPLC analysis procedure [5–8]. Reflectance spectroscopic techniques could be considered as an alternative method due to their nondestructive and simple procedure, quick response characteristics, and the relatively small amount of samples required [9].

The spectral reflectance technique is applied to estimate the functional components and nutritional status of different vegetables, where significant differences in results were observed based on the vegetable species, cultivation methods, and environmental parameters [10–14]. The application of the reflectance spectroscopy technique was introduced in the early 20th century [15]. Results showed that the wavelengths in the infrared region were suitable for the rapid assessment of forage quality. Recently, researchers [9,16] applied and analyzed the reflectance spectra to estimate the nutritional and functional components in various leafy vegetables and medicinal plants. Generally, leaf reflectance spectra were low in the visible region (from 400 to 700 nm) due to light absorption by photosynthesis pigments (e.g., chlorophyll and carotenoids) [8], and sometimes physiological structure also affects the magnitude of reflectance [17]. Leaf pigments such as chlorophyll, carotenoids, and anthocyanins in higher plants could easily be detected in the reflectance range of 400 to 800 nm. More specifically, total chlorophyll content was identified either at 540 to 560 nm (green region) or 700 to 730 nm (red region) and also at 760 nm in beech leaves [18]; 540 to 560 nm and 700 to 705 nm, depending on detection point of lettuce; and 510 to 540 nm for spinach [19]. The carotenoid contents of maple, chestnut, and beech leaves were identified, and it was reported that 510 to 550 nm are closely related to total pigment content [20].

As the consumption of certain cruciferous vegetables (i.e., kale, cabbage) is more strongly related to health benefits, the estimation and study of health-promoting components, specifically glucosinolates and anthocyanins, through spectral reflectance are quite popular among researchers and growers [21–24]. The functional components of Chinese cabbage were estimated using diffuse reflectance spectroscopy, and the sugar, amino acid, glucosinolate and carotenoid contents were modelled using wavelengths of 317, 390, 888, 940 nm; 520, 960 nm; 385, 860, 945 nm; and 454, 472, 530 nm, respectively [9]. The effects of various fertilizer treatments and light intensity on functional components in white head cabbage and Chinese kale were also inspected, respectively [25,26]. A comparison was also carried out to show the similarities of wavelength ranges for estimating the functional components of kale and Chinese cabbage cultivated in plant factory and reported that leading wavelengths were found under 470~1050 nm and 317~960 nm, respectively [27].

Traditionally, kale is cultivated in the open fields using different bio-extracts or bio-decomposed matter with moderate fertilizer and pesticides to reduce impacts on soils. However, maintenance of quantity and nutrient contents cannot be ensured [28–30]. Kim and Chung [30] showed that plant growth and glucosinolate contents were greater in protected cultivation facilities, such as greenhouses and plant factories, than open field cultivation. Moreover, kale production using hydroponic systems has gained popularity in recent years due to uniform controlled environments, sustainable growth, efficient use of nutrients, lower rates of diseases, and year-round high-quality production with minimum influence of geological and climatic conditions [31–33]. Controlled environmental factors, namely temperature, humidity, carbon dioxide, light, and soil fertility, have significant effects on the concentration of health-promoting components, specifically glucosinolates in growing plants and distribution among plant organs. Determination of the effects of controlled environmental parameters and soil properties on the growth, formation, release, germination, yield, and quality of crops have been the focus of many studies worldwide [34–36]. Analysis of the functional components (e.g., glucosinolates, antho-

cyanins, carotenoids, amino acids, and sugars) is also essential for ensuring nutritional levels and determining proper harvesting schedules [9,37,38].

Although controlled-environment cultivation increases the concentration of the functional components of crops, the accumulation rate of every component is not the same. Quantification of the functional components of various vegetables using the spectral reflectance technique was reported in some studies, but significant differences in results were observed, even among species of the same crop family, due to cultivation methods and environmental parameters. As research related to the quantification of functional components on kale leaves is limited, the objective of this study was to estimate the contents of glucosinolates and anthocyanins in leaves of kale grown in a plant factory through UV/VIS/NIR-diffuse reflectance spectroscopy data.

2. Materials and Methods

2.1. Kale Cultivation in a Plant Factory

A plant factory is a closed crop cultivation facility used to grow high-value crops of a high quality throughout the year by utilizing artificially controlled environmental parameters. In this study, a plant factory was used to maintain the desired levels of the controlled environmental parameters precisely and evaluate their effect on the accumulation of the nutritional components of kale. For this purpose, three experiments for the ambient environmental factors (temperature, humidity, and carbon dioxide (CO_2)), three experiments for the light conditions (light type, light intensity, and light photoperiod), and one experiment for the electrical conductivity (EC) were implemented. In each experiment, five different levels for each environmental factor were implemented. All the treatments were prepared following the guidelines of the horticultural crop cultivation process and summarized in Table 1. A wireless sensor and control networks were used for monitoring and controlling the ambient environmental conditions, as detailed by Chung et al. [39].

Kale was selected because it is a nutrient-dense vegetable and is considered a healthy and popular food in many countries for its powerful medicinal properties. A commercial kale variety with smooth, green leaves and a hard and fibred stem was cultivated in the plant factory. Kale seeds were sown in a hydroponic germination sponge, and three weeks after germination, healthy seedlings with true leaves were transplanted into the plant beds. Twenty-four plants were placed in each plant bed. Seedlings were grown using a recycle-type aeroponic nutrient management system over a period of about 40 to 50 days (Figure 1). Commercial nutrient solutions A and B (Daeyu Co., Ltd., Seoul, Korea) were used, and the target nutrient level was monitored and managed once a day using the electrical conduction (EC) and pH sensors. The nutrient solution was sprayed onto the plant root zone for two minutes at 15-min intervals. The unused solution was returned to the nutrient mixing tank, and filtration and sterilization were performed using a commercial UV-sterilizer and filter (HY-600F, Haiyang, China). Additional distilled water and stock solutions were added during nutrient replenishment to prepare the target nutrient solution.

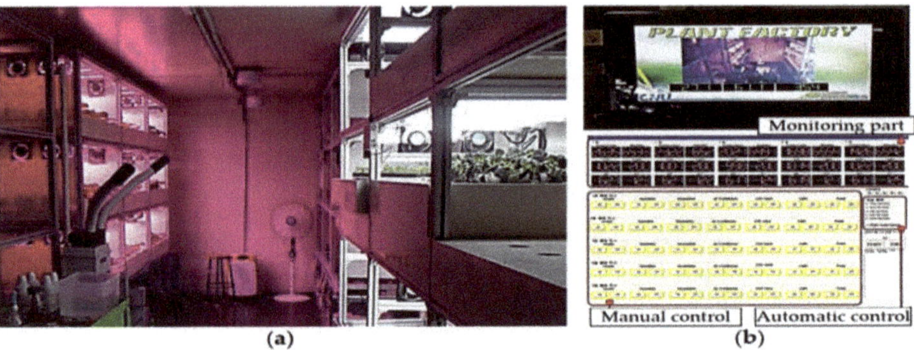

Figure 1. (**a**) Photo of the experimental plant factory and (**b**) ambient environment monitoring and control system.

Table 1. Summary of the target and obtained levels of each treatment along with the specification of sensors used during the cultivation period of kale plants in the plant factory.

Environmental Factors	Experimental Treatments		Specification of the Used Sensors
	Target Levels	**Obtained Levels**	
Temperature (°C)	14 ± 1	14.58 ± 0.74	Model: ETH-01DV
	17 ± 1	17.34 ± 1.8	Range: $-40 \sim 125\,°C$
	20 ± 1	20.25 ± 0.69	Resolution: 14 bit
	23 ± 1	23.26 ± 0.52	Accuracy: $\pm 1.3\,°C$
	25 ± 1	25.97 ± 1.64	
Humidity (%)	45 ± 5	45.78 ± 6.23	Model: ETH-01DV
	55 ± 5	58.06 ± 4.35	Range: $0 \sim 100\%$
	65 ± 5	67.66 ± 4.67	Resolution: 14bit
	75 ± 5	72.66 ± 4.49	Accuracy: $\pm 4.5\%$
	85 ± 5	83.85 ± 4.65	
CO_2 (ppm)	400 ± 100	475.62 ± 106.3	Model: SH-300-DX
	700 ± 100	723.9 ± 140.6	Range: $0 \sim 5000$ ppm
	1000 ± 100	1008.75 ± 175.36	Response time: < 30 s
	1300 ± 100	1375.5 ± 125.11	Accuracy: $\pm 2\%$
	1600 ± 100	1693.21 ± 137.2	
Light source (light emitting diode color ratio)	[1] R:B:W, R:B, R:W, Fluorescent	-	-
Photosynthetic photon flux density ($\mu mol\,m^{-2}s^{-1}$)	100, 130, 160, 190, 220	-	Model: GY-30 Range: 1–65,535 lux Resolution: 16-bit Accuracy: $\pm 3\%$
Photoperiod (day/night hours)	12/12, 14/10, 16/8, 18/6, 20/4	-	Time switch: MaxiRex 5QT Prds rating: AC 230 V, 60 Hz Load capacity: 16 A
Electrical conduction ($\mu S\,cm^{-1}$)	0.80 ± 0.2	0.86 ± 0.3	Model: conductivity probe
	1.00 ± 0.2	1.02 ± 0.34	Range: $2 \sim 20,000\,\mu Scm^{-1}$
	1.20 ± 0.2	1.28 ± 0.22	Resolution: $10\,\mu Scm^{-1}$
	1.40 ± 0.2	1.39 ± 0.24	Accuracy: $\pm 4\%$
	1.60 ± 0.2	1.63 ± 0.25	

[1] R, red; B, blue; W, white.

2.2. Leaf Sample Collection

Two and four weeks after transplanting, leaf sample collection was performed in three steps. Mature and healthy leaf samples were visually selected and collected according to their color and size condition for spectral reflectance measurement and analysis of functional components. Three normal-sized, matured, and healthy leaves were harvested from each plant, and total nine leaves were collected from three plants in each plant bed. Three replications were applied. In total, 27 (3 leaves × 3 plants × 3 replications) kale leaves were harvested. The reflectance spectra were measured first, and the leaves were transferred to the chemical laboratory immediately (to minimize the degradation of nutrient contents) for functional component analyses using a commercial high-performance liquid chromatography (HPLC) machine (model: 1200 series, Agilent Technologies, Santa Clara, CA, USA). Figure 2 shows the different growth stages of kale after different periods of transplantation, a selected plant, and a harvested leaf sample for reflectance acquisition and functional components analysis. The measured reflectance spectra and functional components from 9 leaves were averaged to represent one data point. In total, 204 data points (34 treatments × 2 sampling time × 3 replications) for glucosinolates contents and 90 data points (15 treatments of temperature, humidity, and CO_2 × 2 sampling time × 3 replications) for anthocyanins contents were obtained.

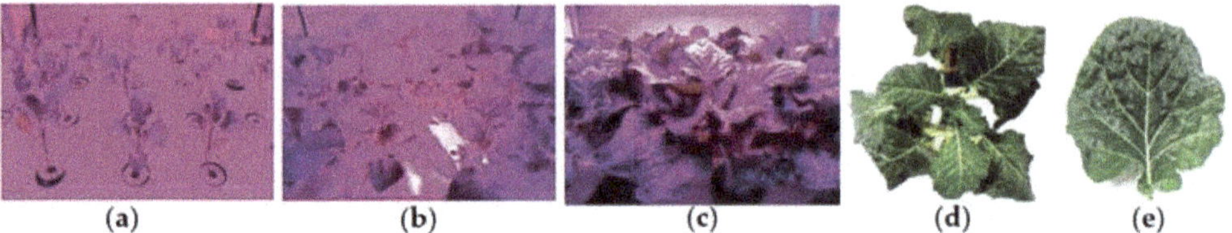

Figure 2. Kale growth after different periods of transplantation for (**a**) 0 day, (**b**) 2 weeks, and (**c**) 4 weeks after transplantation. (**d**) Selected plant for sampling and (**e**) harvested leaf sample for reflectance acquisition and functional components analysis.

2.3. Reflectance Spectra Acquisition

Right after the leaf sample collection, the reflected spectra were measured from each sample leaf using a spectrometer (model: Jaz-Combo-2, Ocean Optics, FL, USA). The applied wavelength range was 190 to 1130 nm with an interval of 0.37 nm, where one detector device provided 190 to 890 nm (UV/VIS) and other detector devices provided 470 to 1130 nm (NIR). However, the wavelength range from 300 to 1050 nm was considered during analysis to avoid excessive noise at the edge of the wavelengths. The received spectra from the mentioned detectors were joined, centering at about 720 nm. The reflected spectra were collected in the dark to minimize the noise caused by background effects and other circumstances. The spectrometer was operated by software provided by the manufacturer. Following the methodology of the previous studies of [9,37,40], the spectral reflectance data were measured from 9 sampling points over the blade part of each sampling leaf, as shown in Figure 3.

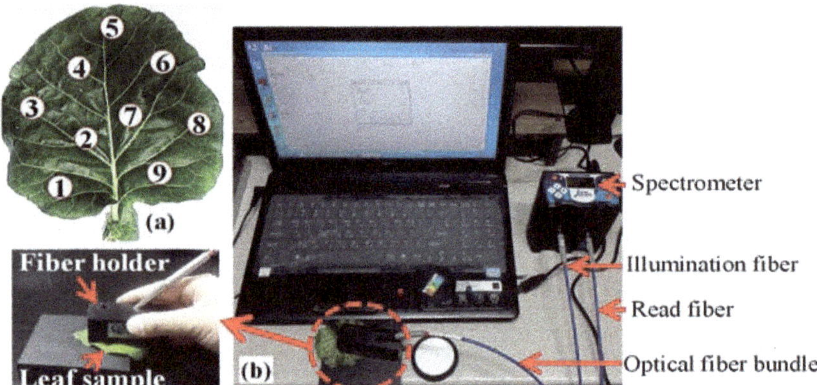

Figure 3. Reflectance spectra acquisition: (**a**) locations of spectral reflectance measurement and (**b**) spectral data acquisition system.

2.4. Extraction of Glucosinolates and Anthocyanins

The freshly harvested leaf samples were freeze-dried using liquid nitrogen for 48 h and ground into a fine powder using a pestle and mortar. Part of the freeze-dried samples (100 mg) was separated, and the crude glucosinolates were extracted with 70% boiling methanol (4.5 mL). The diethyl-aminoethyl (DEAE) anion exchange columns were used to obtain delsulphated glucosinolates. Distilled water (1.5 mL) was used to eluate delsulphated glucosinolates. The prepared eluates were analyzed using an HPLC (model: 1200 series, Agilent Technologies, CA, USA) after filtering the delsulphated glucosinolates through a 0.45 μm polytetrafluoroethylene (PTFE) syringe filter. HPLC conditions were set as follows: a C18 column (150 × 3.0 mm, 3 μm, Inertsil ODS-3, GL Sciences, Tokyo, Japan) was used, the elution solution included ultra-pure water (solvent A) and 100%

acetonitrile (solvent B), flow rate was 0.4 mL min^{-1}, detector wavelength was set at 227 nm, and sinigrin was used as an external standard [41]. The results of glucosinolates are given as mmol g^{-1} dried weight of the samples.

The content of anthocyanin in the leaf samples was analyzed as mentioned by [42,43]. In brief, anthocyanins were extracted overnight at 22 °C. The freeze-dried products (8–10 mg) were extracted with 1 mL of solvent (MeOH:AcOH:H$_2$O = 80:0.2:19.8). The extracts were filtered through a 0.45 μm filter before analysis with the HPLC system. HPLC conditions were set as follows: the normalized collision energy was set to 30%, a C18 column (150 × 2 m, Imtakt Corporation, Kyoto, Japan) was used, flow rate was 0.3 mL min^{-1}, the elution solution included 0–100% solvent A (CH$_3$CN:H$_2$O:TFA = 7.5:92.5:0.1) and solvent B (CH$_3$CN:H$_2$O:TFA = 55:45:0.1) [44,45].

2.5. Statistical Analyses Procedures

A total of 2027 reflectance values (300 to 1050 nm with a 0.37 nm interval) were obtained from each sampling point by the detectors. Although excessive noise was omitted by considering 300 to 1050 nm reflectance spectra instead of 190 to 1130 nm (default range), smoothing (median filter with 21 smoothing points) and 1st derivative methods were additionally applied to remove outliers and reach the actual spectra, and for resolution enhancement, respectively. This combined application technique was applied following the reference [46]. After the preliminary processing, transformation of the raw reflectance value (Ref) was performed independently to check the non-linear correlation of reflectance spectra to the functional components. The used transformations of variables were squared power (Ref^2), squared root ($\sqrt{\text{Ref}}$), logarithm (Ln(Ref)), exponent (e^{Ref}), and inversion (1/Ref) [47].

Due to the similarity of the wavebands, the diffuse reflectance spectra are highly correlated. To reduce the multicollinearity effects and overfitting, some reflectance values were removed using the default procedures of the software. Then, the relationship between reflectance spectra and the content of functional components was investigated using several regression procedures, namely principal component regression (PCR), partial least squares regression (PLSR), and stepwise multiple linear regression (SMLR), to reduce the multicollinearity effects through MATLAB software (version: vR2013b, The MathWorks Ins, Natick, MA, USA). A ten-fold cross-validation (CV) process was applied to optimize the regression results. A total of ten subsets were generated from the dataset. One subset was used as the testing set, and the rest were used for the training set. The models containing from one to ten variables were determined for the remaining observations. The prediction residuals were determined through comparison with the removed subset. This process was repeated 10 times, with each of the subsets used exactly once as the testing data. Then, the average performance across all 10 trials was combined to find the validation residual variance. The coefficient of determination (R^2) and root mean square error (RMSE) were also calculated in order to select the proper regression model [9,48].

The PCR model is generally used for analyzing multiple regression data to avoid prediction instabilities caused by multicollinearity [49]. The general linear matrix form of the PCR model is shown in Equation (1) with usual notation.

$$\hat{Y} = X\hat{\beta} \tag{1}$$

where $\hat{Y}$: dependent or response variable, X: independent or controlled variables, and $\hat{\beta}$: regression coefficient. In comparison with PCR, PLSR delivers a better predictive linear-relationship, and is computed using a selected number of the latent factors from both datasets [50]. The model is linear for each sample n, and the value Y_{nj} is:

$$Y_{nj} = \Sigma_{i=0}^{k} \beta_i X_i \tag{2}$$

where Y_j: the p dependent variables, X_i: the k explanatory variables, and β_i: the regression coefficient. Here, i indicates the number of variables. The basic model and notations of

PCR and PLSR are similar, but the method of β_i determination is different. In SMLR, a multivariate model is constructed for the dependent variable (Y) considering some selected descriptive parameters (independent variables) [51,52]. The model of SMLR can be expressed as Equation (3).

$$Y = \beta_0 + \beta_1 X_1 + \beta_2 X_2 + \ldots + \beta_i X_i \tag{3}$$

where Y: dependent or response variable, $X_1 \sim X_i$: predictor variables (surface reflectance bands), β_0: constant variable, and $\beta_1 \sim \beta_i$: estimated weighted regression coefficient of $X_1 \sim X_i$, respectively. Additionally, the Pearson correlation coefficient was implemented to assess the relationship between the reflectance spectra and functional components. Standardized beta coefficient analysis was also performed by the PCR and PLSR models using the square power of the reflectance data.

3. Results

3.1. Glucosinolate and Anthocyanin Contents

The yield, total glucosinolates, and anthocyanins of kale under the different treatments of each experiment is summarized in Table 2. Besides this, the contents of each laboratory-analyzed glucosinolate and anthocyanin component are shown in Table 3. The number of samples were different, as some of the glucosinolate and anthocyanin components were not identified in all samples due to negligible (nearly zero) content. The quantity of some functional components was not satisfactory, and these were ignored. The minimum, maximum, and mean (non-normal distribution) concentration are also shown in Table 3. Depending on the leaf samples, five to eight components of glucosinolates and four components of anthocyanins were detected. From the results of HPLC analysis, we found that the highest proportion of glucosinolate content was represented by progoitrin (38.61 ± 46.46 µmol g^{-1} DW), and the most abundant glucosinolate component was neoglucobrassicin (0.40 ± 1.03 µmol g^{-1} DW). Similarly, cyanidin (134.10 ± 92.91 µg g^{-1}) and pelargonidin (0.14 ± 0.11 µg g^{-1}) represented the highest and lowest contents of anthocyanin, respectively. However, most of the glucosinolate and anthocyanin components were below 3 µmol g^{-1} DW.

3.2. Characteristics of Spectral Data

The pre-processed reflectance spectra of kale leaves showing low and high concentrations of glucosinolates and anthocyanins are presented in Figure 4. A median filter with 21 smoothing points along with a 1st derivative was applied to pre-process the reflectance spectral data. Data processing with a greater number of smoothing points reduces the noise of raw data, but valuable information that cannot be identified by visual inspection could be lost. As shown in Figure 4, peaks appeared at about 530–570 nm due to pigment variation [53], specifically foliar chlorophyll content [54], while no major difference was observed from 300 to 500 nm, but a significant difference was found among different concentrations of glucosinolates and anthocyanins from 700 to 1050 nm for strong absorption by carotenoids and chlorophylls. Overall, reflectance percentages were lower and almost similar in the visible region (except the peaks at 550 nm), but higher reflectance was obtained in the early NIR region (700 to 1050 nm), where low and high concentrations of glucosinolate and anthocyanins could be distinguished clearly.

Table 2. Content summary of the glucosinolate and anthocyanin components of the kale sample leaves analyzed in the laboratory using high-performance liquid chromatography (HPLC).

Factors	Treatments									
	2nd Week after Transplantation					4th Week after Transplantation				
Temperature (°C)	**14**	**17**	**20**	**23**	**26**	**14**	**17**	**20**	**23**	**26**
Yield (mg)	11.91 ± 1.2	12.59 ± 0.3	13.31 ± 0.2	13.02 ± 0.3	13.24 ± 0.3	21.30 ± 4.1	23.64 ± 6.2	28.29 ± 4.3	25.38 ± 3.2	22.15 ± 4.8
Total GLSs [1] (μmol g^{-1} DW) [1]	86.21 ± 5.9	65.24 ± 13.4	46.32 ± 19.2	29.75 ± 14.1	13.81 ± 7.7	37.12 ± 17.4	33.80 ± 5.1	23.94 ± 5.5	22.42 ± 20.1	12.42 ± 0.5
ATCs [2] (μg g^{-1})	8.9 ± 6.2	47.9 ± 4.3	65.4 ± 2.5	100.5 ± 8.4	90.8 ± 18.2	29.2 ± 23.4	51.1 ± 48.1	126.4 ± 110.9	180.5 ± 57.4	133.6 ± 56.7
Humidity (%)	**45**	**55**	**65**	**75**	**85**	**45**	**55**	**65**	**75**	**85**
Yield (mg)	5.5 ± 0.1	8.22 ± 1.2	6.33 ± 1.4	7.76 ± 1.1	8.89 ± 1.5	23.42 ± 9.2	29.39 ± 3.5	33.01 ± 4.9	25.67 ± 0.9	34.76 ± 6.8
Total GLSs (μmol g^{-1} DW) [1]	1.54 ± 1.54	1.62 ± 1.6	1.17 ± 1.2	1.18 ± 1.2	1.17 ± 1.2	0.86 ± 0.3	0.99 ± 0.2	1.12 ± 0.3	0.88 ± 0.1	0.87 ± 0.2
ATCs (μg g^{-1})	97.1 ± 37.4	116.9 ± 47.2	164.8 ± 40.7	107.1 ± 89.9	146.1 ± 29.6	75.9 ± 67.0	85.7 ± 23.9	174 ± 17.5	133.6 ± 115.7	117 ± 105.9
CO$_2$ (ppm)	**400**	**700**	**1000**	**1300**	**1600**	**400**	**700**	**1000**	**1300**	**1600**
Yield (mg)	10.01 ± 1.8	11.33 ± 1.7	10.38 ± 1.4	8.91 ± 0.4	7.49 ± 0.6	22.15 ± 4.8	28.38 ± 3.2	18.29 ± 4.3	23.64 ± 6.2	25.30 ± 4.1
Total GLSs (μmol g^{-1} DW) [1]	3.06 ± 0.3	2.37 ± 0.4	3.71 ± 0.8	8.08 ± 4.1	3.87 ± 0.8	3.64 ± 1.5	2.42 ± 0.9	2.20 ± 0.3	5.08 ± 0.8	4.48 ± 2.5
ATCs (μg g^{-1})	258.5 ± 96.9	285.7 ± 74.7	249.4 ± 24.4	279.4 ± 124.2	220.1 ± 19.9	224.2 ± 79.8	287.8 ± 18.0	313.16 ± 112.1	322.70 ± 45.1	171.5 ± 47.70
Light source [3]	**R:B:W**	**R:B**	**R:W**	**F**		**R:B:W**	**R:B**	**R:W**	**F**	
				-					-	
Yield (mg)	12.7 ± 0.39	14.1 ± 0.77	11.9 ± 0.16	10.3 ± 0.27		23 ± 1.7	21.9 ± 1.7	20.7 ± 2.4	17.3 ± 2.3	
Total GLSs (μmol g^{-1} DW) [1]	49.33 ± 66.80	14.27 ± 5.31	15.42 ± 13.30	63.38 ± 68.02		79.05 ± 25.37	118.51 ± 16.33	99.04 ± 63.11	103.14 ± 13.42	
Intensity (μmol m^{-2}s^{-1})	**100**	**130**	**160**	**190**	**220**	**100**	**130**	**160**	**190**	**220**
Yield (mg)	9.45 ± 0.82	8.37 ± 0.2	16.23 ± 0.36	12.72 ± 0.27	8.3 ± 0.25	17.24 ± 2.3	18.5 ± 1.74	24.48 ± 1.6	27.57 ± 0.74	18.07 ± 0.66
Total GLSs (μmol g^{-1} DW) [1]	9.08 ± 6.8	15.18 ± 0.6	11.16 ± 5.8	10.81 ± 5.1	10.93 ± 7.7	33.39 ± 2.22	93.17 ± 0.9	11.21 ± 3.7	20.54 ± 12.1	24.75 ± 14.9
Photoperiod (h)	**12/12**	**14/10**	**16/8**	**18/6**	**20/4**	**12/12**	**14/10**	**16/8**	**18/6**	**20/4**
Yield (mg)	9.5 ± 0.91	8.6 ± 0.40	12.23 ± 0.36	10.21 ± 0.47	7.6 ± 0.5	28.45 ± 2.18	18.71 ± 0.48	24.48 ± 1.58	19.21 ± 0.49	17.75 ± 0.53
Total GLSs (μmol g^{-1} DW) [1]	9.22 ± 4.8	9.01 ± 4.0	11.16 ± 5.8	21.21 ± 2.7	6.50 ± 0.9	19.35 ± 9.9	28.23 ± 16.9	11.21 ± 3.7	13.78 ± 12.1	5.46 ± 2.1
EC (μS cm^{-1})	**0.80**	**1.00**	**1.20**	**1.40**	**1.60**	**0.80**	**1.00**	**1.20**	**1.40**	**1.60**
Yield (mg)	14.0 ± 0.9	16.8 ± 1.5	17.4 ± 0.3	15.4 ± 1.6	14.4 ± 1.1	33.0 ± 2.2	45.6 ± 1.5	34.9 ± 0.4	24.0 ± 1.2	30.3 ± 1.8
Total GLSs (μmol g^{-1} DW) [1]	91.50 ± 30.77	82.89 ± 52.70	166.93 ± 23.98	92.75 ± 11.08	169.99 ± 14.59	88.01 ± 33.40	42.59 ± 23.70	76.03 ± 12.98	96.15 ± 27.35	92.41 ± 33.37

[1] GLSc, glucosinolates, [2] ATCs, anthocyanins, [3] R, red; B, blue; W, white.

Table 3. Content summary of the glucosinolate and anthocyanin components of the kale sample leaves analyzed in the laboratory using HPLC.

Component		No. of Samples	Min	Max	Mean ± STD
Glucosinolates (μmol g^{-1} DW) [1]	Progoitrin	55	1.43	96.32	64.57 ± 44.06
	Sinigrin	157	0.05	14.52	6.62 ± 9.19
	Glucoalyssin	22	0.74	6.03	2.77 ± 4.49
	Glucobrassicanapin	7	1.85	4.23	2.52 ± 1.20
	Glucobrassicin	187	0.05	16.77	8.29 ± 12.63
	4-methoxyglucobrassicin	149	0.02	2.58	1.06 ± 1.72
	Gluconasturtiin	23	1.26	3.35	0.75 ± 1.05
	Neoglucobrassicin	146	0.03	1.56	0.81 ± 1.06
	Glucoraphanin	39	0.29	5.2	3.25 ± 5.51
	Gluconapin	86	0.13	1.69	0.78 ± 0.90
Anthocyanins (μg g^{-1})	Cyanidin	72	0.02	217.56	135.98 ± 93.91
	Pelargonidin	71	0.02	0.25	0.13 ± 0.11
	Delphinidin	33	0.03	0.22	0.07 ± 0.10
	Malvidin	68	0.07	11.82	7.39 ± 4.83

[1] DW: dry weight.

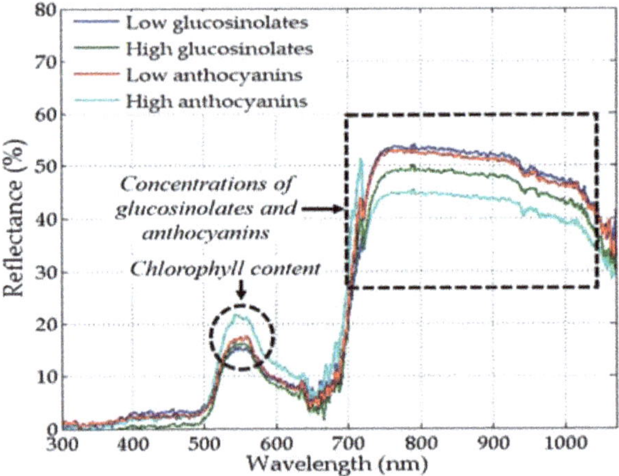

Figure 4. Reflectance spectra for low and high glucosinolates and anthocyanins contents in the kale leaves.

Pearson correlation coefficients between reflectance and an example glucosinolate component (sinigrin) are shown in Figure 5. These correlation coefficients provide transparent views of the relationships between spectra and functional components. The absolute values of correlations were higher for some transformations (e.g., raw spectra, square root, square power, and logarithm) than for exponent and inversion in the near-infrared regions. Wavelengths above 700 nm were highly correlated ($|r| \geq 0.5$) with the concentration of functional components as functional components were highly sensitive to spectra in the NIR regions [23,26].

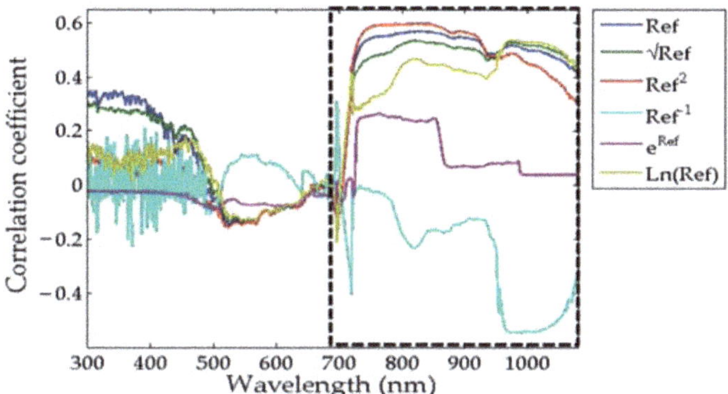

Figure 5. Representation of the correlation coefficients between the spectral reflectance and sinigrin content over the entire wavelength range.

3.3. Estimation Models of the Functional Components

3.3.1. Performance of PCR, PLSR, and SMLR Using Raw Data

A summary of the raw reflectance data analysis using the PCR, PLSR, and SMLR models over the wavelength region (from 300 to 1050 nm) is presented in Table 4. The performance of the PCR model was not satisfactory compared with the PLSR and SMLR models. The coefficients of determination (R^2) of the PCR models were in the range of 0.4 to 0.5 for 4-methoxyglucobrassicin, neoglucobrassicin, malvidin, and pelargonidin, whereas the other components showed worse results using the PCR model (R^2: < 0.4). Using the PLSR model, a strong correlation (R^2: $\geq$ 0.90, RMSE: $\approx$1 μmol g^{-1} DW) was found for glucobrassicin and malvidin; a fair correlation (R^2: $\geq$ 0.60, RMSE: < 0.5 μmol g^{-1} DW) was obtained for sinigrin, 4-methoxyglucobrassicin, neoglucobrassicin, and pelargonidin; and a poor correlation (R^2: $\leq$ 0.60) was observed for gluconapin and cyanidin contents. The SMLR model performed well (R^2: $\geq$ 0.80, RMSE: $\approx$1 μmol g^{-1} DW) for all glucosinolate components, whereas poor performance was shown for anthocyanins (R^2: 0.14, RMSE: 60 μg g^{-1}) and cyanidin, and fair performance (R^2: 0.46, RMSE: 0.04 μg g^{-1}) for pelargonidin and malvidin (R^2: 0.67, RMSE: 2.09 μg g^{-1}). Although good results were obtained for some specific functional components, the transformation of spectral reflectance data was used to include possible non-linear relationships.

Table 4. Summary of raw reflectance data analysis using PCR, PLSR, and SMLR models over the wavelength region (from 300 to 1050 nm).

Components		PCR		PLSR		SMLR	
		R^2	RMSE	R^2	RMSE	R^2	RMSE
Glucosinolates (μmol g^{-1} DW)	Sinigrin	0.38	3.02	0.63	2.33	0.84	1.44
	Glucobrassicin	0.23	3.69	0.90	1.30	0.92	0.90
	4-methoxyglucobrassicin	0.44	0.49	0.62	0.41	0.84	0.24
	Neoglucobrassicin	0.46	0.25	0.61	0.21	0.85	0.12
	Gluconapin	0.35	0.35	0.53	0.29	0.89	0.12
Anthocyanins (μg g^{-1})	Cyanidin	0.28	54.63	0.56	42.54	0.14	60.37
	Malvidin	0.44	2.55	0.91	1.04	0.67	2.09
	Pelargonidin	0.42	0.04	0.74	0.03	0.46	0.04

3.3.2. Performance of PCR, PLSR, and SMLR Using Transformed Data

A summary of the transformed reflectance data analysis using the PCR, PLSR, and SMLR models over the wavelength region (from 300 to 1050 nm) is shown in Table 5. The proper model was identified based on the R^2 and RMSE values of each transformation

of variables, i.e., square (Ref^2) square root ($\sqrt{\text{Ref}}$), exponent (e^{Ref}), inverse (1/Ref), and base 10 logarithmic scale (Ln(Ref)) of the reflectance data. For the PCR model, a poor performance was observed (R^2: < 0.6) for the glucosinolate and anthocyanin components. The results ranged from 0.4 to 0.5 for some components such as sinigrin, neoglucobrassicin, and malvidin using the Ref^2, e^{Ref}, and Ln(Ref) transformation and 4-methoxyglucobrassicin using the Ref^2, $\sqrt{\text{Ref}}$, e^{Ref}, and Ln(Ref) transformation. However, the results of the PCR model were higher than 0.5 for gluconapin (R^2: 0.50 using $\sqrt{\text{Ref}}$ and R^2: 0.51 using Ln(Ref) and pelargonidin (R^2: 0.50 with $\sqrt{\text{Ref}}$).

Table 5. Summary of transformed reflectance data analysis using PCR, PLSR, and SMLR models over the wavelength region from 300 to 1050 nm.

Components		Transformation	PCR		PLSR		SMLR	
			R^2	RMSE	R^2	RMSE	R^2	RMSE
Glucosinolate (μmol g^{-1} DW)	Sinigrin	Ref^2	0.35	3.11	0.64	2.30	0.84	1.41
		$\sqrt{\text{Ref}}$	0.41	2.95	0.65	2.27	0.85	1.56
		e^{Ref}	0.40	2.97	0.50	2.72	0.84	1.43
		1/Ref	0.01	3.82	0.09	3.66	0.85	1.36
		Ln(Ref)	0.41	2.96	0.59	2.45	0.86	1.32
	Glucobrassicin	Ref^2	0.29	3.55	0.35	3.38	0.91	1.04
		$\sqrt{\text{Ref}}$	0.23	3.69	0.96	0.78	0.91	1.06
		e^{Ref}	0.28	3.56	0.60	2.67	0.88	1.30
		1/Ref	0.02	4.18	0.002	4.21	0.92	0.88
		Ln(Ref)	0.24	3.67	0.86	1.57	0.92	0.88
	4-ethoxyglucobrassicin	Ref^2	0.40	0.51	0.60	0.41	0.83	0.25
		$\sqrt{\text{Ref}}$	0.46	0.48	0.65	0.39	0.86	0.22
		e^{Ref}	0.42	0.50	0.62	0.40	0.82	0.26
		1/Ref	0.04	0.65	0.04	0.65	0.87	0.21
		Ln(Ref)	0.44	0.49	0.80	0.29	0.86	0.23
	Neoglucobrassicin	Ref^2	0.38	0.27	0.62	0.21	0.86	0.12
		$\sqrt{\text{Ref}}$	0.42	0.26	0.63	0.20	0.85	0.12
		e^{Ref}	0.43	0.25	0.62	0.21	0.82	0.14
		1/Ref	0.06	0.33	0.10	0.32	0.86	0.12
		Ln(Ref)	0.42	0.26	0.77	0.16	0.87	0.11
	Gluconapin	Ref^2	0.20	0.38	0.68	0.24	0.88	0.12
		$\sqrt{\text{Ref}}$	0.50	0.30	0.86	0.16	0.89	0.12
		e^{Ref}	0.37	0.34	0.61	0.27	0.89	0.12
		1/Ref	0.14	0.40	0.20	0.39	0.88	0.13
		Ln(Ref)	0.51	0.30	0.88	0.15	0.89	0.12
Anthocyanins (μg g^{-1})	Cyanidin	Ref^2	0.02	64.38	0.004	64.26	0.16	60
		$\sqrt{\text{Ref}}$	0.24	56.18	0.50	45.54	0.14	60.58
		e^{Ref}	0.27	54.86	0.46	47.29	0.14	60.78
		1/Ref	0.01	64.15	0.01	64.09	0.12	61.29
		Ln(Ref)	0.20	57.39	0.68	36.59	0.13	60.81
	Malvidin	Ref^2	0.03	3.41	0.36	2.73	0.22	3.08
		$\sqrt{\text{Ref}}$	0.42	2.60	0.96	0.63	0.71	1.97
		e^{Ref}	0.42	2.60	0.86	1.30	0.12	3.28
		1/Ref	0.02	3.41	0.02	3.38	0.48	2.55
		Ln(Ref)	0.41	2.62	0.97	0.61	0.74	1.90
	Pelargonidin	Ref^2	0.01	0.06	0.002	0.06	0.40	0.04
		$\sqrt{\text{Ref}}$	0.50	0.04	0.69	0.03	0.45	0.04
		e^{Ref}	0.36	0.04	0.66	0.03	0.08	0.05
		1/Ref	0.02	0.06	0.03	0.06	0.09	0.05
		Ln(Ref)	0.36	0.04	0.91	0.02	0.38	0.04

During the glucosinolates analysis, good performance of the PLSR method was observed for glucobrassicin (R^2: 0.96 using $\sqrt{Ref}$, R^2: 0.86 using Ln(Ref)), 4-methoxyglucobrassicin (R^2: 0.80 using Ln(Ref)) and gluconapin (R^2: 0.86 using $\sqrt{Ref}$, R^2: 0.88 with Ln(Ref)). Fair results for the PLSR models ($0.60 \leq R^2 \leq 0.79$) were obtained for sinigrin (R^2: 0.64 using Ref^2, R^2: 0.65 using $\sqrt{Ref}$), glucobrassicin (R^2: 0.60 using e^{Ref}), 4-methoxyglucobrassicin (R^2: 0.60, 0.65, and 0.62 using Ref^2, $\sqrt{Ref}$, and e^{Ref}, respectively), neoglucobrassicin (R^2: 0.62, 0.63, 0.62, and 0.77 using Ref^2, $\sqrt{Ref}$, e^{Ref}, and Ln(Ref), respectively), and gluconapin (R^2: 0.68 using Ref^2, R^2: 0.61 using e^{Ref}). Poor performance (R^2: < 0.60) was observed for sinigrin using e^{Ref}, $1/Ref$, and Ln(Ref); glucobrassicin using Ref^2 and $1/Ref$; and 4-methoxyglucobrassicin, neoglucobrassicin, and gluconapin using $1/Ref$. For anthocyanins, good performance of the PLSR models (R^2: ≥ 0.80) were obtained for malvidin (R^2: 0.86, 0.96, and 0.97 using $\sqrt{Ref}$, e^{Ref}, and Ln(Ref)) and pelargonidin (R^2: 0.91 using Ln(Ref)). Fair PLSR model results ($0.60 \leq R^2 \leq 0.79$) were obtained for cyanidin (R^2: 0.68 using Ln(Ref)) and pelargonidin (R^2: 0.66 using e^{Ref}; R^2: 0.69 using the $\sqrt{Ref}$).

The SMLR procedure showed very good performance (R^2: ≥ 0.82) for estimating the glucosinolate components (e.g., sinigrin, glucobrassicin, 4-methoxyglucobrassicin, neoglucobrassicin, gluconapin) with Ref^2, $\sqrt{Ref}$, $1/Ref$, and Ln(Ref) transformation, whereas only the malvidin content of anthocyanins was estimated well by the SMLR model (R^2: 0.71 using $\sqrt{Ref}$, and R^2: 0.74 using Ln(Ref)), and poor performance was observed for the other components ($0.09 \leq R^2 \leq 0.48$).

The B-coefficient values obtained from the PCR and PLSR models reveal important wavelengths for the quantification of functional components. Figure 6 shows the B-coefficient values using square reflectance data for the PCR and PLSR models for glucobrassicin calibration. Wavelengths at 761, 890, 933, and 1000 nm were identified by the PCR model, and the wavelengths determined by the PLSR model were 742, 761, 787, 796, 805, 833, 855, 932, 947, and 1000 nm. The peak in the VIS range at about 579 nm is associated with the green region due to electronic transition. Thus, the wavelength at 544 nm was assigned to chlorophyll, while the wavelengths at 742, 761, 787, 796, 805, 833, 855, 932, 947, and 1000 nm had a strong correlation with glucosinolate content.

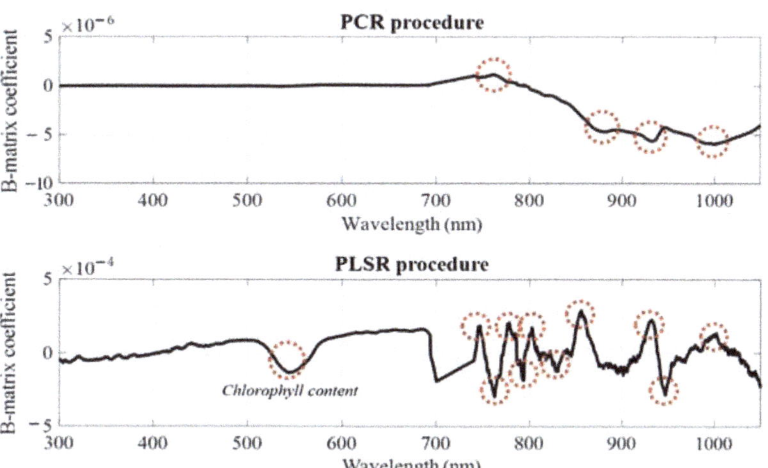

Figure 6. B-coefficient analysis performed by the PCR (top) and PLSR (bottom) models using square power of reflectance data for glucobrassicin.

4. Discussion

According to the laboratory (HPLC) analysis, progoitrin was the most dominant glucosinolate component in kale leaves, and the most abundant components were glu-

coalyssin, gluconasturtiin, and 4-methoxyglucobrassicin. On the other hand, cyanidin was the main component of anthocyanins followed by malvidin. The abundant components were pelargonidin and delphinidin. This variation in functional components depends on crop species, growth stage, cultivation methods, and ambient environment or climatic conditions [23,38,55–58]. For example, glucoiberin, glucoraphanin, and sinigrin were observed as dominating components, sequentially, in kale plants cultivated in a closed chamber and soil-based system [57]. Besides this, glucobrassicin and sinigrin were identified as dominating components in kale when cultivated in open-field conditions [55]. The authors of [44] found sinigrin to be a dominant component in green kale and progoitrin in red kale. A difference in concentrations was also observed among studies based on the cultivation period. However, most of the studies identified almost similar types of abundant components [5,43,44].

The reflected spectra from the kale leaf surface represent the status of leaf photochemical and morphological properties. Minimum reflectance indicates a higher concentration and maximum reflectance indicates a lower concentration of glucosinolates and anthocyanins. In this study, no meaningful variation was observed in the visible region, except the peak around 530 to 570 nm (Figure 4). This sudden higher reflectance (peak) was observed due to pigment variation [53], specifically in foliar chlorophyll content [54], which was very familiar in the species of Brassicaceae family and indicates a lower glucosinolate content level in that region. A significant difference in glucosinolate content was detected for the 700 to 1050 nm wavelengths. Reflectance variation or overlapping during functional components analysis could occur in different cultivation facilities (i.e., greenhouse, plant factory, or open field), covering material types, radiation intensities or artificial light types and even the color or pigment properties of crop species (i.e., red and green kale or lettuce) [59–61]. The reflectance variation might occur in the visible region or NIR region based on the cultivation method, such as hydroponic or organic systems. A strong correlation between the pigment contents of red and green lettuces under different light intensities was also observed by the authors of [59,60].

Besides this, preprocessing of the reflectance spectral is necessary to minimize unwanted background information, along with accentuating the absorption features of the spectra. It also helps to attain accurate models and reduces the number of latent variables [53]. Smoothing and 1st derivative methods were applied in this study. Neto et al. [53] also applied smoothing, as well as the 1st and 2nd derivatives, and observed that the 1st derivative was the best pretreatment process for predicting anthocyanin content. They also mentioned that this pretreatment removed the non-chemical effects, resolved the overlapped bands, and provided a better version of the target data. In addition, PCR, PLSR, and SMLR models were applied in this study, and a significant correlation between spectral reflectance and glucosinolate and anthocyanin concentrations was observed in the NIR region through the SMLR model and Pearson's correlation coefficients test. A similar result was observed during functional components analysis on Chinese cabbage leaves in our previous study [9] and also in the literature [21,62].

Several studies were conducted to identify the variation of crops' physical, chemical, and biological properties based on crop species, cultivation methods, environmental conditions, and fertilization [30,63]. Among these, effects of cultivation systems, such as plant factory, greenhouse, or open field, and cultivation methods, such as soil-based and hydroponics, play a vital role in the variation of the glucosinolate and anthocyanin contents of crops (Table 6). The authors of [30] compared the functional components of kale (*Brassica oleracea var. alboglabra*) grown in three different conditions and found glucobrassicin, sinigrin, neoglucobrassicin, and progoitrin to be dominant components, sequentially, and 4-methoxyglucobrassicin to be an abundant component. They reported that the functional components of kale were higher when grown in the plant factory than in the greenhouse and open field due to potential cultivation conditions. Glucoiberin, glucoraphanin, sinigrin, and glucobrassicin were detected as major components of kale (*Brassica oleracea* var.

acephala) when cultivated using soil-based methods in a closed chamber and an open field, respectively [38,55,57].

Table 6. Dominating glucosinolate components of different plants cultivated under different conditions and analyzed by HPLC and reflectance spectroscopy methods.

Plant	Cultivation Method	Dominating Component (Analyzed by HPLC)	Dominating Component (Analyzed by Spectroscopy)	Wavelengths (nm)
Kale	Aeroponic (Plant factory)	Progoitrin, Sinigrin	Sinigrin, Glucobrassicin	742, 761, 787, 796, 805, 833, 855, 932, 947, 1000
Kale	Soil-based (Closed-chamber)	Glucoiberin, Glucoraphanin, Sinigrin	-	-
Kale	Open field	Glucobrassicin, Sinigrin	-	-
Chinese cabbage	Aeroponic (Plant factory)	Neoglucobrassicin, 4-methoxyglucobrassicin	Glucobrassicin, 4-methoxyglucobrassicin	365, 388, 440, 545, 607, 651, 798, 838, 860, 870, 932, 950
Mustard leaf	Open field	Sinigrin, Glucoiberverin, Gluconasturtiin	-	-
Rocket Leaf	Soil-based (Greenhouse)	-	Glucoerucin, Gluconasturtiin,4-hydroxyglucobrassicin	548, 610, 680, 1432, 1696, 1730, 1920, 2054

5. Conclusions

This study focused on the determination of glucosinolate and anthocyanin contents in kale leaves using the diffuse reflectance spectroscopy technique, where kale plants were cultivated in a plant factory under different levels of environmental factors. The results showed that progoitrin and glucobrassicin, as well as cyanidin and malvidin, were found to be dominating components in glucosinolates and anthocyanins, respectively, in laboratory analysis. Among the applied regression methods, SMLR showed better performance compared with the PCR and PLSR models. Important wavelengths for estimating glucosinolates and anthocyanins were laid between 700 to 1050 nm. Although a similar methodology was applied in some studies on other crops, very little research has been conducted on kale plants. The components of glucosinolates and anthocyanins and their related reflectance spectrum vary based on the crop species, cultivation methods, and environmental parameters, so crop-specific accurate model development is essential. The findings of this study would be useful for designing or improving any multiple-wavelength property sensor applicable for on-site functional components determination of any crops.

Author Contributions: Conceptualization, S.-O.C. and M.C., V.-D.N.; methodology, S.-O.C.; software, V.-D.N.; validation, V.-D.N., M.C.; formal analysis, M.C., V.-D.N., M.N.I., M.A., S.I., K.R., S.-U.P. and S.-O.C.; investigation, S.-O.C. and S.-U.P.; resources, S.-O.C.; data curation, M.C., V.-D.N., M.N.I., M.A., S.I. and K.R.; writing—original draft preparation, M.C.; writing—review and editing, S.-O.C., and S.-U.P.; visualization, M.C., M.N.I., M.A., S.I. and K.R.; supervision, S.-O.C.; project administration, S.-O.C.; funding acquisition, S.-O.C. All authors have read and agreed to the published version of the manuscript.

Funding: This work was supported by the Korea Institute of Planning and Evaluation for Technology in Food, Agriculture and Forestry (IPET) through the Agriculture, Food and Rural Affairs Convergence Technologies Program for Educating Creative Global Leaders, funded by Ministry of Agriculture, Food and Rural Affairs (MAFRA) (project No. 320001-4), Korea.

Institutional Review Board Statement: Not applicable.

Informed Consent Statement: Not applicable.

Data Availability Statement: The data presented in this study are available within the article.

Conflicts of Interest: The authors declare no conflict of interest.

References

1. Wu, X.; Sun, J.; Haytowitz, D.B.; Harnly, J.M.; Chen, P.; Pehrsson, P.R. Challenges of developing a valid dietary glucosin-olate database. *J. Food Compos. Anal.* **2017**, *64*, 78–84. [CrossRef]
2. Esteve, M. Mechanisms Underlying Biological Effects of Cruciferous Glucosinolate-Derived Isothiocyanates/Indoles: A Focus on Metabolic Syndrome. *Front. Nutr.* **2020**, *7*, 111. [CrossRef] [PubMed]
3. Block, G.; Patterson, B.; Subar, A. Fruit, vegetables, and cancer prevention: A review of the epidemiological evidence. *Nutr. Cancer* **1992**, *18*, 1–29. [CrossRef]
4. Kassie, F.; Uhl, M.; Rabot, S.; Grasl-Kraupp, B.; Verkerk, R.; Kundi, M.; Chabicovsky, M.; Schulte-Hermann, R.; Knasmüller, S. Chemoprevention of 2-amino-3-methylimidazo[4,5-f]quinoline (IQ)-induced colonic and hepatic preneoplastic lesions in the F344 rat by cruciferous vegetables administered simultaneously with the carcinogen. *Carcinogenesis* **2003**, *24*, 255–261. [CrossRef] [PubMed]
5. Sun, B.; Liu, N.; Zhao, Y.; Yan, H.; Wang, Q. Variation of glucosinolates in three edible parts of Chinese kale (Brassica alboglabra Bailey) varieties. *Food Chem.* **2011**, *124*, 941–947. [CrossRef]
6. Vale, A.; Santos, J.; Brito, N.; Fernandes, D.; Rosa, E.; Oliveira, M.B.P. Evaluating the impact of sprouting conditions on the glucosinolate content of Brassica oleracea sprouts. *Phytochemistry* **2015**, *115*, 252–260. [CrossRef]
7. Hahn, C.; Müller, A.; Kuhnert, N.; Albach, D. Diversity of Kale (Brassica oleraceavar.sabellica): Glucosinolate Content and Phylogenetic Relationships. *J. Agric. Food Chem.* **2016**, *64*, 3215–3225. [CrossRef]
8. Yu, T.H.; Hsieh, S.P.; Su, C.M.; Huang, F.J.; Hung, C.C.; Yiin, L.M. Analysis of leafy vegetable nitrate using a modified spectrometric method. *Int. J. Anal. Chem.* **2018**, *2018*, 1–6. [CrossRef]
9. Ngo, V.-D.; Jang, B.-E.; Park, S.-U.; Kim, S.-J.; Kim, Y.-J.; Chung, S.-O. Estimation of functional components of Chinese cabbage leaves grown in a plant factory using diffuse reflectance spectroscopy. *J. Sci. Food Agric.* **2019**, *99*, 711–718. [CrossRef]
10. Dechant, B.; Cuntz, M.; Vohland, M.; Schulz, E.; Doktor, D. Estimation of photosynthesis traits from leaf reflectance spectra: Correlation to nitrogen content as the dominant mechanism. *Remote Sens. Environ.* **2017**, *196*, 279–292. [CrossRef]
11. Kataria, S.; Baghel, L.; Guruprasad, K.N. Alleviation of Adverse Effects of Ambient UV Stress on Growth and Some Potential Physiological Attributes in Soybean (Glycine max) by Seed Pre-treatment with Static Magnetic Field. *J. Plant. Growth Regul.* **2017**, *36*, 550–565. [CrossRef]
12. Neto, A.J.S.; Lopes, D.C.; Toledo, J.V.; Zolnier, S.; Silva, T.G.F. Classification of sugarcane varieties using visible/near infrared spectral reflectance of stalks and multivariate methods. *J. Agric. Sci.* **2018**, *156*, 537–546. [CrossRef]
13. Scotford, I.; Miller, P. Applications of Spectral Reflectance Techniques in Northern European Cereal Production: A Review. *Biosyst. Eng.* **2005**, *90*, 235–250. [CrossRef]
14. García-Sánchez, F.; Galvez-Sola, L.; Martínez-Nicolás, J.J.; Muelas-Domingo, R.; Nieves, M. Using Near-Infrared Spectroscopy in Agricultural Systems. In *Developments in Near-Infrared Spectroscopy*; Konstantinos, G., Kyprianidis, Eds.; Intech: Janeza Trdine 9, Croatia, 2017; pp. 98–127.
15. Norris, K.H.; Barnes, R.F.; Moore, J.E.; Shenk, J.S. Predicting Forage Quality by Infrared Replectance Spectroscopy. *J. Anim. Sci.* **1976**, *43*, 889–897. [CrossRef]
16. Mawlong, I.; Kumar, M.S.; Gurung, B.; Singh, K.; Singh, D. A simple spectrophotometric method for estimating total glucosinolates in mustard de-oiled cake. *Int. J. Food Prop.* **2017**, *20*, 3274–3281. [CrossRef]
17. Bowie, B.T.; Chase, D.B.; Griffiths, P.R.; Pont, D. Factors Affecting the Performance of Bench—Top Raman Spectrometers. Part II: Effect of Sample. *Appl. Spectrosc.* **2000**, *54*, 200A–207A. [CrossRef]
18. Gitelson, A.A.; Keydan, G.P.; Merzlyak, M.N. Three-band model for noninvasive estimation of chlorophyll, carotenoids, and anthocyanin contents in higher plant leaves. *Geophys. Res. Lett.* **2006**, *33*. [CrossRef]
19. Xue, L.; Yang, L. Deriving leaf chlorophyll content of green-leafy vegetables from hyperspectral reflectance. *ISPRS J. Photogramm. Remote Sens.* **2009**, *64*, 97–106. [CrossRef]
20. Gitelson, A.A.; Zur, Y.; Chivkunova, O.B.; Merzlyak, M.N. Assessing Carotenoid Content in Plant Leaves with Reflectance Spectroscopy. *Photochem. Photobiol.* **2007**, *75*, 272–281. [CrossRef]
21. Chen, X.; Wu, J.; Zhou, S.; Yang, Y.; Ni, X.; Yang, J.; Zhu, Z.; Shi, C. Application of near-infrared reflectance spectroscopy to evaluate the lutein and β-carotene in Chinese kale. *J. Food Compos. Anal.* **2009**, *22*, 148–153. [CrossRef]
22. Chen, J.; Li, L.; Wang, S.; Tao, X.; Wang, Y.; Sun, A.; He, H. Assessment of Glucosinolates in Chinese Kale by Near-Infrared Spectroscopy. *Int. J. Food Prop.* **2014**, *17*, 1668–1679. [CrossRef]
23. Kim, H.W.; Ko, H.C.; Baek, H.J.; Cho, S.M.; Jang, H.H.; Lee, Y.M.; Kim, J.B. Identification and quantification of glucosinolates in Korean leaf mustard germplasm (Brassica juncea var. integrifolia) by liquid chromatography–electrospray ionization/tandem mass spectrometry. *Eur. Food Res. Technol.* **2016**, *242*, 1479–1484. [CrossRef]
24. Sahamishirazi, S.; Zikeli, S.; Fleck, M.; Claupein, W.; Graeff-Hoenninger, S. Development of a near-infrared spectroscopy method (NIRS) for fast analysis of total, indolic, aliphatic and individual glucosinolates in new bred open pollinating genotypes of broccoli (Brassica oleracea convar. botrytis var. italica). *Food Chem.* **2017**, *232*, 272–277. [CrossRef]
25. Agati, G.; Tuccio, L.; Kusznierewicz, B.; Chmiel, T.; Bartoszek, A.; Kowalski, A.; Grzegorzewska, M.; Kosson, R.; Kaniszewski, S. Nondestructive Optical Sensing of Flavonols and Chlorophyll in White Head Cabbage (Brassica oleraceaL. var.capitatasubvar.alba) Grown under Different Nitrogen Regimens. *J. Agric. Food Chem.* **2016**, *64*, 85–94. [CrossRef]

26. Lin, K.-H.; Shih, F.-C.; Huang, M.-Y.; Weng, J.-H. Physiological Characteristics of Photosynthesis in Yellow-Green, Green and Dark-Green Chinese Kale (*Brassica oleracea* L. var. *alboglabra* Musil.) under Varying Light Intensities. *Plants* **2020**, *9*, 960. [CrossRef]

27. Chung, S.O.; Chowdhury, M.; Ngo, V.D.; Jang, B.E.; Han, M.W.; Ko, H.J. Comparing Models and Wavelength Bands for Es-Timation of Functional Components of Kale and Chinese Cabbage. Available online: https://www.researchgate.net/profile/Milon_Chowdhury2/publication/343704587_Comparing_models_and_wavelength_bands_for_estimation_of_functional_components_of_Kale_and_Chinese_cabbage/links/5f3b24c5299bf13404cd4863/Comparing-models-and-wavelength-bands-for-estimation-of-functional-components-of-Kale-and-Chinese-cabbage.pdf (accessed on 25 January 2021).

28. Noisopa, C.; Prapagdee, B.; Navanugraha, C.; Hutacharoen, R. Effects of bio-extracts on the growth of chinese kale. *Agric. Nat. Resour.* **2010**, *44*, 808–815.

29. Filho, A.B.C.; Bianco, M.S.; Tardivo, C.F.; Pugina, G.C. Agronomic viability of New Zealand spinach and kale intercropping. *An. Acad. Bras. Ciências* **2017**, *89*, 2975–2986. [CrossRef] [PubMed]

30. Kim, K.H.; Chung, S.O. Comparison of plant growth and glucosinolates of Chinese cabbage and kale crops under three cul-tivation conditions. *J. Biosyst. Eng.* **2018**, *43*, 30–36. [CrossRef]

31. Kozai, T. *Smart Plant Factory: The Next Generation Indoor Vertical Farms*; Springer: Berlin/Heidelberg, Germany, 2018.

32. Chowdhury, M.; Jang, B.; Kabir, M.; Kim, Y.; Na, K.; Park, S.; Chung, S. Factors affecting the accuracy and precision of ion-selective electrodes for hydroponic nutrient supply systems. *Acta Hortic.* **2020**, 997–1004. [CrossRef]

33. Kozai, T.; Niu, G.; Takagaki, M. *Plant Factory: An Indoor Vertical Farming System for Efficient Quality Food Production*; Academic Press: San Diego, CA, USA, 2015.

34. Yi, G.-E.; Robin, A.H.K.; Yang, K.; Park, J.-I.; Kang, J.-G.; Yang, T.-J.; Nou, I.-S. Identification and Expression Analysis of Glucosinolate Biosynthetic Genes and Estimation of Glucosinolate Contents in Edible Organs of Brassica oleracea Subspecies. *Molecules* **2015**, *20*, 13089–13111. [CrossRef]

35. Yoon, H.I.; Kim, J.-S.; Kim, D.; Kim, C.Y.; Son, J.E. Harvest strategies to maximize the annual production of bioactive compounds, glucosinolates, and total antioxidant activities of kale in plant factories. *Hortic. Environ. Biotechnol.* **2019**, *60*, 883–894. [CrossRef]

36. Carter, S.; Shackley, S.; Sohi, S.; Suy, T.B.; Haefele, S. The Impact of Biochar Application on Soil Properties and Plant Growth of Pot Grown Lettuce (Lactuca sativa) and Cabbage (Brassica chinensis). *Agronomy* **2013**, *3*, 404–418. [CrossRef]

37. Ngo, V.-D.; Kang, S.-W.; Ryu, D.-K.; Chung, S.-O.; Park, S.-U.; Kim, S.-J.; Park, J.-T. Location of Sampling Points in Optical Reflectance Measurements of Chinese Cabbage and Kale Leaves. *J. Biosyst. Eng.* **2015**, *40*, 115–123. [CrossRef]

38. Lefsrud, M.; Kopsell, D.; Wenzel, A.; Sheehan, J. Changes in kale (Brassica oleracea L. var. acephala) carotenoid and chlorophyll pigment concentrations during leaf ontogeny. *Sci. Hortic.* **2007**, *112*, 136–141. [CrossRef]

39. Chung, S.-O.; Kang, S.-W.; Bae, K.-S.; Ryu, M.-J.; Kim, Y.-J. The potential of remote monitoring and control of protected crop production environment using mobile phone under 3G and Wi-Fi communication conditions. *Eng. Agric. Environ. Food* **2015**, *8*, 251–256. [CrossRef]

40. Chung, S.-O.; Ngo, V.-D.; Kabir, S.N.; Hong, S.-J.; Park, S.-U.; Kim, S.-J.; Park, J.-T. Number of sampling leaves for reflectance measurement of Chinese cabbage and kale. *Korean J. Agric. Sci.* **2014**, *41*, 169–175. [CrossRef]

41. Kim, S.-J.; Ishii, G. Glucosinolate profiles in the seeds, leaves and roots of rocket salad (Eruca sativaMill.) and anti-oxidative activities of intact plant powder and purified 4-methoxyglucobrassicin. *Soil Sci. Plant. Nutr.* **2006**, *52*, 394–400. [CrossRef]

42. Park, N.I.; Li, X.; Suzuki, T.; Kim, S.-J.; Woo, S.-H.; Park, C.H.; Park, S.U. Differential Expression of Anthocyanin Biosynthetic Genes and Anthocyanin Accumulation in Tartary Buckwheat Cultivars 'Hokkai T8' and 'Hokkai T10'. *J. Agric. Food Chem.* **2011**, *59*, 2356–2361. [CrossRef]

43. Park, W.T.; Kim, J.K.; Park, S.; Lee, S.-W.; Li, X.; Kim, Y.B.; Uddin, R.; Park, N.I.; Kim, S.-J.; Park, S.U. Metabolic Profiling of Glucosinolates, Anthocyanins, Carotenoids, and Other Secondary Metabolites in Kohlrabi (Brassica oleraceavar.gongylodes). *J. Agric. Food Chem.* **2012**, *60*, 8111–8116. [CrossRef]

44. Jeong, K.Y.; Kim, J.K.; Kim, H.; Kim, Y.J.; Park, Y.J.; Kim, S.J.; Kim, C.; Park, S.U. Transcriptome analysis and metabolic profiling of green and red kale (Brassica oleracea var. acephala) seedlings. *Food Chem.* **2018**, *241*, 7–13. [CrossRef]

45. Brooks, M.S.L.; Celli, G.B. *Anthocyanins From Natural Sources: Exploiting Targeted Delivery for Improved Health*; Royal Society of Chemistry: London, UK, 2019; Volume 12.

46. Luthria, D.L.; Mukhopadhyay, S.; Lin, L.-Z.; Harnly, J.M. A Comparison of Analytical and Data Preprocessing Methods for Spectral Fingerprinting. *Appl. Spectrosc.* **2011**, *65*, 250–259. [CrossRef]

47. Osborne, J.W. Improving your data transformations: Applying the Box-Cox transformation. *Pract. Assess. Res. Eval.* **2010**, *15*, 1–9.

48. Kawamura, K.; Tsujimoto, Y.; Rabenarivo, M.; Asai, H.; Andriamananjara, A.; Rakotoson, T. Vis-NIR Spectroscopy and PLS Regression with Waveband Selection for Estimating the Total C and N of Paddy Soils in Madagascar. *Remote Sens.* **2017**, *9*, 1081. [CrossRef]

49. Mahesh, S.; Jayas, D.S.; Paliwal, J.; White, N.D.G. Comparison of Partial Least Squares Regression (PLSR) and Principal Components Regression (PCR) Methods for Protein and Hardness Predictions using the Near-Infrared (NIR) Hyperspectral Images of Bulk Samples of Canadian Wheat. *Food Bioprocess. Technol.* **2015**, *8*, 31–40. [CrossRef]

50. Yaroshchyk, P.; Death, D.L.; Spencer, S.J. Comparison of principal components regression, partial least squares regression, multi-block partial least squares regression, and serial partial least squares regression algorithms for the analysis of Fe in iron ore using LIBS. *J. Anal. At. Spectrom.* **2012**, *27*, 92–98. [CrossRef]

51. Ghani, I.M.M.; Ahmad, S. Stepwise Multiple Regression Method to Forecast Fish Landing. *Procedia-Soc. Behav. Sci.* **2010**, *8*, 549–554. [CrossRef]
52. Zhan, X.; Liang, X.; Xu, G.; Zhou, L. Influence of plant root morphology and tissue composition on phenanthrene uptake: Stepwise multiple linear regression analysis. *Environ. Pollut.* **2013**, *179*, 294–300. [CrossRef]
53. Neto, A.J.S.; Moura, L.D.O.; Lopes, D.D.C.; Carlos, L.D.A.; Martins, L.M.; Ferraz, L.D.C.L. Non-destructive prediction of pigment content in lettuce based on visible-NIR spectroscopy. *J. Sci. Food Agric.* **2016**, *97*, 2015–2022. [CrossRef]
54. Kira, O.; Linker, R.; Gitelson, A. Non-destructive estimation of foliar chlorophyll and carotenoid contents: Focus on informative spectral bands. *Int. J. Appl. Earth Obs. Geoinf.* **2015**, *38*, 251–260. [CrossRef]
55. Velasco, P.; Cartea, M.E.; González, M.E.C.; Vilar, M.; Ordás, A. Factors Affecting the Glucosinolate Content of Kale (Brassica oleraceaacephala Group). *J. Agric. Food Chem.* **2007**, *55*, 955–962. [CrossRef]
56. Toledo-Martín, E.M.; Font, R.; Obregón-Cano, S.; De Haro-Bailón, A.; Villatoro-Pulido, M.; Del Río-Celestino, M. Rapid and Cost-Effective Quantification of Glucosinolates and Total Phenolic Content in Rocket Leaves by Visible/Near-Infrared Spectroscopy. *Molecules* **2017**, *22*, 851. [CrossRef]
57. Steindal, A.L.H.; Rødven, R.; Hansen, E.; Mølmann, J. Effects of photoperiod, growth temperature and cold acclimatisation on glucosinolates, sugars and fatty acids in kale. *Food Chem.* **2015**, *174*, 44–51. [CrossRef]
58. Dou, H.; Niu, G.; Gu, M.; Masabni, J. Morphological and Physiological Responses in Basil and Brassica Species to Different Proportions of Red, Blue, and Green Wavelengths in Indoor Vertical Farming. *J. Am. Soc. Hortic. Sci.* **2020**, *145*, 267–278. [CrossRef]
59. Pérez-López, U.; Miranda-Apodaca, J.; Muñoz-Rueda, A.; Mena-Petite, A. Interacting effects of high light and elevated CO_2 on the nutraceutical quality of two differently pigmented Lactuca sativa cultivars (Blonde of Paris Batavia and Oak Leaf). *Sci. Hortic.* **2015**, *191*, 38–48. [CrossRef]
60. Marin, A.; Ferreres, F.; Barberá, G.G.; Gil, M.I. Weather Variability Influences Color and Phenolic Content of Pigmented Baby Leaf Lettuces throughout the Season. *J. Agric. Food Chem.* **2015**, *63*, 1673–1681. [CrossRef] [PubMed]
61. Alrifai, O.; Hao, X.; Marcone, M.F.; Tsao, R. Current Review of the Modulatory Effects of LED Lights on Photosynthesis of Secondary Metabolites and Future Perspectives of Microgreen Vegetables. *J. Agric. Food Chem.* **2019**, *67*, 6075–6090. [CrossRef]
62. Nosenko, T.; Hutsalo, I.; Nosenko, V.; Levchuk, I.; Litvynchuk, S. Analysis of near infrared reflectance spectrum of rape seed with different content of erucic acid. *Ukr. J. Food Sci.* **2013**, 94–99.
63. Challinor, A.J.; Parkes, B.; Ramírez-Villegas, J. Crop yield response to climate change varies with cropping intensity. *Glob. Chang. Biol.* **2015**, *21*, 1679–1688. [CrossRef] [PubMed]

 *horticulturae*

Article

Iodine Biofortification Counters Micronutrient Deficiency and Improve Functional Quality of Open Field Grown Curly Endive

Leo Sabatino [1,*,†], Francesca Di Gaudio [2,†], Beppe Benedetto Consentino [1], Youssef Rouphael [3], Christophe El-Nakhel [3], Salvatore La Bella [1,*], Sonya Vasto [4], Rosario Paolo Mauro [5], Fabio D'Anna [1], Giovanni Iapichino [1], Rosalia Caldarella [6] and Claudio De Pasquale [1]

[1] Dipartimento Scienze Agrarie, Alimentari e Forestali (SAAF), University of Palermo, viale delle Scienze, ed. 5, 90128 Palermo, Italy; beppebenedetto.consentino@unipa.it (B.B.C.); fabio.danna@unipa.it (F.D.); giovanni.iapichino@unipa.it (G.I.); claudio.depasquale@unipa.it (C.D.P.)

[2] Department of Promoting Health, Maternal-Infant, Excellence and Internal and Specialized Medicine (ProMISE) G. D'Alessandro, University of Palermo, 90127 Palermo, Italy; francesca.digaudio@unipa.it

[3] Department of Agricultural Sciences, University of Naples Federico II, 80055 Portici, Italy; youssef.rouphael@unina.it (Y.R.); christophe.elnakhel@unina.it (C.E.-N.)

[4] Department of Biological, Chemical and Pharmaceutical Sciences and Technologies, University of Palermo, 90128 Palermo, Italy; sonya.vasto@unipa.it

[5] Dipartimento di Agricoltura, Alimentazione e Ambiente (Di3A), University of Catania, Via Valdisavoia, 5, 95123 Catania, Italy; rosario.mauro@unict.it

[6] Department of Laboratory Medicine, "P. Giaccone" University Hospital, 90128 Palermo, Italy; liacaldarella@virgilio.it

* Correspondence: leo.sabatino@unipa.it (L.S.); salvatore.labella@unipa.it (S.L.B.); Tel.: +39-09123862252 (L.S.); +39-09123862231 (S.L.B.)

† These authors are equally contributed.

Citation: Sabatino, L.; Di Gaudio, F.; Consentino, B.B.; Rouphael, Y.; El-Nakhel, C.; La Bella, S.; Vasto, S.; Mauro, R.P.; D'Anna, F.; Iapichino, G.; et al. Iodine Biofortification Counters Micronutrient Deficiency and Improve Functional Quality of Open Field Grown Curly Endive. *Horticulturae* **2021**, 7, 58. https://doi.org/10.3390/horticulturae7030058

Academic Editor: Elazar Fallik

Received: 19 February 2021
Accepted: 19 March 2021
Published: 21 March 2021

Publisher's Note: MDPI stays neutral with regard to jurisdictional claims in published maps and institutional affiliations.

Abstract: Human iodine (I) shortage disorders are documented as an imperative world-wide health issue for a great number of people. The World Health Organization (WHO) recommends I consumption through ingestion of seafood and biofortified food such as vegetables. The current work was carried out to appraise the effects of different I concentrations (0, 50, 250, and 500 mg L^{-1}), supplied via foliar spray on curly endive grown in the fall or spring–summer season. Head fresh weight, stem diameter, head height, and soluble solid content (SSC) were negatively correlated to I dosage. The highest head dry matter content was recorded in plants supplied with 250 mg I L^{-1}, both in the fall and spring–summer season, and in those cultivated in the fall season and supplied with 50 mg I L^{-1}. The highest ascorbic acid concentration was recorded in plants cultivated in the spring–summer season and biofortified with the highest I dosage. The highest fructose and glucose concentrations in leaf tissues were obtained in plants cultivated in the spring–summer season and treated with 250 mg I L^{-1}. Plants sprayed with 250 mg I L^{-1} and cultivated in the fall season had the highest I leaf concentration. Overall, our results evidently suggested that an I application of 250 mg L^{-1} in both growing seasons effectively enhanced plant quality and functional parameters in curly endive plants.

Keywords: growing season; *Cichorium endivia* L. var. *crispum* Hegi; yield; sugars; mineral profile; iodine concentration; functional compounds

1. Introduction

Iodine (I) is a crucial trace element for the biosynthesis of thyroid hormones in humans [1]. Iodine deficiency illnesses are caused by unsatisfactory dietary iodine consumption [2] and associated with inadequate thyroid hormone synthesis, which, in turn, produces deleterious effects on the human organism, such as goiters, reproductive failure, hearing loss, growth impairment, cretinism, and numerous kinds of brain injury [3–6].

The World Health Organization (WHO) [7] highlights that approximately 45% of European inhabitants are distressed by I deficiency. As declared by the European Food Safety Authority [8], the recommended daily allowance (RDA) for I is estimated as follows: 90–120 µg for children, 150 µg for adults, and 290 µg for pregnant or breastfeeding women. However, according to the WHO [7], human I content, determined via urinary I concentration in spot urine samples, is considered insufficient at less than 100 µg L^{-1}, moderately deficient at 20–49 µg L^{-1}, and severely deficient at less than 20 µg L^{-1}.

Although Zimmermann [9] and Gonzali et al. [10] suggest that the primary approach to overcome low I assumption is the conventional iodination of table salt, Mottiar and Altosaar [11] point out that salt iodination alone is unsatisfactory to cover the entire human necessity of I. In addition, inorganic I is volatile, and therefore, its loss is difficult to control during storage, transport, and during cooking, particularly in the presence of high-temperature oils [12]. Moreover, the use of table salt is not recommended for people affected by cardiovascular disorders [13]. The WHO [7] suggests I intake via consumption of seafood and biofortified food such as fruiting and leafy green vegetables.

From an environmental and economic point of view, crop biofortification is recognized as a feasible strategy to combat human mineral malnourishment [14]. Although I is an imperative trace element for humans and animals [15], it is not essential for plants. As indicated by Tschiersch et al. [16], higher plants can absorb I by the roots or by the shoot and leaves via the stomata and/or the cuticular waxes. Additionally, Lawson et al. [17] reported that I supply through foliar sprays is more effective than soil application to enhance I concentration in butterhead lettuce. It is known that plant response to I enrichment is related to various factors, such as the chemical adopted form, the concentration in the nutrient solution, and the cultivation system [18,19]. A number of research initiatives have been aimed at enriching I concentration in various fruit and leafy vegetable crops, such as lettuce, spinach, *Brassica* genotypes, and tomato [13,20–24].

Curly endive (*Cichorium endivia* L. var. *crispum* Hegi) is widely grown all over the world and appreciated as a constituent of mixed salads. Furthermore, curly endive encloses a considerable level of bioactive constituents, such as ascorbic acid, phenolics, glucosinolates, sesquiterpene lactones, and minerals, especially potassium and calcium [25–27]. There is also evidence that the concentration of these compounds is significantly influenced by the growing season [28]. To the best of our knowledge, there is a lack of scientific literature on I biofortification of curly endive. Starting from the aforesaid evidence, the aim of the current study was to evaluate the effects of four levels of I supply on yield, bioactive compounds, sugars, and mineral profile of curly endive cultivated in the spring–summer or fall season.

2. Materials and Methods

2.1. Trial Setup, Plant Materials, and Crop Management

A two-year trial (2018 and 2019) was carried out in open field conditions in two consecutive growing seasons (spring–summer and fall). The research was performed at Blufi, Palermo Province (longitude 14°04′ E, latitude 37°45′ N, altitude 500 m) Sicily (Italy) in an experimental field of the Department of Agricultural, Food and Forest Sciences (SAAF) of the University of Palermo. Daily temperature (maximum and minimum) and rainfall throughout the plant cultivation cycles were collected (Figures 1 and 2).

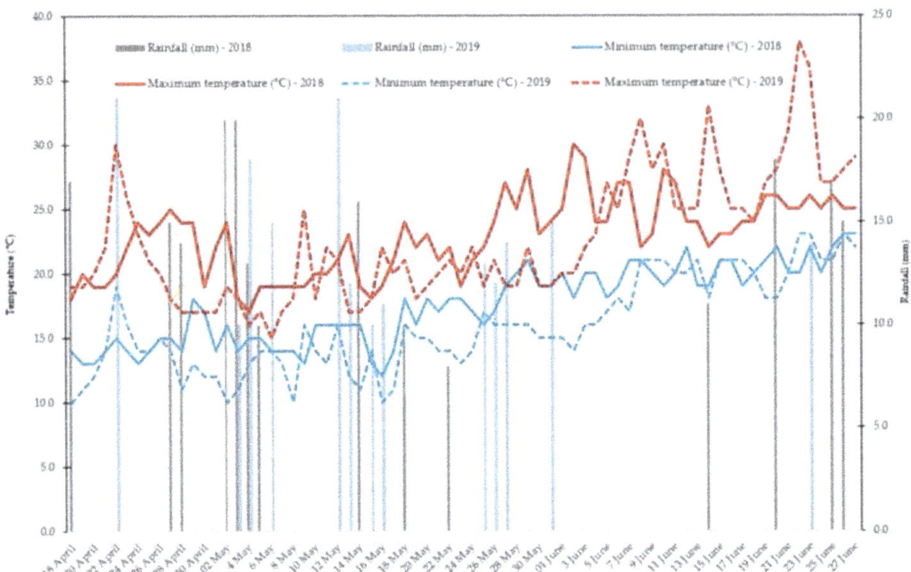

Figure 1. Daily temperature (maximum and minimum) and rainfall from 18 April to 27 June (2018 and 2019).

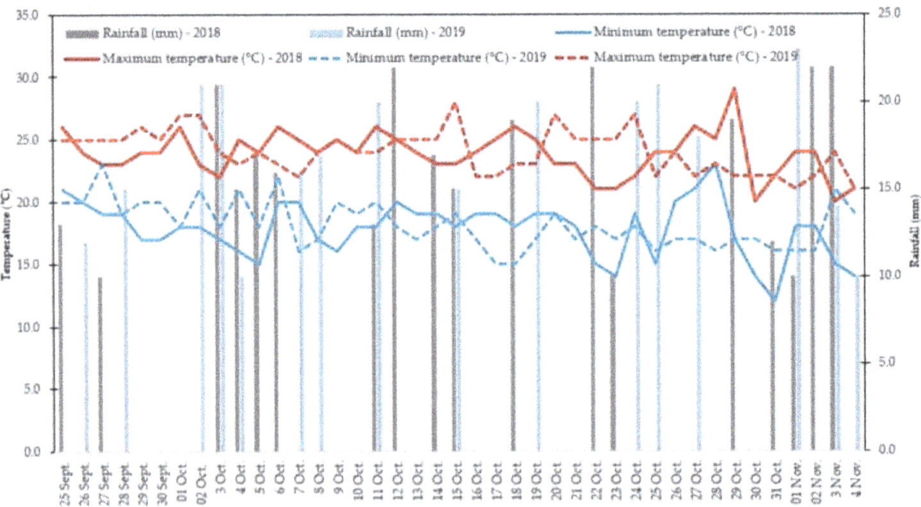

Figure 2. Daily temperature (maximum and minimum) and rainfall from 25 September to 4 November (2018 and 2019).

On 18 April (2018 and 2019) and 25 September (2018 and 2019), plug plants of curly endive (*Cichorium endivia* L., var. *crispum* Hegi) (var. Trusty, HM Clause, France) were grown with 0.33 m between rows and 0.30 m apart within the row, rendering 10 plants m^{-2}. Experimental soil was essentially sandy clay loam, characterized by a total nitrogen of 1.5% and organic matter of 3.0%.

Iodine-enrichment was made by supplying I in form of potassium iodate (KIO$_3$, Sigma-Aldrich ACS reagent, purity 99.5%). Four concentrations of I (0, 50, 250, and 500 mg L^{-1}) were provided through foliar spray. The foliar applications were carried out every 14 days, beginning on 2 May and 9 October (2018 and 2019) and finishing on 13 June and 20 October (2018 and 2019) for the spring–summer and fall seasons, respectively. In sum, for each

growing cycle, four foliar applications were performed. For each foliar spray application, 1.5 L m^{-2} of solution was distributed. Curly endive plants belonging to the plots maintained at 0 mg L^{-1} of I (control) received 1.5 L m^{-2} of water foliar spray. Fertilization was managed via drip irrigation during the cultivation cycle and comprised 100 kg nitrogen ha^{-1}, 60 kg phosphorous pentoxide ha^{-1}, and 180 kg potassium oxide ha^{-1} [29].

2.2. Yield and Biometric Parameters

Endive plants were harvested 70 days after plug transplant. All plants were taken into consideration for yield assessment and biometric traits determination. Biometric traits, consisting of head fresh weight, head height, stem diameter, and number of leaves, were recorded for all curly endive plants. To appraise dry matter content, five casually designated plants from each replicate were dehydrated in an oven (Memmert, Serie standard, Venice, Italy) set at 105 °C until constant weight.

2.3. Nutraceutical Features

Samples dedicated to nutraceutical quality investigation were collected immediately after harvest. Five casually designated plants from each replicate were taken into consideration for the functional property determinations. Soluble solid content (SSC) was assessed via a refractometer (MTD-045nD, Three-In-161 One Enterprises Co. Ltd., New Taipei City, Taiwan). Prior to SSC determination, curly endive samples were juiced and filtered. Titratable acidity (TA) was evaluated as reported by Han et al. [30]. Briefly, 10 g aliquots of curly endive were mixed in 50 mL of distilled water and titrated with 0.1 N NaOH to an end-point of pH 8.1. TA was expressed as percentage of malic acid. Ascorbic acid content was appraised by reflectometer Merck RQflex* 10 m using Reflectoquant Ascorbic Acid Test Strips, as reported by Sabatino et al. [27]. Concisely, 1 g of leaf juice sample was mixed with distilled water till reaching a final volume of 10 mL. Ascorbic acid was expressed as mg of ascorbic acid kg^{-1} fresh weight. To measure total phenolics, the methodology reported by Rivero et al. [31] was adopted. Briefly, 5 g of leaf sample were used for the extraction procedure using methanol as solvent and evaluated quantitatively by A765. Total phenolics concentration was appraised using Folin–Ciocalteu reagent and the outcomes were shown as mg of caffeic acid g^{-1} fresh weight.

2.4. Sugars Assessment

Five plants from each replicate were considered for sugars investigation. Sugars were appraised as described by Serna et al. [32]. Thus, leaf samples of 3 g were homogenised with 10 mL of deionized water and centrifuged at 15,000× g for 20 min at 4 °C. For sugars quantification, high-performance liquid chromatography (HPLC) was used and 10 µL of the supernatant was employed. Standard curves for pure standards of sugars (glucose, fructose, and sucrose) (Sigma, Poole, UK) were utilized for quantification. Findings were communicated as g 100 g^{-1} of fresh weight.

2.5. Mineral Profile

Five plants from each replicate were considered for minerals determination. Leaves Nitrogen (N) concentration was evaluated using the Kjeldahl method. The procedure described by Morand and Gullo [33] was followed for calcium (Ca), magnesium (Mg), and potassium (K) determination. Thus, atomic absorption spectroscopy (SavantAA, 200 ERRECI, Milan, Italy) was used. Phosphorus concentration was appraised using colorimetry, as reported by Fogg and Wilkinson [34].

With regard to I determination, the total I content in leaves tissues was assessed via inductively coupled plasma mass spectrometry (ICP-MS). In line with the official methodology for I evaluation (European Standard BS EN 15111:2007), an alkaline extraction was performed using the tetramethylammonium hydroxide. Afterwards, all the samples were filtered and analyzed via ICP-MS. The I content was expressed as mg kg^{-1} of dry weight.

2.6. Experimental Design and Statistics

Two different growing seasons (fall and spring–summer) were combined with four I enrichment levels (0, 50, 250, and 500 mg L^{-1}) in a two factorial experimental design rendering a total of eight treatments, two growing seasons (G) times four I doses. Every treatment constituted of three replicates, each containing 15 plants, for a total of 360 plants. The same experiment was performed in two consecutive years (2018 and 2019) following the same experimental scheme. All data sets were subjected to two-way analysis of variance (ANOVA), setting growing season and I dosage as source of variation. To appraise the influence of the year, a preliminary ANOVA analysis was performed. Percentage data were subjected to the arcsin transformation prior ANOVA analysis ($\varnothing$ = arcsin(p/100)$^{1/2}$). Tukey honestly significant difference (HSD) test was applied to separate mean values ($p < 0.05$). The statistical analyses were accomplished using the SPSS software version 20 (StatSoft, Inc., Chicago, IL, USA). Principal component analysis (PCA) was provided for the agronomical dataset (yield and biometric traits, nutraceutical features, sugars, and mineral profile) to evaluate any underlying relationships among the diverse I dosages and growing seasons of curly endive. Principal components with eigenvalues higher than 1.0 were considered for the individuation of the principal factors numbers (PCs). As a result, the PCs permit the investigation of relationships between the variables of the data set. Thus, the original variables were planned into the space demarcated by the PC1 and PC2, and connected variables were acknowledged. SPSS version 20.0 (StatSoft, Inc., Chicago, IL, USA) was utilized to accomplish PCA analysis.

3. Results

The trial was reiterated a second year using the identical experimental design and attaining analogous outcomes (Table S1). Thus, data from 2018 are shown.

3.1. Plant Performance and Quality

ANOVA analysis for head fresh weight, stem diameter, and head height did not display a significant influence on the interaction G × I (Table 1).

Table 1. Effect of the growing season (fall or spring–summer) and iodine biofortification supply (0, 50, 250 or 500 mg L^{-1}) on head fresh weight, stem diameter, and head height of curly endive.

Treatments	Head Fresh Weight (g)		Stem Diameter (mm)		Head Height (cm)	
Growing season						
Fall	734.07	a	22.94	a	31.00	a
Spring–summer	797.81	a	23.02	a	32.08	a
Iodine Biofortification (mg L^{-1})						
0	1125.45	a	26.28	a	36.78	a
50	853.02	b	25.30	a	32.53	b
250	697.25	c	23.15	b	32.53	b
500	388.05	d	17.18	c	24.32	c
Significance						
Growing season (G)	NS		NS		NS	
Iodine biofortification (I)	***		***		***	
G × I	NS		NS		NS	

Values within a column followed by different letters are significantly different at $p \leq 0.05$. *** significant at 0.001, respectively. NS, not significant.

Regardless of the biofortification, growing season did not significantly affect head fresh weight and stem diameter (Table 1). Conversely, non-biofortified plants showed the highest head fresh weight, followed by those biofortified with 50 mg I L^{-1}. The lowest

head fresh weight was recorded in plants treated with 500 mg I L^{-1}. Irrespective of the growing season, control plants and plants supplied with 50 mg I L^{-1} had the biggest stem diameter, whereas the smallest stem diameter was recorded in plants biofortified with 500 mg I L^{-1}. Aside from the biofortification, head height was greater in plants cultivated during the spring–summer season (Table 1). Irrespective of the growing season, control plants had the highest head height, followed by those treated with 50 or 250 mg I L^{-1}, whereas plants treated with a dosage of 500 mg I L^{-1} had the lowest height.

ANOVA for the number of leaves and head dry matter revealed a significant effect of the interaction G × I (Figure 3).

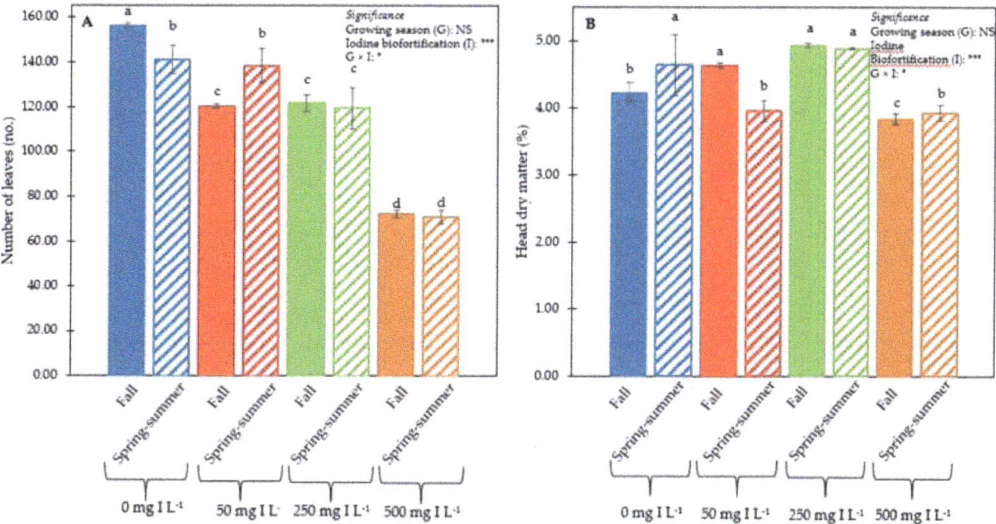

Figure 3. Number of leaves (**A**) and head dry matter (**B**) as affected by combining the growing season (fall or spring–summer) and iodine supply (0, 50, 250, or 500 mg L^{-1}) in curly endive. Different letters indicate significant differences at $p \leq 0.05$. *, *** significant at 0.05 and 0.001, respectively. NS, not significant.

Control plants cultivated in the fall season had the highest number of leaves, followed by spring–summer grown control plants and by plants biofortified with 50 mg I L^{-1} cultivated in the spring–summer season (Figure 3A). Plants subjected to the highest I-biofortification dosage grown both in the fall and spring–summer season displayed the lowest number of leaves. Dry matter percentage was the highest in the combinations 0 mg I L^{-1} × spring–summer season, 50 mg I L^{-1} × fall season, and 250 mg I L^{-1} × fall or spring–summer season (Figure 3B), whereas the lowest value was observed in plants fed with the highest biofortification dosage (500 mg I L^{-1}) grown in the fall.

ANOVA for titratable acidity, soluble solid content, and total phenolics did not exhibit a significant effect of the main treatments and of their interaction. (Table 2).

Regardless of I-biofortification, the growing season did not have a significant effect on SSC (Table 2). Irrespective of the growing season, the highest SSC values were detected in control plants and in those treated with 250 mg I L^{-1}. Plants supplied with an I-biofortification dosage of 50 mg I L^{-1} did not significantly differ neither from control plants nor from those biofortified with 250 mg I L^{-1}. The lowest SSC was recorded in plants biofortified with the highest dosage. Notwithstanding the I-biofortification, plants grown in the spring–summer season had a higher total phenolic concentration than those cultivated during the fall season (Table 2). Disregarding the growing season, plants biofortified with the highest dosage of I displayed the highest total phenolics concentration, followed by those biofortified with 250 mg I L^{-1}. The lowest total phenolics concentration was observed in control plants.

Table 2. Effect of growing season (fall or spring–summer) and iodine biofortification supply (0, 50, 250 or 500 mg L^{-1}) on TA, SSC, and total phenolics of curly endive.

Treatments	TA (%)		SSC (°Brix)		Total Phenolics (mg of Caffeic Acid g^{-1} FW)	
Growing season						
Fall	0.667		2.71		0.701	b
Spring–summer	0.667		2.64		0.786	a
Iodine Biofortification (mg L^{-1})						
0	0.683		2.90	a	0.580	d
50	0.683		2.87	ab	0.700	c
250	0.663		2.62	b	0.790	b
500	0.667		2.32	c	0.910	a
Significance						
Growing season (G)	NS		NS		***	
Iodine biofortification (I)	NS		***		***	
G × I	NS		NS		NS	

Values within a column followed by different letters are significantly different at $p \leq 0.05$. *** significant at 0.001. NS, not significant. TA: titratable acidity, SSC: soluble solid content.

ANOVA for ascorbic acid showed a significant effect of the interaction G × I (Figure 4); plants from the combination spring–summer cycle × 500 mg I L^{-1} had the highest ascorbic acid value followed by those from the combination fall × 500 mg I L^{-1}, which, in turn, revealed a higher ascorbic acid content than those grown during the spring–summer season and supplied with 250 mg I L^{-1} (Figure 4). The lowest ascorbic acid concentration was exhibited by control plants grown in the fall season.

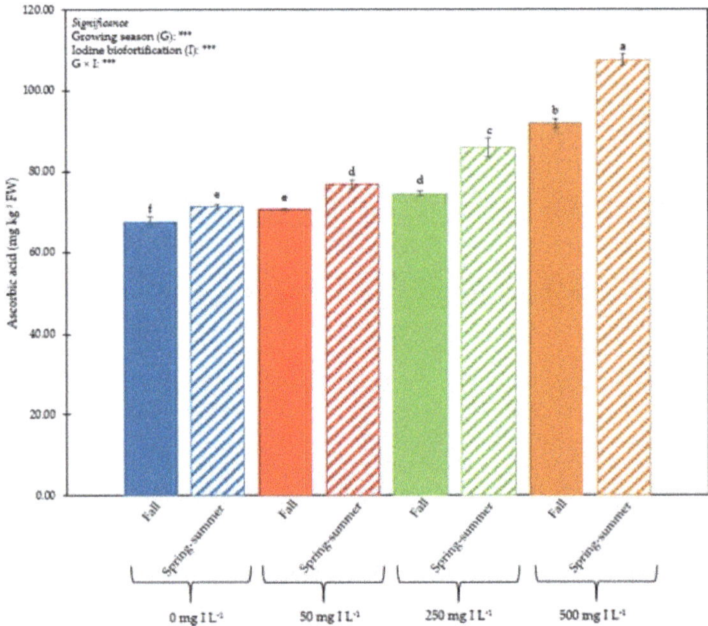

Figure 4. Ascorbic acid concentration as affected by combining growing season (fall or spring–summer) and iodine supply (0, 50, 250 or 500 mg L^{-1}) in curly endive. Different letters indicate significant differences at $p \leq 0.05$. *** significant at 0.001.

Concerning fructose concentration, a significant effect of the interaction G × I was detected (Figure 5A). Plants grown in the spring–summer season and treated with 250 mg I L^{-1} had the highest fructose concentration, followed by those biofortified with the same I dosage but grown during the fall season (Figure 5A). The lowest plant fructose concentration was recorded in control plants and in those biofortified with the highest I dosage.

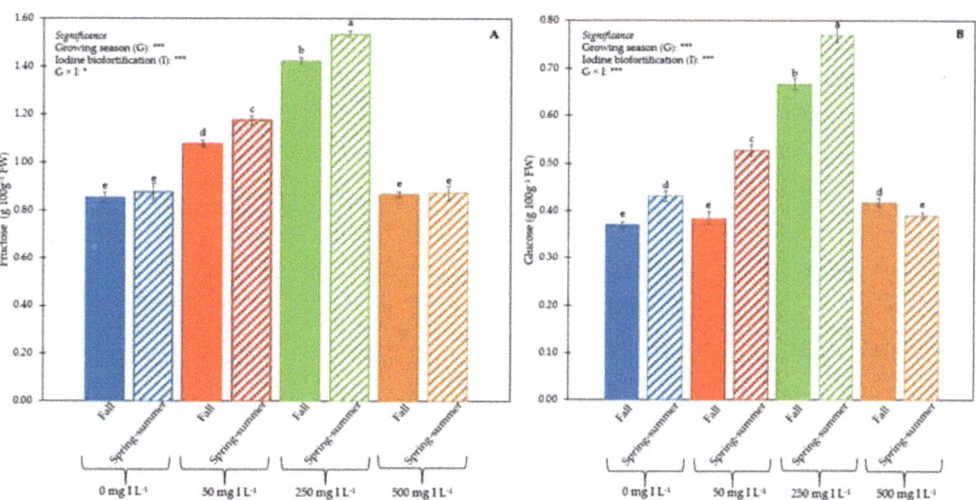

Figure 5. Fructose (**A**) and glucose (**B**) concentration as influenced by combining the growing season (fall or spring–summer) and iodine supply (0, 50, 250 or 500 mg L^{-1}) in curly endive. Different letters indicate significant differences at $p \leq 0.05$. *, *** significant at 0.05 and 0.001, respectively.

Regarding glucose concentration, ANOVA revealed a significant effect of the interaction between the growing season and I-biofortification (Figure 5B); plants from the combination spring–summer × 250 mg I L^{-1} had the highest glucose concentration, followed by plants exposed to the same I concentration but grown during fall (Figure 5). The lowest glucose concentration was observed in control plants cultivated in the fall and in plants grown during spring–summer and treated with 500 mg I L^{-1}. The treatments had no effect on sucrose concentration (Figure 6).

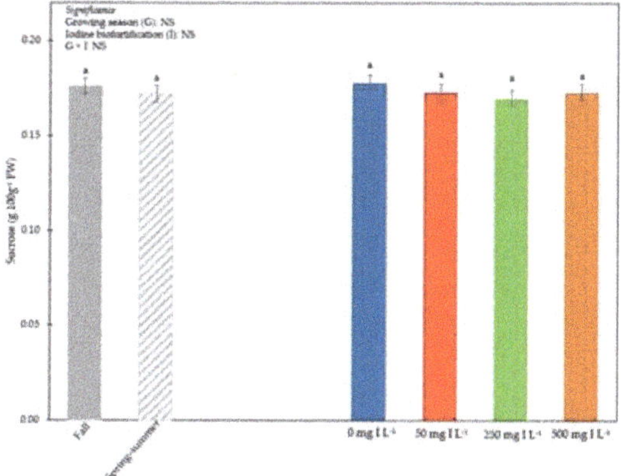

Figure 6. Effect of the growing season (fall or spring–summer) and iodine supply (0, 50, 250 or 500 mg L^{-1}) on sucrose concentration in curly endive. Different letters indicate significant differences at $p \leq 0.05$. NS, not-significant.

ANOVA for N, P, K, and Mg concentration did not show a significant effect of the interaction G × I (Table 3).

Table 3. Effect of the growing season (fall or spring–summer) and iodine supply (0, 50, 250 or 500 mg L^{-1}) on N, P, K, and Mg concentration in curly endive.

Treatments	N (mg g^{-1} DW)		P (mg g^{-1} DW)		K (mg g^{-1} DW)	Mg (mg g^{-1} DW)
Growing season						
Fall	5.77	a	0.61	a	3.23	0.34
Spring–summer	5.59	b	0.57	b	3.23	0.33
Iodine Biofortification (mg L^{-1})						
0	5.73		0.58		3.25	0.30
50	5.73		0.60		3.25	0.35
250	5.68		0.59		3.23	0.33
500	5.68		0.59		3.19	0.34
Significance						
Growing season (G)	*		***		NS	NS
Iodine biofortification (I)	NS		NS		NS	NS
G × I	NS		NS		NS	NS

Values within a column followed by different letters are significantly different at $p \leq 0.05$. *, *** significant at 0.05 and 0.001, respectively. NS, not significant. DW: dry weight.

Independently of the I-biofortification, plants grown in the fall showed a higher N concentration than plants cultivated in the spring–summer season (Table 3). On the contrary, regardless of the growing season, ANOVA analysis did not reveal a significant influence of the I-biofortification. Data on P concentration maintained the trend recognised for N concentration (Table 3). For K and Mg, ANOVA analysis did not display a significant effect of the treatments (Table 3).

ANOVA for Ca concentration revealed a significant influence of the interaction between growing season and I-biofortification (Figure 7).

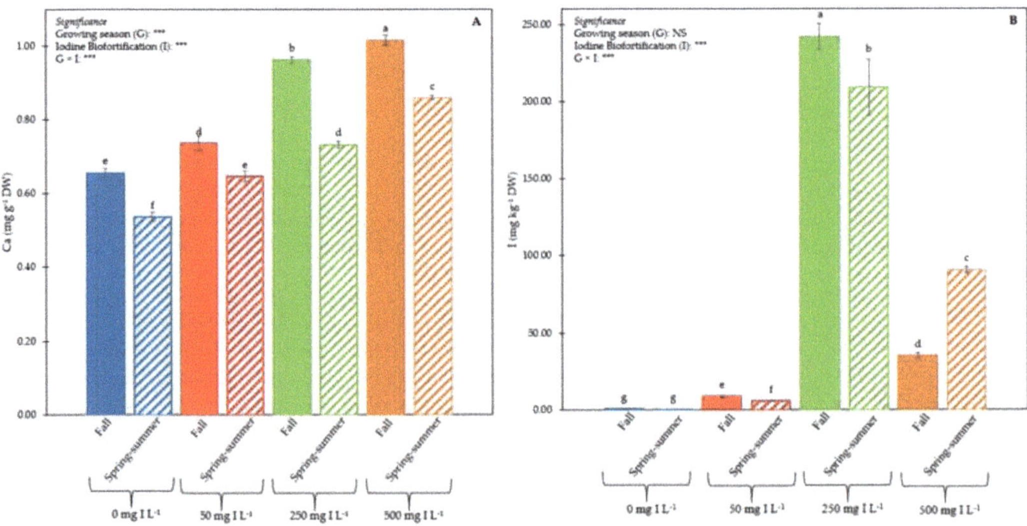

Figure 7. Ca (**A**) and I (**B**) concentration as affected by combining growing season (fall or spring–summer) and iodine supply (0, 50, 250, or 500 mg L^{-1}) in curly endive. Different letters indicate significant differences at $p \leq 0.05$. *** significant at 0.001. NS, not significant.

Plants grown in the fall season and treated with the highest I dosage had the highest Ca concentration, followed by those grown in the fall season and supplied with 250 mg I L^{-1} (Figure 7). The lowest Ca concentration was recorded in control plants cultivated in the spring–summer season. However, curly endive plants cultivated in the fall season revealed a higher Ca concentration than plants cultivated in the spring–summer season at the same I dosage (Figure 7).

3.2. Iodine Concentration in Leaf Tissues

ANOVA analysis showed a significant effect of the interaction G × I in terms of plant I concentration (Figure 7). Plants enriched with 250 mg I L^{-1} and grown in the fall season had the highest I leaf tissue concentration, followed by plants grown in the spring–summer season and biofortified with the same dosage (Figure 7). These plants, in turn, displayed a higher I concentration than plants cultivated in the spring–summer season and treated with 500 mg I L^{-1}. The lowest leaf I concentration was observed in non-biofortified plants.

3.3. Principal Component Analysis of all Plant Traits (PCA)

Principal component analysis (PCA) was conducted on all agronomical datasets. The loading plot and scores are presented in Figure 8.

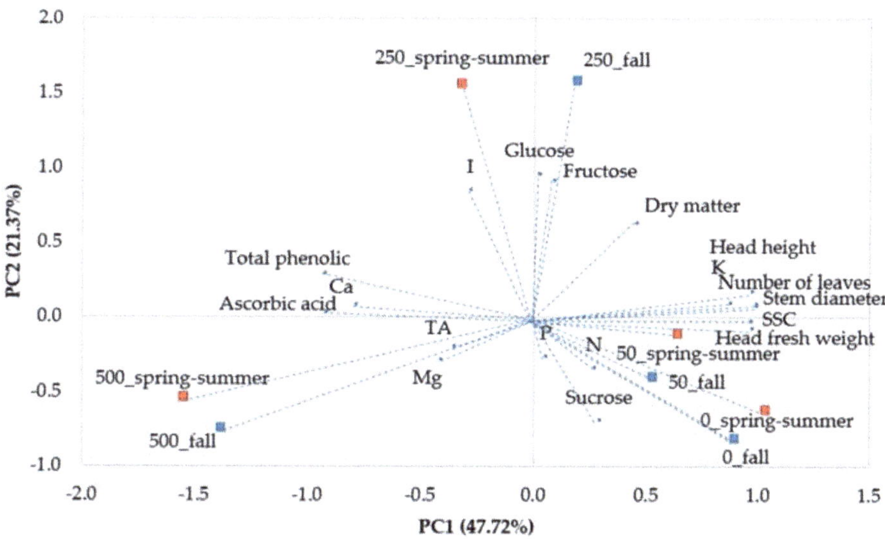

Figure 8. Scores and loading plots of PCA of curly endive plant performance and quality traits as affected by combining the growing season [fall (blue squares) or spring–summer (red squares)] and iodine dosage (0, 50, 250 or 500 mg L^{-1}).

As presented in Table S2, the outcomes of the PCA revealed four main factors (PCs) with eigenvalues higher than 1.00, representing 47.72%, 21.37%, 15.46%, and 8.38% of the total variance, respectively, and, consequently, clarifying 92.94% of the entire variance. PC1 was predominantly positively correlated to head fresh weight, head height, stem diameter, number of leaves, and SSC and negatively correlated to ascorbic acid, total phenolics, and Ca; PC2 was mostly positively correlated to fructose, glucose, and I; PC3 was mainly positively correlated to N and P; PC4 was essentially positively correlated to TA (Table S2). The PC1-PC2 graphic representation can be assumed in Figure 8. The 250_fall is placed on the top-right side of the plot of loading; the 0_fall, 0_spring–summer, 50_fall, and 50_spring–summer are located in the bottom-right side of the plot of loading; the 250_spring–summer is allocated in the top-left side of the loading plot; and finally, 500_spring–summer and 500_fall are placed in the bottom-left side of the plot of loading (Figure 8).

4. Discussion

Mineral malnutrition can be controlled via an appropriate dietary diversification, mineral increase consumption, foodstuff fortification, and by enhancing the bio-available mineral content in edible crops (a procedure named biofortification) [27,35,36]. Accordingly, functional food is very interesting and promising to prevent and cure diverse human disorders. Iodine is an imperative trace element for human and can be mainly assimilated via seafood and/or biofortified food intake, such as vegetables [7]. Indeed, there are reports that I shortage determines a number of negative effects on human health related to insufficient thyroid hormone production [3–6]. The current study highlighted that I supply and growing season can significantly influence plant performance and quality of curly endive grown in an open field. Our results showed that increasing I dosage in the nutrient solution resulted in significant decrease in yield (head fresh weight), head height, stem diameter, number of leaves, and percentage of head dry matter compared to the control. Our findings are in line with those obtained by Blasco et al. [37], who, studying the interactive effect between I and mineral nutrients in lettuce plants, found a reduction in the biomass of I-biofortified plants. Our outcomes concur with those reported by Smoleń et al. [20], who tested the effect of selenium and I biofortification on lettuce grown in a NFT hydroponic system and found a decrease of the biomass of I-enriched plants. Our results are also supported by those attained by Incrocci et al. [38], who reported a significant decrease in plant height, total dry matter, and leaf area of sweet basil I-enriched plants. A reduction in plant biomass was also reported in tomato and potato [22]. Conversely, other authors [39–41] reported a stimulating plant growth effect of I supply in barley, tomato, spinach, and strawberry. Blasco et al. [42] observed injurious effects in lettuce when I concentration in the nutrient solutions was higher than 10–40 μM or 100–200 μM. However, Signore et al. [43] reported that I biofortification does not have significant effect on leaves and roots biomass in a carrot Italian landrace. Thus, considering our results and those reported by other authors, it seems that the lowest I concentration tested in the current study (50 mg L^{-1}) is excessive for curly endive.

Our outcomes revealed that head fresh weight and stem diameter were not affected by the growing season. On the contrary, Sabatino et al. [28] reported that the number of leaves is positively influenced by the fall season in control plants, whereas spring–summer plants treated with 50 mg I L^{-1} performed better than those cultivated in the fall. Moreover, the number of leaves in spring–summer grown plants treated with 250 or 500 mg I L^{-1} did not significantly differ from that recorded in the fall grown plants supplied with the same I dosage. Thus, considering that the optimum growing temperature for curly endive is the 15–18 °C range [44] and since such temperatures occur in Sicily in the fall, we may assume that curly endive mineral absorption is more efficient during this season. Consequently, the toxic I threshold value was reached in the fall at a lower I supply dosage than in the spring–summer season.

Our results showed that neither the growing season nor I biofortification affected TA. These findings are in accordance with those reported by Islam et al. [45] in cherry tomatoes. Our findings, also, showed that I-biofortification significantly decreased SSC in curly endive. However, this is in contrast with the results by Golubkina et al. [46] in Indian mustard and by Islam et al. [45] in cherry tomato. Our outcomes, also, highlighted that the growing season did not affect SSC. This result is in accordance with the finding of Sabatino et al. [28], who did not determine differences in terms of SSC between curly endive plants grown in the fall and those grown in the spring–summer season. Moreover, our outcomes showed that, regardless of the growing season, total phenolics increased as I concentration in the nutrient solution increased. Our results are supported by those of Blasco et al. [42], who showed an increase of total phenolic in lettuce plants biofortified with an I dosage ranging from 0 to 240 μM. Furthermore, our results are in line with those reported by Kiferle et al. [47], who declared that KIO$_3$ treatments enhance phenolic concentration in basil. However, our findings did not concur with those of Incrocci et al. [38], who reported that, unlike I-biofortification via KI, I-biofortification via KIO$_3$ does not affect total phenolic

concentration in sweet basil. Our results pointed out, also, that the spring–summer season promoted total phenolic concentration in curly endive plants. Hence, considering that: (i) stress conditions promote phenolic synthesis [48–52]; (ii) I-biofortification is a stressful treatment for plants because iodide might be oxidized to elemental I and, consequently, can irreversibly damage the root cell membranes and oxidize chlorophylls and carotenoids, resulting in chlorosis of leaf and decreased CO_2 assimilation [17,18,53]; (iii) the optimum growth temperature for curly endive is 15–18 °C [44], we may speculate that the higher total phenolic concentration detected in curly endive plants grown in the spring–summer season and treated with higher I dosages could be positively correlated to the stressful conditions previously reported. Focusing on ascorbic acid, we found that curly endive cultivated during the spring–summer season and treated with a higher I-dosage revealed a higher ascorbic acid concentration. Our findings concur with those of Blasco et al. [42] and Blasco et al. [53], who found that I supply increases ascorbate concentration in lettuce plants. Our results also agree with those of Lester [54], who reported that a higher light intensity promotes ascorbic acid synthesis in green mustard. Furthermore, there are reports that radiation stress elicits plant ascorbic acid concentration [55,56]. Thus, considering that, in the Mediterranean region, curly endive is generally grown during the fall, we assume that the higher ascorbic acid concentration found in our study could be related to unfavourable spring–summer light intensity and photoperiod.

Regarding fructose, we found that I supply positively affected fructose concentration up to 250 mg L^{-1}. Conversely, I-biofortification at 500 mg L^{-1} reduced fructose concentration to a level similar to the control. Furthermore, plants grown in the spring–summer season and treated with 500 or 250 mg I L^{-1} had a higher fructose concentration than those grown during the fall. Data on glucose supported the tendency established for fructose. Our results concur with those by Blasco et al. [57], who, studying the effect of I on photosynthesis and metabolism of sugars in lettuce plants, found that increasing I supply results in an increase of fructose and glucose leaf concentration. Thus, in our study, the decrease in fructose and glucose concentration in the plants treated with 500 mg I L^{-1} could be due to a toxic effect of high I dosages. Our results are in agreement with those of Weston and Barth [58] and Caruso et al. [59], who reported that strawberry and tomato plants grown in full sunlight contain more sugar than those cultivated in the shade and suggested that a lower light intensity can significantly reduce sugar accumulation in vegetables.

Medrano-Macias et al. [19] reported that I supply has a relevant effect on the redox state of the system that absorbs elements; thus, it interrelates also with metal ions, altering the oxidation state and bioavailability.

Independently of the season, our data on mineral profile are fully in agreement with those reported by Islam et al. [45], who found that I-implementation does not influence N, P, K, and Mg fruit concentration compared to the control. Our findings are also in accordance with those by Incrocci et al. [38], who found that I supplied by KIO_3 does not significantly affect N, P, K, and Mg content in sweet basil. However, irrespective of I-biofortification, our results agree with those by Sabatino et al. [27], who reported that plants fertilized via standard nitrogen source and cultivated during the fall have a N leaf concentration higher than plants grown in the spring–summer. Concerning Ca content, we found that a higher I-dosage stimulated Ca leaf concentration. Additionally, fall grown plants displayed a higher Ca concentration than plants grown in the spring–summer season. This is in accord with the results of Incrocci et al. [38], who found a positive correlation between I-dosage and Ca leaf content. Our results partially concur with those by Blasco et al. [37], who found no significant effect in terms of N, P, and K when I was supplied via IO_3^-. As reported by Kato et al. [60], plant I-enrichment via IO_3^- form might elicit reductase activity in the roots. This could have an impact on the mineral nutrients bioavailability and on the iodate reductase and it may induce a redox signalling, resulting in a plant response to counter the I effect. Our data on mineral concentration showed that the nutrients fluctuated within the optimal range for curly endive [44]. Thus, as observed by other authors [18,38], the reduction in plant growth cannot be linked to I-induced mineral deficiencies.

Plants can absorb I by the root and, also, by epigeal organs such as stem and leaves via stomata and cuticular waxes [16]. As stated by White and Broadley [61], plants absorb I via ionic channels and chloride transporters. Moreover, there is evidence that I supplied by foliar spray is more effective than by soil applications for enhancing I concentration in plant tissues [17]. Additionally, Voogt et al. [62] established that the differences in iodine allocation among genotypes and seasons can be elucidated by the change in the transpiration rate. Our outcomes on leaf I concentration revealed that plants treated with 250 mg I L^{-1} had the highest I concentration in leaf tissues. Furthermore, plants enriched with a dosage of 50 or 250 mg I L^{-1} during the fall had a higher I concentration than those treated with the highest I-dosage.

Our results are in agreement with those of Incrocci et al. [38], who evidenced that I supplied as KI or KIO_3 causes an increase in I plant tissue concentration. Our findings also confirm those reported by Blasco et al. [37], who claimed that an I supply higher than 120 μM determines a reduction in I leaf concentration.

According to our results and taking into consideration the plant I uptake capacity and tolerance, we may suggest that both in the fall and in the spring–summer season, 250 mg I L^{-1} represents the best dosage to improve curly endive functional and nutraceutical traits.

5. Conclusions

In the current study, growing season combined with I-enrichment significantly affected yield and plant biometric traits, functional features, sugars, and mineral profile in curly endive. Overall, I-biofortification improved total phenolic and ascorbic acid, especially in plants cultivated in the spring–summer season. Furthermore, I-enrichment enhanced fructose and glucose concentration in curly endive up to the dose of 250 mg I L^{-1}, particularly in plants grown during the spring–summer season. The I concentration in the spraying solution was positively related to the Ca concentration in plant tissues. Our results also displayed that fall season increased Ca concentration as compared to the spring–summer season. Plants cultivated in the fall and I-biofortified at 250 mg L^{-1} had the highest I concentration in leaf tissues. Finally, our outcomes suggested that a combination of fall or spring–summer growing season and an I-dosage of 250 mg L^{-1} may effectively improve plant functional and nutritional quality of curly endive.

Supplementary Materials: The following are available online at https://www.mdpi.com/2311-7524/7/3/58/s1, Table S1: Significance of three-way ANOVA analysis (growing season × iodine biofortification × year). Table S2: Eigenvalues, proportion of total variability and correlation between the 28 variables and the first four principal components (PCs).

Author Contributions: Conceptualization, L.S., G.I., F.D., and B.B.C.; methodology, L.S. and G.I.; software, L.S. and B.B.C.; validation, L.S., G.I., B.B.C., Y.R., and C.E.-N.; formal analysis, L.S., B.B.C., C.D.P., and F.D.G.; investigation, L.S., S.L.B., B.B.C., C.E.-N., S.V., R.P.M., and R.C.; resources, L.S., F.D., and C.D.P.; data curation, L.S., Y.R., B.B.C., and C.E.-N.; writing—original draft preparation, L.S., G.I., Y.R., and S.L.B.; writing—review and editing, L.S., G.I., and Y.R.; visualization, L.S., G.I., and Y.R.; supervision, L.S., F.D., G.I., S.L.B., and Y.R; project administration, L.S., F.D., G.I., S.L.B., and Y.R; funding acquisition, F.D. and C.D.P. All authors have read and agreed to the published version of the manuscript.

Funding: This research received no external funding.

Institutional Review Board Statement: Not applicable.

Informed Consent Statement: Not applicable.

Acknowledgments: Authors are grateful to the farm Di Gioia Salvatore for hosting our experiment.

Conflicts of Interest: The authors declare no conflict of interest.

References

1. Grasberger, H.; Refetoff, S. Genetic causes of congenital hypothyroidism due to dyshormonogenesis. *Curr. Opin. Pediatr.* **2011**, *23*, 421–428. [CrossRef] [PubMed]
2. Zimmermann, M.B. Iodine deficiency. *Endocr. Rev.* **2009**, *30*, 376–408. [CrossRef] [PubMed]
3. Andersson, M.; Takkouche, B.; Egli, I.; Allen, H.E.; de Benoist, B. Current global iodine status and progress over the last decade towards the elimination of iodine deficiency. *Bull. World Health Organ.* **2005**, *83*, 518–525.
4. Delange, F. The role of iodine in brain development. *Proc. Nutr. Soc.* **2000**, *59*, 75–79. [CrossRef] [PubMed]
5. Dillon, J.C.; Milliez, J. Reproductive failure in women living in iodine deficient areas of West Africa. *Br. J. Obstet. Gynaecol.* **2000**, *107*, 631–636. [CrossRef] [PubMed]
6. Haddow, J.E.; Palomaki, G.E.; Allan, W.C.; Williams, J.R.; Knight, G.J.; Gagnon, J.; O'Heir, C.E.; Mitchell, M.L.; Hermos, R.J.; Waisbren, S.E.; et al. Maternal thyroid deficiency during pregnancy and subsequent neuropsychological development of the child. *N. Engl. J. Med.* **1999**, *341*, 549–555. [CrossRef] [PubMed]
7. World Health Organization. UNICEF, International Council for the Control of Iodine Deficiency Disorders. In *Assessment of Iodine Deficiency Disorders and Monitoring Their Elimination*, 3rd ed.; World Health Organization: Geneva, Switzerland, 2007.
8. EFSA Panel on Dietetic Products, Nutrition and Allergies (NDA). Scientific opinion on dietary reference values for iodine. *EFSA J.* **2014**, *12*, 3660.
9. Zimmermann, M.B. Iodine. In *Nutrition and Health in a Developing World*; de Pee, D., Taren, M., Bloem, S., Eds.; Springer International Publishing: Cham, Switzerland, 2017; pp. 287–295.
10. Gonzali, S.; Kiferle, C.; Perata, P. Iodine biofortification of crops: Agronomic biofortification, metabolic engineering and iodine bioavailability. *Curr. Opin. Biotechnol.* **2017**, *44*, 16–26. [CrossRef]
11. Mottiar, Y.; Altosaar, I. Iodine sequestration by amy lose to combat iodine deficiency disorders. *Trends Food Sci. Technol.* **2011**, *22*, 335–340. [CrossRef]
12. Tonacchera, M.; Dimida, A.; De Servi, M.; Frigeri, M.; Ferrarini, E.; De Marco, G.; Grasso, L.; Agretti, P.; Piaggi, P.; Alghino-Lombardi, F.; et al. Iodine fortification of vegetables improves human iodine nutrition: In vivo evidence for a new model of iodine prophylaxis. *J. Clin. Endocrinol. Metab.* **2013**, *98*, E694–E697. [CrossRef]
13. Kiferle, C.; Gonzali, S.; Holwerda, H.T.; Real Ibaceta, R.; Perata, P. Tomato fruits: A good target for iodine biofortification. *Front. Plant Sci.* **2013**, *4*, 205. [CrossRef] [PubMed]
14. Marschner, P. *Mineral Nutrition of Higher Plants*, 3rd ed.; Elsevier: Amsterdam, The Netherlands; Academic Press: Amsterdam, The Netherlands, 2012.
15. Stein, A.J. Global impacts of human mineral nutrition. *Plant Soil* **2010**, *335*, 133–154. [CrossRef]
16. Tschiersch, J.; Shinonaga, T.; Heuberger, H. Dry deposition of gaseous radioiodine and particulate radiocaesium onto leafy vegetables. *Sci. Total Environ.* **2009**, *407*, 5685–5693. [CrossRef] [PubMed]
17. Lawson, P.G.; Daum, D.; Czauderna, R.; Meuser, H.; Härtling, J.W. Soil versus foliar iodine fertilization as a biofortification strategy for field-grown vegetables. *Front. Plant Sci.* **2015**, *6*, 450. [CrossRef] [PubMed]
18. Duborská, E.; Urík, M.; Kubová, J. Interaction with soil enhances the toxic effect of iodide and iodate on barley (*Hordeum vulgare* L.) compared to artificial culture media during initial growth stage. *Arch. Agron. Soil Sci.* **2018**, *64*, 46–57. [CrossRef]
19. Medrano-Macías, J.; Leija-Martínez, P.; González-Morales, S.; Juárez- Maldonado, A.; Benavides Mendoza, A. Use of iodine to biofortify and promote growth and stress tolerance in crops. *Front. Plant Sci.* **2016**, *7*, 1146. [CrossRef] [PubMed]
20. Smoleń, S.; Kowalska, I.; Sady, W. Assessment of biofortification with iodine and selenium of lettuce cultivated in the NFT hydroponic system. *Sci. Hortic.* **2014**, *166*, 9–16. [CrossRef]
21. Smoleń, S.; Sady, W. Influence of iodine form and application method on the effectiveness of iodine biofortification, nitrogen metabolism as well as the content of mineral nutrient sand heavy metals in spinach plants (*Spinacia oleracea* L.). *Sci. Hortic.* **2012**, *143*, 176–183. [CrossRef]
22. Caffagni, A.; Arru, L.; Meriggi, P.; Milc, J.; Perata, P.; Pecchioni, N. Iodine fortification plant screening process and accumulation in tomato fruits and potato tubers. *Commun. Soil Sci. Plant Anal.* **2011**, *42*, 706–718. [CrossRef]
23. Smoleń, S.; Sady, W.; Kołton, A.; Wiszniewska, A.; Liszka-Skoczylas, M. Iodine biofortification with additional application of salicylic acid affects yield and selected parameters of chemical composition of tomato fruits (*Solanum lycopersicum* L.). *Sci. Hortic.* **2015**, *188*, 89–96. [CrossRef]
24. Gonnella, M.; Renna, M.; D'Imperio, M.; Santamaria, P.; Serio, F. Iodine Biofortification of Four *Brassica* Genotypes is Effective Already at Low Rates of Potassium Iodate. *Nutrients* **2019**, *11*, 451. [CrossRef] [PubMed]
25. Koudela, M.; Petříková, K. Nutritional composition and yield of endive cultivars—*Cichorium endivia* L. *Hort. Sci.* **2007**, *34*, 6–10. [CrossRef]
26. D'Antuono, L.F.; Ferioli, F.; Manco, M.A. The impact of sesquiterpene lactones and phenolics on sensory attributes: An investigation of a curly endive and escarole germplasm collection. *Food Chem.* **2016**, *199*, 238–245. [CrossRef] [PubMed]
27. Sabatino, L.; Ntatsi, G.; Iapichino, G.; D'Anna, F.; De Pasquale, C. Effect of selenium enrichment and type of application on yield, functional quality and mineral composition of curly endive grown in a hydroponic system. *Agronomy* **2019**, *9*, 207. [CrossRef]
28. Sabatino, L.; Iapichino, G.; La Bella, S.; Tuttolomondo, T.; D'Anna, F.; Cardarelli, M.; Rouphael, Y. An Appraisal of Calcium Cyanamide as Alternative N Source for Spring-Summer and Fall Season Curly Endive Crops: Effects on Crop Performance, NUE and Functional Quality components. *Agronomy* **2020**, *10*, 1357. [CrossRef]

29. Pardossi, A.; Prosdocimi Gianquinto, G.; Santamaria, P.; Incrocci, L. *Orticoltura Principi e Pratica*; Edagricole: Milano, Italy, 2018.
30. Han, C.; Zhao, Y.; Leonard, S.W.; Traber, M. Edible coatings to improve storability and enhance nutritional value of fresh and frozen strawberries (*Fragaria* × *ananassa*) and raspberries (*Rubus ideaus*). *Postharvest Biol. Technol.* **2008**, *33*, 67–78. [CrossRef]
31. Rivero, R.M.; Ruiz, J.M.; Garcia, P.C.; López-Lefebre, L.R.; Sánchez, E.; Romero, L. Resistance to cold and heat stress: Accumulation of phenolic compounds in tomato and watermelon plants. *Plant Sci.* **2001**, *160*, 315–321. [CrossRef]
32. Serna, M.; Hernández, F.; Coll, F.; Coll, Y.; Amorós, A. Effects of brassinosteroid analogues on total phenols, antioxidant activity, sugars, organic acids and yield of field grown endive (*Cichorium endivia* L.). *J. Sci. Food Agric.* **2013**, *93*, 1765–1771. [CrossRef]
33. Morand, P.; Gullo, J.L. Mineralisation des tissus vegetaux en vue du dosage de P, Ca, Mg, Na, K. *Ann. Agron.* **1970**, *21*, 229–236.
34. Fogg, D.N.; Wilkinson, N.T. The colorimetric determination of phosphorus. *Analist* **1958**, *83*, 406–414. [CrossRef]
35. Puccinelli, M.; Landi, M.; Maggini, R.; Pardossi, A.; Incrocci, L. Iodine biofortification of sweet basil and lettuce grown in two hydroponic systems. *Sci. Hortic.* **2021**, *276*, 109783. [CrossRef]
36. Sabatino, L.; D'Anna, F.; Iapichino, G.; Moncada, A.; D'Anna, E.; De Pasquale, C. Interactive effects of genotype and molybdenum supply on yield and overall fruit quality of tomato. *Front. Plant Sci.* **2019**, *9*, 1922. [CrossRef]
37. Blasco, B.; Ríos, J.J.; Sánchez-Rodríguez, E.; Rubio-Wilhelmi, M.M.; Leyva, R.; Romero, L.; Ruiz, J.M. Study of the interactions between iodine and mineral nutrients in lettuce plants. *J. Plant Nutr.* **2012**, *35*, 1958–1969. [CrossRef]
38. Incrocci, L.; Carmassi, G.; Maggini, R.; Poli, C.; Saidov, D.; Tamburini, C.; Pardossi, A. Iodine accumulation and tolerance in sweet basil (*Ocimum basilicum* L.) with green or purple leaves grown in floating system technique. *Front. Plant Sci.* **2019**, *10*, 1494. [CrossRef]
39. Borst Pauwels, G.W.F.H. Iodine as a micronutrient for plants. *Plant Soil* **1961**, *14*, 377–392. [CrossRef]
40. Zhu, Y.G.; Huang, Y.-Z.; Hu, Y.; Liu, Y.-X. Iodine uptake by spinach (*Spinacia oleracea* L.) plants grown in solution culture: Effects of iodine species and solution concentrations. *Environ. Int.* **2003**, *29*, 33–37. [CrossRef]
41. Li, R.; Liu, H.P.; Hong, C.L.; Dai, Z.X.; Liu, J.W.; Zhou, J.; Hu, C.Q.; Weng, H.X. Iodide and iodate effects on the growth and fruit quality of strawberry. *J. Sci. Food Agric.* **2017**, *97*, 230–235. [CrossRef] [PubMed]
42. Blasco, B.; Rios, J.J.; Cervilla, L.M.; Sánchez-Rodrigez, E.; Ruiz, J.M.; Romero, L. Iodine biofortification and antioxidant capacity of lettuce: Potential benefits for cultivation and human health. *Ann. Appl. Biol.* **2008**, *152*, 289–299. [CrossRef]
43. Signore, A.; Renna, M.; D'Imperio, M.; Serio, F.; Santamaria, P. Preliminary evidences of biofortification with iodine of "Carota di Polignano", an Italian carrot landrace. *Front. Plant Sci.* **2018**, *9*, 170. [CrossRef]
44. Tesi, R. *Orticoltura Mediterranea Sostenibile*; Pàtron Editore: Bologna, Italy, 2010.
45. Islam, M.Z.; Ho-Min, K.A.N.G. Iron, iodine and selenium effects on quality, shelf life and microbial activity of cherry tomatoes. *Not. Bot. Horti Agrobot. Cluj Napoca* **2018**, *46*, 388–392. [CrossRef]
46. Golubkina, N.; Kekina, H.; Caruso, G. Yield, quality and antioxidant properties of Indian mustard (*Brassica juncea* L.) in response to foliar biofortification with selenium and iodine. *Plants* **2018**, *7*, 80. [CrossRef]
47. Kiferle, C.; Ascrizzi, R.; Martinelli, M.; Gonzali, S.; Mariotti, L.; Pistelli, L.; Flamini, G.; Perata, P. Effect of Iodine treatments on Ocimum basilicum L.: Biofortification, phenolics production and essential oil composition. *PLoS ONE* **2020**, *15*, e0229016. [CrossRef]
48. Dixon, R.A.; Paiva, N.L. Stress-induced phenylpropanoid metabolism. *Plant Cell* **1995**, *7*, 1085–1097. [CrossRef]
49. Moglia, A.; Lanteri, S.; Comino, C.; Acquadro, A.; de Vos, R.; Beekwilder, J. Stress induced biosynthesis of dicaffeoylquinic acids in globe artichoke. *J. Agric. Food Chem.* **2008**, *56*, 8641–8649. [CrossRef]
50. Gisbert, C.; Prohens, J.; Raigón, M.D.; Stommel, J.R.; Nuez, F. Eggplant relatives as sources of variation for developing new rootstocks: Effects of grafting on eggplant yield and fruit apparent quality and composition. *Sci. Hortic.* **2011**, *128*, 14–22. [CrossRef]
51. Maršić, N.K.; Mikulič-Petkovšek, M.; Štampar, F. Grafting influences phenolic profile and carpometric traits of fruits of greenhouse-grown eggplant (*Solanum melongena* L.). *J. Agric. Food Chem.* **2014**, *62*, 10504–10514. [CrossRef]
52. Sabatino, L.; Iapichino, G.; Rotino, G.L.; Palazzolo, E.; Mennella, G.; D'Anna, F. *Solanum aethiopicum* gr. gilo and its interspecific hybrid with *S. melongena* as alternative rootstocks for eggplant: Effects on vigor, yield, and fruit physicochemical properties of cultivar 'Scarlatti'. *Agronomy* **2019**, *9*, 223. [CrossRef]
53. Blasco, B.; Rios, J.J.; Cervilla, L.M.; Sánchez-Rodríguez, E.; Rubio-Wilhelmi, M.M.; Rosales, M.A.; Romero, L. Photorespiration process and nitrogen metabolism in lettuce plants (*Lactuca sativa* L.): Induced changes in response to iodine biofortification. *J. Plant Growth Regul.* **2010**, *29*, 477–486. [CrossRef]
54. Lester, G.E. Environmental regulation of human health nutrients (ascorbic acid, β-carotene, and folic acid) in fruits and vegetables. *HortScience* **2006**, *41*, 59–64. [CrossRef]
55. Eskling, M.; Akerlund, H.E. Changes in the quantities of violaxanthin de-epoxidase, xanthophylls and ascorbate in spinach upon shift from low to high light. *Photosynth. Res.* **1998**, *57*, 41–50. [CrossRef]
56. Schonhof, I.; Kläring, H.P.; Krumbein, A.; Claußen, W.; Schreiner, M. Effect of temperature increase under low radiation conditions on phytochemicals and ascorbic acid in greenhouse grown broccoli. *Agric. Ecosyst. Environ.* **2007**, *119*, 103–111. [CrossRef]
57. Blasco, B.; Rios, J.J.; Cervilla, L.M.; Sánchez-Rodríguez, E.; Rubio-Wilhelmi, M.M.; Rosales, M.A.; Ruiz, J.M. Iodine application affects nitrogen-use efficiency of lettuce plants (*Lactuca sativa* L.). *Acta Agric. Scand. Sect. B Soil Plant Sci.* **2011**, *61*, 378–383.
58. Weston, L.A.; Barth, M.M. Preharvest factors affecting postharvest quality of vegetables. *HortScience* **1997**, *32*, 812–816. [CrossRef]

59. Caruso, G.; Villari, A.; Villari, G. Quality characteristics of *Fragaria vesca* L. fruits influenced by NFT solution EC and shading. In *South Pacific Soilless Culture Conference-SPSCC 648*; ISHS: Belgium, Leuven, February 2003; pp. 167–175.
60. Kato, S.; Wachi, T.; Yoshihira, K.; Nakagawa, T.; Ishikawa, A.; Takagi, D.; Takahashi, M. Rice (*Oryza sativa* L.) roots have iodate reduction activity in response to iodine. *Front. Plant Sci.* **2013**, *4*, 227. [CrossRef]
61. White, P.J.; Broadley, M.R. Biofortification of crops with seven mineral elements of ten lacking in human diets–iron, zinc, copper, calcium, magnesium, selenium, and iodine. *New Phytol.* **2009**, *182*, 49–84. [CrossRef]
62. Voogt, W.; Steenhuizen, J.W.; Eveleens, B.A. *Uptake and Distribution of Iodine in Cucumber, Sweet Pepper, Round, and Cherry Tomato*; Wageningen UR Greenhouse Horticulture: Bleiswijk, The Netherlands, 2014; pp. 1–69.

 *horticulturae*

 MDPI

Article

Chemical Composition, Antioxidant, and α-Glucosidase-Inhibiting Activity of Aqueous and Hydroethanolic Extracts of Traditional Antidiabetics from Croatian Ethnomedicine

Kristina Bljajić [1], Andrea Brajković [1,2], Ana Čačić [1,2], Lovorka Vujić [1], Jasna Jablan [1], Isabel Saraiva de Carvalho [2] and Marijana Zovko Končić [1,*]

[1] Faculty of Pharmacy and Biochemistry, University of Zagreb, 10000 Zagreb, Croatia; bljajickristina@gmail.com (K.B.); abrajkovic@pharma.hr (A.B.); anacacicka@hotmail.com (A.Č.); lvujic@pharma.hr (L.V.); jjablan@pharma.hr (J.J.)
[2] Food Science Laboratory, MED-Mediterranean Institute for Agriculture, Environment and Development, University of Algarve, 8005-139 Faro, Portugal; icarva@ualg.pt
* Correspondence: mzovko@pharma.hr; Tel.: +385-1-6394-792

Abstract: Type 2 diabetes (T2D) is a chronic disease with a growing prevalence worldwide. In addition to the conventional therapy, many T2D patients use phytotherapeutic preparations. In the present study, chemical composition, antioxidant, and α-glucosidase inhibiting activity of traditional antidiabetics from Croatian ethnomedicine (*Achillea millefolium, Artemisia absinthium, Centaurium erythraea, Morus alba, Phaseolus vulgaris, Sambucus nigra,* and *Salvia officinalis*) were assessed. The efficacy of water and 80% ethanol as extraction solvents for bioactive constituents was compared. HPLC analysis revealed that the prepared extracts were rich in phenols, especially rutin, ferulic, and chlorogenic acid. Antiradical (against DPPH and ABTS radicals), reducing (towards Mo^{6+} and Fe^{3+} ions), and enzyme inhibiting properties were in linear correlation with the content of phenolic constituents. Ethanolic extracts, richer in phenolic substances, showed dominant efficacy in those assays. Aqueous extracts, on the other hand, were better Fe^{2+} ion chelators and more active in the β-carotene linoleic acid assay. Extracts from *S. officinalis* and *A. millefolium* were particularly active antioxidants and α-glucosidase inhibitors. *A. absinthium,* another potent α-glucosidase inhibitor, contained chromium, a mineral that promotes insulin action. The investigated plants contained significant amounts of minerals useful in management of T2D, with negligible amounts of heavy metals deeming them safe for human use.

Keywords: chromium; ethnopharmacology; flavonoids; glucose-lowering activity; HPLC; natural antioxidants; polyphenol

Citation: Bljajić, K.; Brajković, A.; Čačić, A.; Vujić, L.; Jablan, J.; Saraiva de Carvalho, I.; Zovko Končić, M. Chemical Composition, Antioxidant, and α-Glucosidase-Inhibiting Activity of Aqueous and Hydroethanolic Extracts of Traditional Antidiabetics from Croatian Ethnomedicine. *Horticulturae* **2021**, *7*, 15. https://doi.org/10.3390/horticulturae7020015

Academic Editor: Christophe El-Nakhel

Received: 30 December 2020
Accepted: 20 January 2021
Published: 26 January 2021

Publisher's Note: MDPI stays neutral with regard to jurisdictional claims in published maps and institutional affiliations.

1. Introduction

Unhealthy dietary choices and sedentary living style have led to an increased incidence of different chronic diseases including metabolic syndrome and type 2 diabetes (T2D) [1]. Those diseases are primarily characterized by insulin resistance resulting not only in hyperglycemia but reactive oxygen species (ROS) production and accumulation [2]. While ROS, in physiological concentrations, play an important role in healthy metabolism, excessive ROS production can lead to oxidative stress, a state characterized by damage of cellular macromolecules, impaired protein function, and, eventually, cell death [3]. A constant state of enhanced oxidative stress leads to the development of diabetic complications including nephropathy, neuropathy, and retinopathy, as well as to liver damage [3,4]. The consequences of hyperglycemia-induced oxidative stress may be ameliorated using various endo- and exogenous antioxidants. Among the latter, phenolic natural substances

have an especially prominent place due to their notable antioxidant, but also other biological activities. Numerous studies indicate that they can modulate carbohydrate and lipid metabolism, attenuate hyperglycemia, dyslipidemia, and insulin resistance, as well as alleviate oxidative stress and inflammatory processes [3,5]. Secondary metabolites can target the key enzymes of carbohydrate metabolism and retard the postprandial increase of glucose concentration. For instance, they can inhibit the activity of α-glucosidase, the enzyme that degrades the oligosaccharides to glucose. Plants used in traditional medicine are an especially rich source of α-glucosidase inhibitors making them valuable nutritional and therapeutic tool for prevention of onset and long-term T2D complications [5,6]. Recent studies have shown that most patients use medicinal plants as a complementary therapy for T2D [7] and a potential value of herbal products for T2D may lay primarily in the area of prevention of diabetic complications [8]. In order to develop such products from plants, it is important to investigate the influence of different types of extraction, as they may radically affect the composition and, as a result, the biological activity of the prepared extracts [9].

Traditional medicine is an accessible, affordable, and culturally acceptable form of healthcare trusted by large numbers of people worldwide. Ethnobotanical data indicate that more than 800 plants around the world are used as traditional remedies for the treatment of T2D, often with efficacy comparable to conventional drugs [10]. However, the use of medicinal plants is not without its risks because they often contain harmful substances, such as pesticides and heavy metals [11]. On the other hand, the presence of some minerals may positively contribute to the antidiabetic effects of plants e.g., by enhancing insulin effects [12].

Medicinal plants are a rich source of bioactive phenols that may affect multiple targets and act beneficially in treatment of T2D. Thus, the aim of this work was evaluation and comparison of the phenolic composition, antioxidant, and α-glucosidase-inhibiting activity of aqueous and ethanolic extracts of the plants traditionally used in Croatia for management of diabetes: *Achillea millefolium*, *Artemisia absinthium*, *Centaurium erythraea*, *Morus alba*, *Phaseolus vulgaris*, *Sambucus nigra*, and *Salvia officinalis*. In order to assess their safety, but also potential favorable effects of the minerals present in the plant material, the content of heavy and other metals was determined.

2. Materials and Methods

2.1. Plant Material, Chemicals, and Apparatus

Herbal material included in this study was bought from the herbalists from the local markets. Using the appropriate monographs of European pharmacopoeia [13] and other relevant literature [14], the identity was confirmed by the authors. Vouchers are deposited at the University of Zagreb Faculty of Pharmacy and Biochemistry (Department of Pharmacognosy). Extraction was performed using Bandelin SONOREX® Digital 10 P DK 156 BP ultrasonic bath (Berlin, Germany). For total reflection X-ray fluorescence (TXRF) determination of metal content in the plant material, the bench top TXRF spectrometer "S2 Picofox" (Bruker Nano GmbH, Berlin, Germany), equipped with a Mo target micro focus tube, a multilayer monochromator (80% reflectivity), as well as the liquid nitrogen-free XFlash® silicon drift detector and with energy resolution of <150 eV (Mn Kα) was used. The instrument was operated at 50 kV/750 µA. A microplate reader Stat Fax 3200 (Awareness Technologies, Palm City, FL, USA) was used for UV-VIS spectroscopy. An Agilent 1200 series HPLC instrument with an autosampler, DAD detector, and Zorbax Eclipse XDB-C18 (5 µm, 12.5 mm × 4.6 mm) (Agilent Technologies, Santa Clara, CA, USA) column and guard column was used for determination of individual flavonoids and phenolic acids. HPLC standards (purity $\geq$ 97%) and α-glucosidase (type I from *Saccharomyces cerevisiae*) were purchased from Sigma-Aldrich (St. Louis, MO, USA). For chromatographic separation, HPLC grade methanol was used, while the other reagents and chemicals were of analytical grade.

2.2. Determination of Metal Content in Plant Material

For TXRF analysis, powdered plant material (0.3 g) was ultrasonicated with 15 mL of HNO$_3$ (0.14 M) for 10 min at 25 °C. Upon filtration, aliquots (1 mL) of each prepared extract were mixed with 10 µL of a yttrium stock solution (1 g/L, internal standard). After homogenization, an aliquot (10 µL) of thus prepared solution was centred at an unsiliconized quartz glass sample carrier. The solvent was removed on a plate heater at the temperature of 50 °C and inserted into a TXRF spectrometer. The measurement time was 1000 s.

2.3. Preparation of the Extracts

Extracts were prepared from aerial parts of *Artemisia absinthium*, *Achillea millefolium*, *Centaurium erythraea*, leaves of *Morus alba*, and *Salvia officinalis*, as well as *Phaseolus vulgaris* pods, and *Sambucus nigra* flowers. Dried plant material was milled and passed through a sieve of 850 µm mesh size. To 20 mL of the appropriate solvent (water or 80% ethanol) in a 50 mL Erlenmeyer flask, 2 g of plant material was added. After 30 min in an ultrasonication bath (80 °C, 720 W), the extracts were subjected to centrifugation for 30 min at 3400 rpm. Supernatant was collected and subjected either to freeze-drying (aqueous extracts), or evaporation at 30 °C in rotavapor (hydroethanolic extracts).

2.4. Spectrophotometric Determination of Phenolic Compounds

Total phenols (TP) content was determined using a Folin–Ciocalteau reagent [15]. For the determination of total flavonoids (TF), a reaction with aluminum chloride was used [16]. Total phenolic acids (TPA) content was assessed with a nitrite-molybdate reagent [17]. Content of TP, TF and TPA was expressed as mg of gallic acid (GAE), quercetin (QE), and caffeic acid (CAE) equivalents in g of dry weight (DW), respectively.

2.5. HPLC Analysis of Phenolic Constituents

Methanolic solutions of the extracts (2 mg/mL) and the phenolic standards (0.2 mg/mL), filtered through a 0.45 µm PTFE syringe filter, were subjected to HPLC chromatographic separation at a temperature of 40 °C and flow of 1.0 mL/min. The solvents A and B were water, methanol, and formic acid in proportions 93:5:2 (v:v:v) and 3:95:2 (v:v:v), respectively. Hyperoside, kaempferol, protocatechuic acid, and quercetin were quantified at 270 nm, while for caffeic, chlorogenic, ferulic, and *p*-coumaric acid, absorbance at 320 nm was recorded. In order to construct a calibration curve, varying volumes of standard solutions were injected using an autosampler. The peak assignment and identification were based on comparison of retention times of peaks in sample chromatogram and UV spectra with those of the standards. Calibration curve parameters, level of detection (LOD), and level of quantification and (LOQ) are reported in Table 1.

Table 1. Calibration curve parameters for flavonoids and phenolic acid standards.

Standard	Equation	r^2	LOD (µg/mL)	LOQ (µg/mL)
Caffeic acid	$y = 5335.00x - 19.45$	0.9999	0.012	0.041
Chlorogenic acid	$y = 2587.30x + 73.42$	0.9996	0.036	0.11
p-coumaric acid	$y = 6214.63x - 137.38$	0.9973	0.09	0.299
Ferulic acid	$y = 5045.04x - 45.10$	0.9998	0.026	0.088
Gallic acid	$y = 4808.10x + 12.88$	0.9998	0.026	0.087
Hyperoside	$y = 1426.20x + 15.40$	0.9999	0.013	0.04
Luteolin	$y = 3205.25x - 46.60$	0.9998	0.025	0.077
Myricetin	$y = 4341.80x - 691.78$	0.9997	0.315	1.049
Quercetine	$y = 2200.20x - 36.75$	0.9998	0.027	0.083
Rosmarinic acid	$y = 2518.90x + 21.44$	0.9999	0.003	0.01
Rutin	$y = 3197.71x - 478.76$	0.9988	0.289	0.962

LOD = level of detection; LOQ = Level of quantification; y = Area under curve (mAU × s); x = amount of the standard (µg).

2.6. ABTS and DPPH Radical Scavenging Activity

Radical scavenging activity (RSA) of the extracts was evaluated using 2,2′-azino-bis(3-ethylbenzothiazoline-6-sulfonic acid) (ABTS) and 2,2-diphenyl-1-picrylhydrazyl (DPPH) free radicals as described in the references [18,19], using Trolox and butylated hydroxyanisole (BHA) as standard radical scavengers, respectively. The solution of the extract was added to the free radical solution. After 30 min, the absorbance was read either at 734 nm (for ABTS RSA) or 545 nm (for DPPH RSA), respectively. RSA was calculated according to Equation (1):

$$RSA = \frac{A_{control} - A_{sample}}{A_{control}} \times 100 \tag{1}$$

where $A_{control}$ and A_{sample} are the absorbances of the negative control (methanol) and the solution containing the extract, respectively.

2.7. Fe²⁺ Chelating Activity

Chelating activity (ChA) determination was based on complexation of Fe^{2+} ions with ferrozine [19]. To the methanolic solution (150 μL), of either extract or ethylenediaminetetraacetic acid (EDTA) (chelating standard), 50 μL $FeCl_2$ (0.25 mM) was added. After 5 min, ferrozine solution (1.0 mM, 100 μL) was added. After additional 10 min, absorbance was measured at 545 nm. ChA was calculated according to Equation (2):

$$ChA = \frac{A_{control} - A_{sample}}{A_{control}} \times 100 \tag{2}$$

where is $A_{control}$ is absorbance of the negative control (reaction mixture without ion chelators), and A_{sample} is the absorbance of the solution with the extract.

2.8. β-Carotene-Linoleic Acid Assay

The β-carotene-linoleic acid was performed as follows [19]: Tween 40 (200 mg) was added to the mixture of linoleic acid (20 mg) and β-carotene (1.0 mL, γ = 0.2 g/L) in chloroform. After vortexing the mixture, chloroform was removed and 30 mL of water, previously saturated with oxygen, was added. Upon the addition of the aliquots of thus prepared emulsion (200 μL) to the extract solutions (50 μL), the reaction mixture was incubated at 50 °C for 120 min and the absorbance at 450 nm measured every 15-min. antioxidant activity in β-carotene-linoleic acid assay (ANT) was calculated according to Equation (3):

$$ANT = \frac{R_{control} - R_{sample}}{R_{control}} \times 100 \tag{3}$$

where $R_{control}$ and R_{sample} are reaction rates for the water (control) and the extract, respectively. BHA was used as the standard antioxidant. Final concentrations of the extracts and the BHA in the reaction solution were 8 μg/mL.

2.9. Total Antioxidant Activity

For determination of total antioxidant activity (TAA) [20], 0.1 mL of the extract solution was combined with 1 mL of solution consisting of sulfuric acid (0.6 M), sodium phosphate (28 mM), and ammonium molybdate (4 mM). Reaction mixture was incubated for 90 min at 95 °C. The absorbance was measured at 695 nm after cooling the mixture to the room temperature. The activity calculation was based on the calibration curve of ascorbic acid, and expressed as mg ascorbic acid equivalents (AAE) per g of DW.

2.10. Reducing Power

Reducing power (RP) determination was performed as described previously [21]. The extract solution (0.2 mL) was mixed with 2.5 mL phosphate buffer (0.2 M, pH 6.6) and 0.5 mL $K_3[Fe(CN)_6]$ (1.0%). After 20 min at 50 °C, 0.5 mL of trichloroacetic acid (10%)

was added. An aliquot (0.5 mL) of supernatant was mixed with water (0.5 mL) and $FeCl_3$ (0.1 mL, 0.1%). The absorbance was read at 700 nm. The calibration curve of Trolox was constructed and RP expressed as mg Trolox equivalent (TE) per g of DW.

2.11. Ferric Reducing Antioxidant Power

Ferric reducing antioxidant power (FRAP) was performed as described in reference [22]. To 0.9 mL of the FRAP reagent, consisting of acetate buffer (25 mL, 300 mM), of 2,4,6-tripyridyl-2-triazine solution (2.5 mL, 10 mM in in 40 mM HCl) and 2.5 mL ferric chloride solution (20 mM), the extract solution was added (0.1 mL) and left in the dark at 25 °C and the absorbance at 593 nm recorded after 30 min. FRAP was calculated using the Trolox calibration curve and expressed as mg Trolox equivalent (TE) per g of DW.

2.12. Determination of α-Glucosidase Inhibiting Activity

Determination was performed according to Tiwari et al. [23]. The enzyme (1.0 U/mL in 0.1 M phosphate buffer, pH 6.8) was added to the 100 µL of the extract. After 10 min at 37 °C., 50 µL of 5 mM *p*-nitrophenyl-α-D-glucopyranoside, dissolved in the same buffer, was added. After 5 min, the absorbance was measured at 405 nm and the α-glucosidase inhibiting activity (AG) calculated according to Equation (4):

$$AG = \frac{A_{control} - A_{sample}}{A_{control}} \times 100 \tag{4}$$

where $A_{control}$ is the absorbance of the control mixture (mixture with the buffer instead of the inhibitor), and A_{sample} represents the absorbance of samples containing the inhibitor (the extracts or acarbose).

2.13. Statistical Analysis

The experiments were performed in triplicate. The results were expressed as mean $\pm$ SD. Statistical analyses were performed using GraphPad Prism 8.0 (www.graphpad.com). Analyses were performed using one-way ANOVA followed by either Dunnett (comparisons of the individual extracts with the controls) or Tukey (for comparisons between the extracts) post-hoc tests. The differences between ethanolic and aqueous extracts were investigated using a paired *t*-test. Unless otherwise noted, *p*-values < 0.05 were considered statistically significant. IC_{50} values in ABTS RSA, DPPH RSA, ChA, and α-glucosidase assay were calculated using regression analysis.

3. Results
3.1. Plant Included in the Study

A recent ethnopharmacological study has shown that various medicinal plants are used either as complementary or the individual therapy of diabetes in Croatia [7]. In this study, seven plants were selected and their chemical content and biological effects were compared. The plants and their parts included in this study, as well as the abbreviations used throughout the text, are presented in Table 2. One group of the plants was selected according to the prevalence of their use against T2D. For example, most of the plants selected for this study were used by more than 30% of interviewed herbalists (AA, CE, MA, and PV) [7]. The other plants were selected due to their widespread use in Croatia for culinary or other medicinal purposes (AM, SN, and SO). Some of them were better researched in regard to their antidiabetic activity (e.g., MA) [24], while the others were the subject of a fewer studies (e.g., AA). For example, it was interesting to note that, while both beans and pods (pericarp) of *Phaseolus vulgaris* are well-known traditional antidiabetic agents, only the beans are rather well studied and recognized as functional food for diabetics. The pods, on the other hand, are still under-researched and the existing studies failed to establish its efficacy unequivocally [8]. In spite of that, *Phaseolus vulgaris* pods were still one of the most popular remedies in Croatia, recommended by 50% of herbalists, surpassed only by *Urtica dioica* [7].

Plant materials are complex mixtures of phytochemical constituents, which, due to their differing chemical properties (e.g., polarities), may not all be equally efficiently extracted with the same extraction solvents. Thus, different extraction methods may affect the composition and thus biological effects of the prepared extracts. In order to include a wider spectrum of metabolites, as well as to determine the solvent best suited for potential production of food supplements based on the selected plants, the extraction was performed using two solvents of different polarities: water and 80% ethanol (v/v). The prepared extracts and their abbreviations are presented in Table 2.

Table 2. The medicinal plants, extracts, and their abbreviations used in this study.

Plant Species	Type of Material	Abbreviation *
Artemisia absinthium L., Asteraceae (wormwood)	Flowering aerial parts	AA
Achillea millefolium L., Asteraceae (yarrow)	Flowering aerial parts	AM
Centaurium erythraea Rafn., Gentianaceae (centaury)	Flowering aerial parts	CE
Morus alba L., Moraceae (white mulberry)	Leaf	MA
Phaseolus vulgaris L., Fabaceae (common bean)	Fruit (pericarp)	PV
Sambucus nigra L., Caprifoliaceae (elder)	Flower	SN
Salvia officinalis L., Lamiaceae (sage)	Leaf	SO

* = In further text, the suffixes denote the solvents used for extraction: -E = extract prepared using 80% ethanol (v/v); -W = extract prepared using water.

3.2. Content of Metals in Selected Plants

Herbal teas for T2D, similar to other herbal teas at local markets, often sold without appropriate health safety and quality control, may represent a potential health-hazard to the patients who use them. In order to investigate the safety of the plant material, but also to assess the potential contribution of selected minerals to their antidiabetic effects, multielemental analysis was performed using TXRF (Table 3).

Chromium is an essential element necessary for the action of insulin most probably through its complex with a chromium-binding oligopeptide called chromoduline. This complex binds to β-subunit of the insuline receptor, thus activating it and increasing the insulin signal [12]. Even though it has been postulated that PV might exert its potential antidiabetic effect due to its chromium content [8], among the investigated samples, only AA contained a significant amount of this metal. On the other hand, all the investigated plants contained significant amounts of magnesium, zinc, and manganese. Magnesium is a cofactor in more than 300 enzymatic reactions. It is important for maintaining cellular membrane integrity, muscle contraction, nervous system conduction, and vascular tone. Its deficiency is associated with a number of clinical disorders, including insulin resistance, type 2 diabetes, hypertension, and cardiovascular disease. Magnesium supplementation has been reported to improve insulin sensitivity in patients with T2D [25]. Among the investigated samples, PV was the richest in this metal, and its content was two-fold higher than in MA, another sample rich in Mg.

Table 3. Content of investigated metals in dry plant material.

Plant Material	Cr mg/kg	Mg mg/kg	Zn mg/kg	Mn mg/kg	Cu mg/kg	Pb mg/kg	Ni mg/kg	As mg/kg
AA	0.5 ± 0.1	205 ± 80	31.9 ± 0.15	28.4 ± 0.2	25.2 ± 0.1	1.6 ± 0.05	2.0 ± 0.05	0.8 ± 0.05
AM	*n.d.*	655 ± 90	21.4 ± 0.1	24.6 ± 0.2	11.0 ± 0.1	0.8 ± 0.05	0.6 ± 0.05	1.1 ± 0.05
CE	*n.d.*	295 ± 80	29.5 ± 1.3	37.1 ± 2.3	13.6 ± 1.0	0.4 ± 4.25	3.1 ± 0.07	1.3 ± 4.0
MA	*n.d.*	795 ± 110	14.4 ± 0.1	9.5 ± 0.2	5.6 ± 0.1	0.8 ± 0.05	1.5 ± 0.05	1.0 ± 0.05
PV	*n.d.*	1575 ± 110	16.4 ± 0.1	12.5 ± 0.2	11.3 ± 0.1	0.3 ± 0.05	5.8 ± 0.1	1.2 ± 0.05
SN	*n.d.*	715 ± 105	22.5 ± 0.2	11.1 ± 0.2	12.0 ± 0.1	0.4 ± 0.05	0.4 ± 0.05	1.1 ± 0.05
SO	*n.d.*	310 ± 75	16.7 ± 0.1	7.8 ± 0.1	9.7 ± 0.1	0.4 ± 0.05	0.5 ± 0.05	0.6 ± 0.05

The abbreviations for plant material are presented in Table 2; *n.d.* = not detected. Values are average of 3 replications ± SD.

Zinc is another important mineral in the human diet. The interaction of zinc with insulin causes conformational changes and enhances binding to the insulin receptor. Zinc ions possess insulin-mimetic activity, presumably through the ability to inhibit protein tyrosine phosphatases. Zinc is a cofactor of several key enzymes related to glucose metabolism. It is the activator of fructose-1-6-bisphosphate aldolase and the inhibitor of fructose-1-6-bisphosphatase. It can also have antioxidant activity, and is a cofactor in copper/zinc superoxide dismutase, the major antioxidant enzyme [25]. Manganese aids in glucose metabolism and is required for normal synthesis and secretion of insulin. The manganese-activated enzymes play an important role in the metabolism of carbohydrates, amino acids, and cholesterol. Copper deficiency, on the other hand, leads to glucose intolerance, decreased insulin response and increased glucose response. It has been associated with hypercholesterolemia and atherosclerosis [26]. AA was particularly rich in zinc, manganese and copper, followed by CE and SN.

Heavy metals are one of the most important contemporary environmental problems due to their toxic effects and high accumulation capacity. They remain in the soil for long periods of time and can be transferred to the food chain in significant quantities [27]. Heavy metals in food and drinking water adversely affect developmental processes, as well as the nervous, gastrointestinal, immune, urogenital, cardiovascular, and musculoskeletal system [28]. For example, it has been repeatedly shown that there is an association between arsenic exposure and T2D [29]. In the samples investigated herein, the content of of lead, nickel, and arsenic was generally low, with the largest amount of lead contained in AA and nickel in PV, while the content of arsenic was highest in CE. Given that the preparations purchased from herbalists were recommended in the doses of up to three teaspoons per day, it is important to point out that the heavy metals analyzed in the tested samples were below the maximum allowed daily dose [30], regardless if used in a mixture or individually.

3.3. Phenolic Content of the Extracts

The research on natural phenols, including flavonoids and phenolic acids, is gaining momentum due to their effectiveness in prevention and therapy of numerous chronic diseases including insulin resistance, T2D and its complications. Numerous scientific studies indicate that natural phenols can protect cellular targets in the eye, kidney, liver, and other organs affected by T2D complications [3]. Content of TP, TF, and TPA in the prepared extracts is presented in Table 4. In order to estimate the effects and influence of individual flavonoids and phenolic acids on the potential antidiabetic effects of the extracts, HPLC analysis was used (Table 5).

Significant differences in the TP content can be noted among extracts (Table 4). TP in hydroethanolic extracts varied from 54.51 mg GAE/g DW (PV-E) to as much as 423.51 mg GAE/g DW (SN-E), while the aqueous extracts contained much lower amounts of phenolic compounds, and their amounts ranged from 27.59 mg GE/g DW (PV-V) to 147.78 mg GE/g DW (SO-W). TP was statistically higher in the hydroethanolic than in aqueous extracts (paired *t*-test, $p < 0.05$). On average, the amount of total phenols in hydroethanolic extracts was almost threefold higher than their amount in the aqueous extract. This is expected because 80% ethanol is a solvent of relatively low polarity (as compared to water), suitable for dissolving natural phenolic substances.

Flavonoids are a group of phenolic compounds that have a significant potential to combat the consequences and complications of T2D e.g., by attenuating the progression of diabetic retinopathy. It is widely recognized that many of their observed effects are closely related to their antioxidant properties [31]. TF content of the investigated extracts is presented in Table 4. Among the hydroethanolic extracts, MA-E was particularly rich in flavonoids with 154.51 mg QE/g DW, while CE-E contained only 17.43 mg QE/g DW of TF. SO-W, on the other hand, contained 47.85 mg QE/g DW of flavonoids, much more than some hydroethanolic extracts, while the amount of flavonoid in PV-W was almost negligible (6.22 mg QE/g DW). The amount of TF was statistically higher in hydroethanolic than in aqueous extracts (paired *t*-test, $p < 0.05$), and the effect of the solvent on the TF was

even more pronounced than with TP. The average yield of the extracted flavonoids was almost 3.5 lower when water was used instead of 80% ethanol.

Table 4. Content of total phenols (TP), total flavonoids (TF), and total phenolic acids (TPA) in the extracts.

Extract	TP (mg GAE/g DW)	TF (mg QE/g DW)	TPA (mg CAE/g DW)
AA-E	154.37 ±18.95 [CDE]	63.67 ± 2.74 [BCD]	13.46 ± 0.74 [GH]
AA-W	68.62 ± 8.33 [DE]	17.25 ± 1.09 [DE]	42.77 ± 0.8 [D]
AM-E	362.58 ± 36.06 [AB]	43.43 ± 1.6 [CDE]	54.41 ± 2.23 [C]
AM-W	116.67 ± 3.73 [DE]	38.4 ± 0.39 [CDE]	74.7 ± 1.35 [A]
CE-E	111.92 ± 20.24 [DE]	17.43 ± 0.57 [DE]	9.08 ± 0.6 [HI]
CE-W	44.6 ± 4.99 [DE]	13.29 ± 0.85 [DE]	13.04 ± 0.19 [GH]
MA-E	281.29 ± 6.75 [BC]	154.51 ± 22.64 [A]	14.37 ± 0.64 [GH]
MA-W	89.74 ± 7.36 [DE]	31.51 ± 4.25 [CDE]	16.37 ± 1.15 [G]
PV-E	54.51 ± 7.87 [DE]	34.07 ± 5.17 [CDE]	3.8 ± 0.51 [IJ]
PV-W	27.59 ± 1.89 [E]	6.22 ± 1.23 [E]	1.54 ± 0.00 [E]
SN-E	423.51 ± 13.88 [A]	105.55 ± 3.24 [AB]	35.87 ± 1.06 [E]
SN-W	87.08 ± 5.56 [DE]	21.39 ± 1.03 [CDE]	31.78 ± 2.59 [E]
SO-E	271.99 ± 10.66 [BC]	140.03 ± 21.35 [A]	65.67 ± 2.97 [B]
SO-W	147.78 ± 14.85 [CDE]	47.85 ± 4.6 [CDE]	77.47 ± 3.76 [A]

The extracts' abbreviations are presented in Table 2; CAE = caffeic acid equivalents; DW—dry weight of the extract; GAE = gallic acid equivalents; QE = Quercetin equivalents; Values are average of three replications ± SD. [A–J] = differences between the extracts within a column (extracts not connected with the same capital letter are statistically different, Tukey post-test, $p < 0.05$).

The results revealed that the extracts contained several common flavonoids investigated in this work such as hyperoside, luteolin, myricetin, and rutin (Table 5). The presence of other flavonoids used as standards (baicalein, chrysin, hesperetin, and kaempferol) was not detected in the extracts. Rutin was the most common flavonoid in the investigated extracts (AA-E, AM-E, MA-E, SN-E, SN-W, SO-E, and SO-W) (Table 5). It is a flavonoid with the well-documented anti-inflammatory and antioxidant activity. Its antihyperglycemic effects are based on multiple mechanisms such as a decrease of carbohydrates absorption from the small intestine, inhibition of tissue gluconeogenesis, increase of tissue glucose uptake, stimulation of insulin secretion from beta cells, and protecting Langerhans islet against degeneration. Rutin also decreases the formation of sorbitol, reactive oxygen species, advanced glycation end-product precursors, and inflammatory cytokines. These mechanisms are considered to be responsible for the neuroprotective, nephroprotective, and hepatoprotective, as well as the protective effect against cardiovascular disorders in T2D [32].

Another important group of phenolic compounds universally present in plants are phenolic acids. They have a high antioxidant capacity and the ability to remove free radicals [33]. They can also inhibit certain enzymes responsible for the production of ROS [34]. In addition, they can modulate carbohydrate metabolism due to their ability to inhibit α-glucosidase, the enzyme that degrades complex carbohydrates into glucose [35]. The content of TPA is presented in Table 4. While the TP and TF were significantly higher in the hydroethanolic than in the aqueous extracts, TPA extracted with the two solvents did not differ significantly (paired *t*-test, $p > 0.05$). Sage leaf extracts were richest in phenolic acids with 65.67 mg CAE/g DW and 77.47 mg CAE/g DW in hydroethanolic and aqueous extract, respectively. The amount of phenolic acids in PV was again lowest among the investigated extracts being as low as 1.54 mg CAE/g DW in PV-W. In general, the results obtained in this work are in line with some previous studies of phenols and flavonoid content comparison of selected plants e.g., for *A. absinthium* and *A. millefolium* [36].

Using HPLC analysis, numerous phenolic acids were detected in the extracts. The results, presented in Table 5, revealed the presence of caffeic acid, chlorogenic acid, *p*-coumaric acid, ferulic acid, gallic acid, and rosmarinic acid. Ferulic acid was the most

widespread phenolic acid in the investigated extracts (AA-E, AM-E, AM-W, SN-E, SN-W, SO-E, and SO-W). On the other hand, chlorogenic acid, found in AM-E, AM-W, MA-E, SN-E, and SN-W, was the most represented phenolic acid by weight. Ferulic acid may play an important role in the potential antidiabetic activity of the investigated extracts. It improves glucose and lipid profile in diabetic rats by enhancing activities of antioxidant enzymes, superoxide dismutase, and catalase in the pancreatic tissue. In addition, combination of ferulic acid with metformin improved both, in vitro glucose uptake activity and in vivo hypoglycemic activity of the metformin, thus making it possible to reduce the dose of metformin by four folds [37]. Chlorogenic acid is a substance with well researched antidiabetic properties. Besides hypoglycemic, it shows also hypolipidemic, anti-inflammatory, antioxidant, and other activities potentially useful in combating T2D causes and consequences. Regular consumption of coffee rich in chlorogenic acid has been linked to the prevention of T2D in the consumers. Caffeic acid relieves the complications arising from T2D, such as diabetic nephropathy, diabetic retinopathy, and diabetic peripheral neuropathy [5,38].

Table 5. Quantity of individual phenolic acids and flavonoids in the extracts.

Extract	The Identified Flavonoids and Phenolic Acid
AA-E	Chlorogenic acid (12.73 mg/g DW), Ferulic acid (12.07 mg/g DW), Rosmarinic acid (9.32 mg/g DW), Hyperoside (1.62 mg/g DW), Rutin (15.53 mg/g DW)
AA-W	Hyperoside (0.89 mg/g DW)
AM-E	Chlorogenic acid (39.72 mg/g DW), Ferulic acid (2.34 mg/g DW), Luteolin (3.19 mg/g DW), Rutin (30.24 mg/g DW)
AM-W	Chlorogenic acid (11.07 mg/g DW), Ferulic acid (1.2 mg/g DW),
CE-E	*p*-coumaric acid (3.55 mg/g DW), Gallic acid (1.22 mg/g DW)
CE-W	*p*-coumaric acid (2.82 mg/g DW), Gallic acid (3.67 mg/g DW)
MA-E	Caffeic acid (0.53 mg/g DW), Chlorogenic acid (15.16 mg/g DW), Rutin (14.74 mg/g DW)
MA-W	Caffeic acid (2.68 mg/g DW)
PV-E	*p*-coumaric acid (<LOQ)
PV-W	*n.d.*
SN-E	Caffeic acid (0.95 mg/g DW), Chlorogenic acid (25.83 mg/g DW), Ferulic acid (9.38 mg/g DW), Myricetin (16.79 mg/g DW), Rutin (41.9 mg/g DW)
SN-W	Caffeic acid (0.45 mg/g DW), Chlorogenic acid (10.27 mg/g DW), Ferulic acid (1.67 mg/g DW), Myricetin (14.79), Rutin (16.31 mg/g DW)
SO-E	Ferulic acid (5.11), Rosmarinic acid (39.2 mg/g DW), Rutin (22.62 mg/g DW)
SO-W	Ferulic acid (1.19 mg/g DW), Rosmarinic acid (0.45 mg/g DW), Rutin (14.2 mg/g DW)

The extracts' abbreviations are presented in Table 2; DW = dry weight of the extract; <LOQ = below level of quantification, *n.d.* = not detected.

While the presence of some of the selected phenolic standards confirms previously reported findings (e.g., rutin in *S. officinalis* leaf [39] and *S. nigra* flower [40], ferulic acid in *S. nigra* flower [40], and *A. absinthium* aerial parts [41]), the other previously reported compounds were not recorded in this work (e.g., caffeic acid in *A. absinthium* aerial parts [41]). This may be related either to the interspecies variations or the specific condition of extract preparation. This finding further stresses the well-known necessity of the standardization of herbal material used for T2D treatment [42].

3.4. Antioxidant Activity of the Extracts

Antioxidant activity of plant secondary metabolites and other natural antioxidants may take place directly, e.g., through hydrogen atom or single electron transfer, or, indirectly, by their complexation of pro-oxidative metal ions. Thus, it is often necessary to use

several methods to give a comprehensive analysis of the antioxidant activity of complex mixtures such as herbal extracts [43]. In this work, RSA of the extracts was evaluated using ABTS and DPPH free radicals, while TAA, RP, and FRAP were used to explore reducing properties of the investigated extracts. ChA and ANT assay were used to assess chelating and polyunsaturated fatty acid-protecting ability, respectively. BHA, ascorbic acid, Trolox and EDTA, antioxidants, and ion chelator often employed in the food and pharmaceutical industry, as well as common standards for antioxidant and chelating assays, were used as the positive controls [44]. The results of the employed assays are presented in Tables 6 and 7, and Figure 1.

Table 6. Radical scavenging activity for ABTS (IC_{50} ABTS RSA) and DPPH (IC_{50} DPPH RSA) free radical and chelating activity (ChA).

Extract	(IC_{50} ABTS RSA) (μg DW/mL)	IC_{50} DPPH RSA (μg DW/mL)	ChA IC_{50} (μg DW/mL)
AA-E	799.6 ± 24.3 [F]	156.85 ± 8.37 [DEF]	56.04 ± 1.40 [FGX]
AA-W	1066.1 ± 28.2 [E]	206.27 ± 2.32 [D]	18.26 ± 1.50 [GX]
AM-E	569.0 ± 29.3 [G]	68.78 ± 7.63 [FGX]	334.37 ± 17.87 [BCDE]
AM-W	481.6 ± 32.7 [GH]	75.24 ± 4.10 [FGX]	223.59 ± 14.35 [CDEFG]
CE-E	1474.1 ± 57.2 [D]	835.86 ± 64.59 [B]	272.88 ± 7.01 [CDEF]
CE-W	2399.9 ± 160.7 [B]	388.75 ± 32.79 [C]	230.41 ± 30.42 [CDEFG]
MA-E	515.1 ± 41.3 [GH]	177.79 ± 10.87 [DE]	399.40 ± 39.02 [BC]
MA-W	749.7 ± 18.3 [F]	225.50 ± 19.05 [D]	277.29 ± 12.97 [CDEF]
PV-E	1997.5 ± 76.7 [C]	876.82 ± 87.35 [B]	392.19 ± 7.43 [BC]
PV-W	2700.7 ± 72.0 [A]	2127.86 ± 35.89 [A]	142.99 ± 3.72 [DEFGX]
SN-E	370.3 ± 10.5 [HI]	85.91 ± 9.86 [FG]	541.13 ± 61.51 [B]
SN-W	548.1 ± 0.1 [G]	84.07 ± 2.41 [FGX]	364.11 ± 30.72 [BCD]
SO-E	300.4 ± 6.6 [IJ]	68.86 ± 6.15 [FGX]	900.74 ± 265.24 [A]
SO-W	185.2 ± 6.8 [J]	40.70 ± 0.84 [GX]	217.45 ± 12.26 [CDEFG]
Standard	[a] 41.7 ± 1.8 [X]	[b] 8.40 ± 0.19 [X]	[c] 17.76 ± 3.90 [X]

The extracts' abbreviations are presented in Table 2; Standards: [a] = Trolox, [b] = BHA, [c] = EDTA; DW—dry weight of the extract; Values are an average of three replications $\pm$ SD; [A-J] = differences between the extracts within a column (extracts not connected with the same capital letter are statistically different, Tukey post-test, $p < 0.05$); [X] = differences with the positive control within a column (extracts not connected with the same capital letter are statistically different, Dunnet's post-test, $p < 0.05$).

As shown in Table 6, ABTS RSA, DPPH RSA, and ChA differed depending on the extract and the assay. In general, the most efficient ABTS radical scavengers were the extracts prepared from SN, SO, and AM. Those extracts were also the most successful scavengers of DPPH free radicals. In addition, the activity of the SO and AM extracts did not statistically differ from the synthetic antioxidant, butylated hydroxyanisole (BHA) (Dunnet's post-test, $p > 0.05$). Comparison of IC_{50} values shows that neither ABTS RSA nor DPPH RSA assay results differed statistically between the hydroethanolic and the aqueous extracts (paired *t*-test, $p > 0.05$). Among the investigated extracts AA-E and, especially AA-W, were the most efficient Fe^{3+} ion chelators, and its activity was not different from the standard, ethylenediaminetetraacetic acid (EDTA). Interestingly, the activity of PV-W, although seemingly lower, did not differ from the activity of EDTA (Dunnet's post-test, $p > 0.05$), indicating that this extract acted as a relatively good secondary antioxidant. It is interesting to note that ChA IC_{50} values were slightly lower for aqueous extracts than their hydroethanolic counterparts (paired *t*-test, $p < 0.1$), indicating better activity of the aqueous extracts.

The ANT activity is presented in Figure 1. In general, the aqueous extracts were very good inhibitors of β-carotene degradation, and the ANT activity of the majority of the investigated aqueous extracts was equal to the activity of BHA. The most active extracts were AM-W, MA-W, PV-W, and SO-W. The observed high activity of PV in this assay is rather surprising considering its low phenolic content and weak activity in the other assays. Hydroethanolic extracts, on the other hand, displayed significantly lower activity, the most

active among them being MA-E. The observed good activity of the aqueous extract in this assay may be related to the aqueous medium in which the reaction is performed [19].

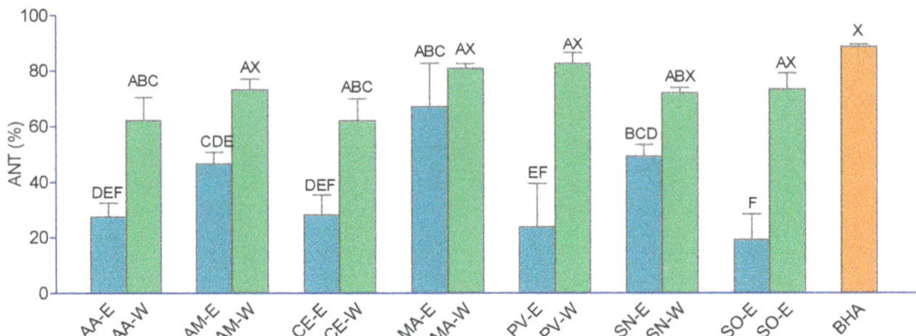

Figure 1. Antioxidant activity of the extracts and BHA (tested in the concentration of 8 µg/mL) in β-carotene-linoleic acid assay (ANT). The extracts' abbreviations are presented in Table 2. Values are the average of three replications ± SD. A–F = differences between the extracts within a column (extracts not connected with the same capital letter are statistically different, Tukey post-test, $p < 0.05$); X = differences with the positive control within a column (extracts not connected with the same capital letter are statistically different, Dunnet's post-test, $p < 0.05$).

Table 7. Total antioxidant activity (TAA), reducing power (RP), and ferric reducing power (FRAP) of the extracts.

Extract	TAA mg AAE/g DW	RP mg TA/g DW	FRAP mg TA/g DW
AA-E	149.0 ± 4.0 [D]	189.5 ± 5.8 [E]	123.3 ± 6.3 [F]
AA-W	81.2 ± 2.7 [H]	50.9 ± 2.6 [IJ]	77.7 ± 3.0 [G]
AM-E	94.7 ± 1.5 [G]	261.6 ± 9.0 [C]	204.4 ± 4.3 [E]
AM-W	170.7 ± 7.2 [C]	81.8 ± 2.1 [H]	288.1 ± 6.4 [D]
CE-E	111.2 ± 1.0 [F]	65.0 ± 6.7 [HI]	56.4 ± 3.8 [H]
CE-W	134.7 ± 3.0 [E]	29.4 ± 1.4 [JK]	42.5 ± 1.8 [H]
MA-E	142.3 ± 4.1 [DE]	140.0 ± 1.8 [F]	109.6 ± 4.4 [F]
MA-W	116.9 ± 1.6 [F]	61.0 ± 1.7 [HI]	86.2 ± 4.6 [G]
PV-E	30.1 ± 1.4 [I]	21.8 ± 0.9 [K]	14.3 ± 0.3 [I]
PV-W	26.4 ± 0.5 [I]	25.3 ± 2.1 [K]	12.6 ± 0.3 [I]
SN-E	137.1 ± 1.7 [DE]	385.5 ± 23.4 [B]	323.4 ± 10.4 [C]
SN-W	141.6 ± 6.5 [DE]	111.6 ± 4.3 [G]	203.4 ± 7.9 [E]
SO-E	199.5 ± 3.4 [B]	501.9 ± 6.9 [A]	435.9 ± 0.2 [A]
SO-W	220.6 ± 10.6 [A]	234.8 ± 12.1 [D]	408.3 ± 5.4 [B]

The extracts' abbreviations are presented in Table 2; AE = ascorbic acid equivalents; TE = Trolox equivalents. DW—dry weight of the extract; values are the average of three replications ± SD; A–K = differences between the extracts within a column (extracts not connected with the same capital letter are statistically different, Tukey post-test, $p < 0.05$); X = differences with the positive control within a column (extracts not connected with the same capital letter are statistically different, Dunnet's post-test, $p < 0.05$).

In Table 7, the results of the three assays based on the reducing abilities of the analytes are presented. The TAA assay measures the ability to reduce Mo^{6+} to Mo^{5+} ions [20]. RP and FRAP assays, on the other hand, are based on the reduction of Fe^{3+} to Fe^{2+}, with the FRAP assay being the more specific of the two. In this work, SO extracts were the most efficient reducing agents. However, the SO-W was more active in TAA assay, while SO-E was more active in the RP and FRAP assays. Other extracts also showed notable reducing properties, such as AM-W in TAA assay, and SN-E in RP and FRAP assay. Hydroethanolic extracts were statistically better Fe^{3+} reducing agents than aqueous extracts in the RP assay (paired t-test, $p < 0.05$), while such differences were not observed in the other two assays.

3.5. α-Glucosidase-Inhibitory Activity of the Extracts

The investigated extracts were tested for their activity against α-glucosidase, an important enzyme that participates in carbohydrate digestion. The results are presented in Figure 2. While the inhibiting activity of most extracts was excellent, MA extracts were particularly good α-glucosidase inhibitors with the IC_{50} value statistically equal to IC_{50} of acarbose, conventional anti-diabetic drug. Two other hydroethanolic extracts also showed the activity equal to the activity of acarbose AM-E and SO-E. These findings are in accordance with previous studies showing good α-glucosidase inhibiting activity of the sage [45] and yarrow [46] extracts. However, good activity of wormwood extracts is an interesting find because, to the best of our knowledge, this is the first report of anti-α-glucosidase activity of this plant commonly used in bitter alcoholic beverages with digestive properties. In this study, hydroethanolic extracts were significantly better α-glucosidase inhibitor than the corresponding aqueous extracts (paired t-test, $p < 0.05$). This is an interesting finding in light of the traditional preparation mode for the investigated plants being in the form of infusions [7] as the results clearly indicate that it may be desirable to use them in the form of hydroalcoholic extracts or preparations thereof.

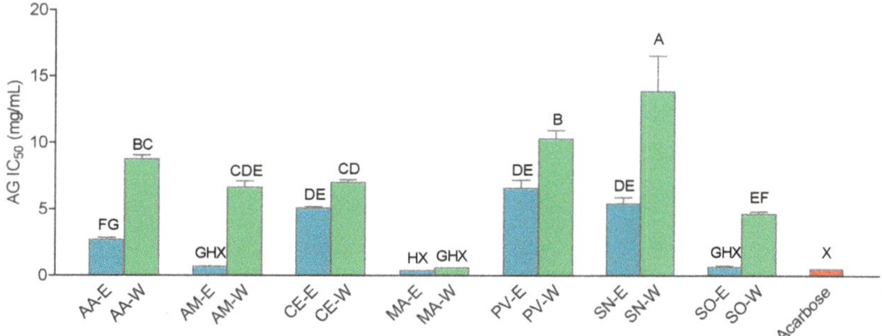

Figure 2. α-glucosidase inhibitory activity (AG IC_{50}) of the extracts. The extracts' abbreviations are presented in Table 2. Values are average of 3 replications $\pm$ SD; [A–H] = differences between the extracts within a column (extracts not connected with the same capital letter are statistically different, Tukey post-test, $p < 0.05$); [X] = differences with the positive control within a column (extracts not connected with the same capital letter are statistically different, Dunnet's post-test, $p < 0.05$).

The observed anti-glucosidase activity may be related to the presence of specific flavonoids, which hydroethanolic extracts were particularly abundant in. It has been found that the α-glucosidase-inhibiting effects of rutin, the most prevalent flavonoid in the hydroethanolic extracts, surpass the inhibiting effects of acarbose [47]. Significant correlation of the activity in this assay with the flavonoids content was also observed in this work, as discussed further in the text. In addition, phenolic acids may also contribute to the observed α-glucosidase inhibiting activity. Chlorogenic acid, the most represented phenolic acid in the investigated extracts, can inhibit α-glucosidase and alleviate postprandial hyperglycemia [48]. Furthermore, aglycones formed by hydrolysis of rutin and chlorogenic acid, quercetin, and caffeic acid, respectively, may also strongly suppress α-glucosidase activity [48,49]. In addition, current studies suggest that combinations of plant phenolics may have an additive effect on α-glucosidase inhibition [50].

3.6. The Relationship between the Investigated Variables

In this work, multivariate analysis was employed in order to investigate possible relationship between the composition, antioxidant, and α-glucosidase inhibiting potential of the investigated plants.

The color maps with the correlations between the measured variables are presented in Figure 3, while the statistically significant coefficients of determinations between the content

of selected phenolics and the performed assays' results are presented in Table 8. In Figure 3, two distinct groups, marked by two dark red zones, where groups of variables are positively (Figure 3a) and significantly (Figure 3a) correlated to one another, can be discerned. The first group consists of four variables TF, TP, RP, and ChA IC_{50}. While it is well known that natural flavonoids and other phenols possess reducing abilities [51], and their correlation with RP was thus expected, it was interesting to see that ChA IC_{50} showed a positive correlation with TP and TF, indicating lower activity of the extracts rich in flavonoids and other phenols. This is rather unexpected because flavonoids are well-known ion chelators [52]. As previously noted, TF and TP were better extracted with hydroethanolic mixtures. However, it seems that the aqueous extraction is better suited for obtaining the extracts with high chelating abilities. Several other studies also demonstrated higher chelating activity of aqueous extracts in comparison with the hydroethanolic extracts [53].

The second group of variables, TPA, TAA, and FRAP, correlated positively (Figure 3a) and significantly (Figure 3b) with one another and, in some cases, with the variables from the first group. Expectedly, the results of the reducing assays (TAA, FRAP from the second, and RP from the first group of assays, respectively) showed a good correlation. It seems that phenolic acids were responsible for most of the observed reducing properties of the extracts as TPA was in a statistically significant positive correlation with all the assays based on this type of activity (Figure 3a,b, Table 8). Interestingly, in addition to the RP assay from the previous group, TP correlated only with the FRAP assay and not with the TAA. On the other hand, TF did not show any connection to any of the assays in the second group. It is not unusual that different antioxidants would give comparable but not identical results in different electron transfer-based assays. This is due to the specific conditions of every reaction such as redox potential, pH, and kinetics [54]. In this particular case, the reason may lay in typically different reduction potentials of phenolic acid and flavonoids, which, in addition to their chemical characteristics, may also influence their biological behavior [55].

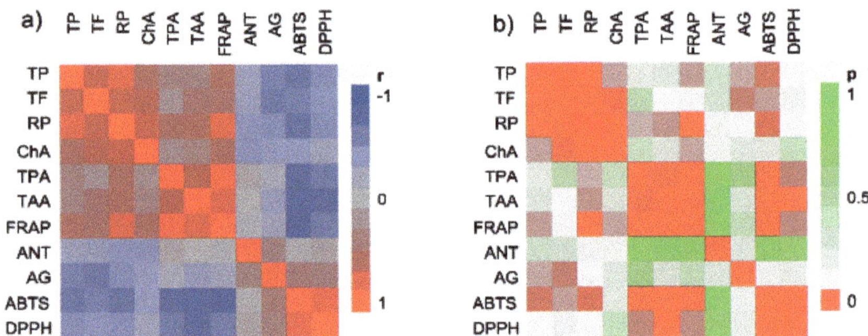

Figure 3. Color maps of correlations between the measured variables (**a**) and corresponding *p*-values (**b**). TP = total phenolic content; TF = total flavonoid content; TPA = total phenolic acid content; ABTS = IC_{50} value of radical scavenging activity for ABTS free radical; DPPH = IC_{50} value of radical scavenging activity for DPPH free radical; ChA = IC_{50} value of chelating activity; ANT = antioxidant activity in β-carotene-linoleate assay; TAA = total antioxidant activity; RP = reducing power; FRAP = ferric reducing power; AG = IC_{50} value of α-glucosidase inhibiting activity.

Table 8. Coefficients of determination for significant ($p < 0$) correlations between the content of phenolic compounds (TP, TF, and TPA) and antioxidant activity.

	ABTS RSA	DPPH RSA	ChA	ANT	TAA	RP	FRAP	AG
TP	$r^2 = 0.39$ (−)	n.s.	$r^2 = 0.31$ (+)		n.s.	$r^2 = 0.64$ (+)	$r^2 = 0.31$ (+)	$r^2 = 0.30$ (−)
TF	$r^2 = 0.31$ (−)	n.s.	$r^2 = 0.46$ (+)		n.s.	$r^2 = 0.51$ (+)	n.s.	$r^2 = 0.36$ (−)
TPA	$r^2 = 0.48$ (−)	$r^2 = 0.35$ (−)	n.s.		$r^2 = 0.45$ (+)	$r^2 = 0.29$ (+)	$r^2 = 0.74$ (+)	n.s.

TP = total phenolic content; TF = total flavonoid content; TPA = total phenolic acid content; ABTS RSA = IC_{50} value of radical scavenging activity for ABTS free radical; DPPH RSA = IC_{50} value of radical scavenging activity for DPPH free radical; ChA = IC_{50} value of chelating activity; ANT = antioxidant activity in β-carotene-linoleate assay; TAA = total antioxidant activity; RP = reducing power; FRAP = ferric reducing power; AG = IC_{50} value of α-glucosidase inhibiting activity; (−) = negative correlation; (+) = positive correlation. n.s.: Not significant.

Among the remaining variables, the IC_{50} value of ABTS RSA and DPPH RSA correlated well with one another (Figure 3, Table 8). In addition, the IC_{50} value of ABTS was negatively affected by TP, TF, and TPA, indicating a positive correlation with the actual activity in the assay (Table 8). The IC_{50} values of DPPH RSA, on the other hand, were affected only by TPA. Knowing that numerous phenolic substances may form resonance-stabilized phenoxyl radicals which scavenge other free radicals and reduce oxidative stress [37], the role of phenolic substances in these assays is not surprising. While ANT did not show any significant interaction with the selected groups of the phenolic compounds, AG was affected by TP and TF content, further confirming the α-glucosidase inhibitory effects of flavonoids [47] and other phenolics [48,49] present in the extracts. It may also be observed that the coefficients of determinations were typically low (Table 8), indicating that TP, TF, and TPA simultaneously influence the results of the performed antioxidant activity assays [56]. While it was not possible to ascertain that phenolic compounds were the only metabolites responsible for all the observed activity in the presented work, the statistically significant correlations between the content of phenolic compounds, and the results of the majority of assays presented herein, show that they do play a pivotal role in the performed assays. The determination of all the pyhtochemical components of the investigated extracts was outside of scope of this work. However, it is also possible that other compounds may be partly involved in the observed activity. This is especially true for the activity in the β-carotene-linoleic acid assay that showed no connection to any of the types of phenolic compounds quantified in this work. It is our hope that comprehensive analytical studies, who will disclose the full spectrum of plant metabolites responsible for the activity in this and the other assays, will be performed in the future.

The results presented here confirm traditional medicine as a starting place for development of new antidiabetics. Croatian ethnomedicine in particular was shown to be a rich source of traditional remedies for T2D whose activity may be based on their bioactive phenolics, antioxidant, and anti-α-glucosidase properties. However, the effects of the investigated plants greatly varied according to the species and the solvent used for the extraction. Some of the plants investigated in this study, like sage and white mulberry, were previously well-researched in relation to their antioxidant and antidiabetic properties. Several studies have noted the abundance of phenolic compounds in sage extracts, connecting them to the potential antidiabetic effects of this plant [45]. Sage phenolics may affect carbohydrate digestion [57], inhibit hepatocyte gluconeogenesis, and decrease insulin resistance. In this study, hydroethanolic sage extract was shown to be an excellent α-glucosidase inhibitor, especially when compared to the aqueous one. Better activity of the sage extracts prepared using moderately polar solvents was also observed in the study of Mahdi et al. [57]. In their study, the extract prepared using ethyl-acetate was better inhibitor of α-amylase and α-glucosidase than the extracts prepared using either hydro-methanol solution or *n*-butanol. The authors related this effect to the presence of flavonoids that ethyl acetate extract was particularly abundant in. The same effect was noted in this study with flavonoid-rich 80% ethanolic extract.

Another excellent α-glucosidase inhibitor in this study was white mulberry. Its leaf is used as a traditional antidiabetic in both Croatian and Chinese ethnomedicine. Superior anti-α-glucosidase activity of mulberry extracts observed in this work is not surprising because several studies have shown its excellent ability to inhibit this enzyme in vitro, and this effect was observed even in clinical setting [58]. A study performed in China demonstrated that flavonoid-rich mulberry leaf extracts may improve skeletal muscle insulin resistance and mitochondrial function [59]. The abundance of flavonoids and other phenolics in the hydroethanolic extract of mulberry leaf from Croatia, as well as its antioxidant and anti-glucosidase effects may well justify the widespread use of this plant. Another good inhibitor of α-glucosidase was yarrow. Previous studies have shown that aqueous extracts of this plant were better inhibitors of this enzyme than ethyl acetate extracts [46]. Since ethanol extract in this study was even more potent than the aqueous one, it seems that it should be the extract of choice for incorporation into antidiabetic formulations.

Unlike yarrow, sage, and mulberry leaf, bean pods are rather under-researched when it comes to their antidiabetic effects even though they are among the most commonly used antidiabetic remedies in Croatia [7]. However, their activity in this study was rather unremarkable, and they demonstrated only a weak activity in the large majority of the performed assays. The exception was excellent activity of the aqueous extract in the β-carotene linoleic acid assay. This activity could not be related to the phenolic compounds due to their low content in the extract. However, knowing that the pods are a source of pectic polysaccharides with antioxidant properties [60], we may postulate that they could be responsible for the observed effect in this assay. In addition to that, the abundance of magnesium in the pods may also contribute to the potential anti-T2D effects in the users of herbal teas containing pods. Similar to bean pods, centaury, another widely used antidiabetic plant, was rather poor in phenolic compounds and displayed modest effects in the majority of the assays in this study. While the anti-α-glucosidase activity of water and hydroethanolic centaury extracts is reported for the first time in this study, the activity of methanol and chloroform extracts was, similar to the results reported herein, moderate [61]. Richest in phenolic compounds in this study were elderflower extracts, material that is not so well researched for its anti-T2D activity in spite of its relatively common use in European cuisine. However, similar to the results presented herein, recent studies indicate that they are excellent antioxidants and rich source of phenolic compounds, richer in fact than elder fruits, part of the plants that are much better researched [62]. On the other hand, there are few studies on the antidiabetic effects of elder flowers. One study demonstrated that elderflowers may positively affect glucose- and fatty acid uptake in human myotubes and HepG2-cells [63]. It seems that phenolic compounds from elderflower are mostly responsible for that activity, which is not surprising given the qualitative and quantitative abundance of various elder flower phenolics detected in this study. One of the most interesting findings was very good anti-α-glucosidase activity of wormwood hydroethanolic extract. It is interesting that, in spite of wormwood widespread use in traditional digestives and alcoholic beverages, this is, to the best of our knowledge, the first report of its anti-α-glucosidase activity. In addition to that, the aqueous extract of this plant was rich in phenolic acids and demonstrated excellent antiradical, chelating, and Fe^{3+} reducing properties, making it one of the most interesting plants in this research. It is important to note that wormwood was the only one among the investigated plants to contain chromium, a mineral that is necessary for insulin action [12].

This study has demonstrated that the selected plants from Croatian ethnomedicine may have a significant potential in complementary therapy of diabetes. The hydroethanolic extracts were especially active, so they seem like the first candidate for development of anti-T2D supplements and herbal drugs. However, it is important to note that, before they could be widely applied, their toxicological profile should also be thoroughly assessed. For example, it is well known that sage and wormwood raw ethanolic extracts contain not only their phenolic components, but also their essential oils rich in thujone, camphor,

and other potentially toxic natural compounds. For example, prolonged use or overdose of sage ethanolic extract or essential oil may produce unwanted effects such as vomiting, tachycardia, vertigo, and even convulsions [45]. It is important to note that the extracts investigated in this study were evaporated to dry prior to performing the antioxidant- and anti-glucosidase assays. In the evaporation process, not only ethanol was removed, but also essential oils, which are known to contain potentially toxic volatile compounds. This rendered the prepared extracts safer for human consumption. Great differences in the observed activity of the extracts prepared by two solvents implicate that more than one extraction method should be considered when developing new herbal treatments from traditional medicine. The displayed antioxidant and anti-glucosidase effects of the investigated plants give hope that they may be a source of new natural and affordable antidiabetics. However, before being widely accepted for use as antidiabetics, detailed studies of pharmacological activity should be performed in the clinical setting.

4. Conclusions

The results presented herein indicate that the investigated extracts were active radical scavengers, ion chelators, and reducing agents with significant α-glucosidase inhibitory activity. Ethanolic extracts from *S. officinalis* leaf and *A. millefolium* aerial parts were the most active radical scavengers, reducing agents and α-glucosidase inhibitors. Notable antioxidant activity was also displayed by *S. nigra*, while *A. absinthium* and *M. alba* extracts were good α-glucosidase inhibitors. *A. absinthium* contained chromium, a mineral that promotes insulin action. The investigated plants contained and low amounts of toxic heavy metals, deeming them safe for human use. The long-standing tradition of their use, as well as their significant antioxidant and α-glucosidase-inhibitory effects, indicate that the investigated plants represent a potentially viable alternative to conventional medicine for complementary treatment of diabetes and its complications.

Author Contributions: Conceptualization, M.Z.K.; methodology, I.S.d.C., L.V., J.J. and M.Z.K.; validation, K.B., I.S.d.C., L.V., J.J. and M.Z.K.; investigation, K.B., A.B., A.Č., I.S.d.C., L.V., J.J. and M.Z.K.; resources, I.S.d.C. and M.Z.K.; writing—original draft preparation, M.Z.K.; writing—review and editing, J.J. and L.V.; supervision, I.S.d.C. and M.Z.K. All authors have read and agreed to the published version of the manuscript.

Funding: This research is funded by the scientific support fund from University of Zagreb and Croatian Science Foundation, project Bioactive plant principles extraction using green solvents—a step towards green cosmeceuticals, Grant No. IP-2018-01-6504. The APC is funded by the authors.

Institutional Review Board Statement: Not applicable.

Informed Consent Statement: Not applicable.

Conflicts of Interest: The authors declare no conflict of interest.

References

1. Esposito, K.; Ciotola, M.; Maiorino, M.I.; Giugliano, D. Lifestyle Approach for Type 2 Diabetes and Metabolic Syndrome. *Curr. Atheroscler. Rep.* **2008**, *10*, 523–528. [CrossRef] [PubMed]
2. Luc, K.; Schramm-Luc, A.; Guzik, T.J.; Mikolajczyk, T.P. Oxidative Stress and Inflammatory Markers in Prediabetes and Diabetes. *J. Physiol. Pharmacol.* **2019**, *70*, 809–824. [CrossRef]
3. Kapoor, R.; Kakkar, P. Protective Role of Morin, a Flavonoid, against High Glucose Induced Oxidative Stress Mediated Apoptosis in Primary Rat Hepatocytes. *PLoS ONE* **2012**, *7*, e41663. [CrossRef]
4. Fiorentino, T.V.; Prioletta, A.; Zuo, P.; Folli, F. Hyperglycemia-Induced Oxidative Stress and Its Role in Diabetes Mellitus Related Cardiovascular Diseases. *Curr. Pharm. Des.* **2013**, *19*, 5695–5703. [CrossRef] [PubMed]
5. Lin, D.; Xiao, M.; Zhao, J.; Li, Z.; Xing, B.; Li, X.; Kong, M.; Li, L.; Zhang, Q.; Liu, Y.; et al. An Overview of Plant Phenolic Compounds and Their Importance in Human Nutrition and Management of Type 2 Diabetes. *Molecules* **2016**, *21*, 1374. [CrossRef] [PubMed]
6. Truba, J.; Stanisławska, I.; Walasek, M.; Wieczorkowska, W.; Woliński, K.; Buchholz, T.; Melzig, M.F.; Czerwińska, M.E. Inhibition of Digestive Enzymes and Antioxidant Activity of Extracts from Fruits of *Cornus alba*, *Cornus sanguinea* Subsp. Hungarica and Cornus florida—A Comparative Study. *Plants* **2020**, *9*, 122. [CrossRef]

7. Zovko Končić, M.; Bljajić, K. Chapter 42—Traditional Herbal Products Used for the Management of Diabetes in Croatia: Linking Traditional Use with α-Glucosidase-Inhibitory Activity. In *Bioactive Food as Dietary Interventions for Diabetes*, 2nd ed.; Watson, R.R., Preedy, V.R., Eds.; Academic Press: Oxford, UK, 2019; pp. 647–664, ISBN 978-0-12-813822-9.
8. Helmstädter, A. Beans and Diabetes: Phaseolus Vulgaris Preparations as Antihyperglycemic Agents. *J. Med. Food* **2010**, *13*, 251–254. [CrossRef]
9. Wen, Y.; He, L.; Peng, R.; Lin, Y.; Zhao, L.; Li, X.; Ye, L.; Yang, J. A Novel Strategy to Evaluate the Quality of Herbal Products Based on the Chemical Profiling, Efficacy Evaluation and Pharmacokinetics. *J. Pharm. Biomed. Anal.* **2018**, *161*, 326–335. [CrossRef]
10. Bilal, M.; Iqbal, M.S.; Shah, S.B.; Rasheed, T.; Iqbal, H.M.N. Diabetic Complications and Insight into Antidiabetic Potentialities of Ethno- Medicinal Plants: A Review. *Recent Pat. Inflamm Allergy Drug Discov.* **2018**, *12*, 7–23. [CrossRef]
11. Dghaim, R.; Al Khatib, S.; Rasool, H.; Ali Khan, M. Determination of Heavy Metals Concentration in Traditional Herbs Commonly Consumed in the United Arab Emirates. *J. Environ. Public Health* **2015**, *2015*, e973878. [CrossRef]
12. Lewicki, S.; Zdanowski, R.; Krzyżowska, M.; Lewicka, A.; Dębski, B.; Niemcewicz, M.; Goniewicz, M. The Role of Chromium III in the Organism and Its Possible Use in Diabetes and Obesity Treatment. *Ann. Agric. Environ. Med.* **2014**, *21*, 331–335. [CrossRef] [PubMed]
13. *European Pharmacopoeia*, 8th ed.; Council of Europe: Strasbourg, France, 2013.
14. Abbasi, A.M.; Khan, M.A.; Khan, N.; Shah, M.H. Ethnobotanical Survey of Medicinally Important Wild Edible Fruits Species Used by Tribal Communities of Lesser Himalayas-Pakistan. *J. Ethnopharmacol.* **2013**, *148*, 528–536. [CrossRef] [PubMed]
15. Singleton, V.L.; Orthofer, R.; Lamuela-Raventós, R.M. Analysis of total phenols and other oxidation substrates and antioxidants by means of folin-ciocalteu reagent. In *Methods in Enzymology*; Elsevier: Amsterdam, The Netherlands, 1999; Volume 299, pp. 152–178, ISBN 978-0-12-182200-2.
16. Kumazawa, S.; Hamasaka, T.; Nakayama, T. Antioxidant Activity of Propolis of Various Geographic Origins. *Food Chem.* **2004**, *84*, 329–339. [CrossRef]
17. Nicolle, C.; Carnat, A.; Fraisse, D.; Lamaison, J.-L.; Rock, E.; Michel, H.; Amouroux, P.; Rémésy, C. Characterisation and Variation of Antioxidant Micronutrients in Lettuce (*Lactuca sativa* Folium). *J. Sci. Food Agric.* **2004**, *84*, 2061–2069. [CrossRef]
18. Re, R.; Pellegrini, N.; Proteggente, A.; Pannala, A.; Yang, M.; Rice-Evans, C. Antioxidant Activity Applying an Improved ABTS Radical Cation Decolorization Assay. *Free Radic. Biol. Med.* **1999**, *26*, 1231–1237. [CrossRef]
19. Končić, M.Z.; Barbarić, M.; Perković, I.; Zorc, B. Antiradical, Chelating and Antioxidant Activities of Hydroxamic Acids and Hydroxyureas. *Molecules* **2011**, *16*, 6232–6242. [CrossRef]
20. Prieto, P.; Pineda, M.; Aguilar, M. Spectrophotometric Quantitation of Antioxidant Capacity through the Formation of a Phosphomolybdenum Complex: Specific Application to the Determination of Vitamin E. *Anal. Biochem.* **1999**, *269*, 337–341. [CrossRef]
21. Bljajić, K.; Petlevski, R.; Vujić, L.; Čačić, A.; Šoštarić, N.; Jablan, J.; Saraiva de Carvalho, I.; Zovko Končić, M. Chemical Composition, Antioxidant and α-Glucosidase-Inhibiting Activities of the Aqueous and Hydroethanolic Extracts of *Vaccinium myrtillus* Leaves. *Molecules* **2017**, *22*, 703. [CrossRef]
22. Benzie, I.F.; Strain, J.J. The Ferric Reducing Ability of Plasma (FRAP) as a Measure of "Antioxidant Power": The FRAP Assay. *Anal. Biochem.* **1996**, *239*, 70–76. [CrossRef]
23. Tiwari, A.K.; Swapna, M.; Ayesha, S.B.; Zehra, A.; Agawane, S.B.; Madhusudana, K. Identification of Proglycemic and Antihyper-glycemic Activity in Antioxidant Rich Fraction of Some Common Food Grains. *Int. Food Res. J.* **2011**, *18*, 915–923.
24. Phimarn, W.; Wichaiyo, K.; Silpsavikul, K.; Sungthong, B.; Saramunee, K. A Meta-Analysis of Efficacy of *Morus alba* Linn. to Improve Blood Glucose and Lipid Profile. *Eur. J. Nutr.* **2017**, *56*, 1509–1521. [CrossRef]
25. Evans, J.L.; Bahng, M. Non-pharmaceutical intervention options for type 2 diabetes: Diets and Dietary Supplements (Botanicals, Antioxidants, and Minerals). In *Endotext*; De Groot, L.J., Chrousos, G., Dungan, K., Feingold, K.R., Grossman, A., Hershman, J.M., Koch, C., Korbonits, M., McLachlan, R., New, M., et al., Eds.; MDText.com, Inc.: South Dartmouth, MA, USA, 2000.
26. Siddiqui, K.; Bawazeer, N.; Joy, S.S. Variation in Macro and Trace Elements in Progression of Type 2 Diabetes. *Sci. World J.* **2014**, *2014*, 461591. [CrossRef] [PubMed]
27. Salamon, I.; Labun, P.; Petruska, P. Occurrence of Heavy Metals, Radioactivity, and Pesticide Residues in Raw Materials of Elderberry and Other Herbs and Fruits in Slovak Republic. *Acta Hortic.* **2015**, *1061*, 259–266. [CrossRef]
28. Krejpcio, Z.; Król, E.; Sionkowski, S. Evaluation of Heavy Metals Contents in Spices and Herbs Available on the Polish Market. *Pol. J. Environ. Stud.* **2007**, *16*, 97–100.
29. Sung, T.-C.; Huang, J.-W.; Guo, H.-R. Association between Arsenic Exposure and Diabetes: A Meta-Analysis. *Biomed. Res. Int* **2015**, *2015*, 368087. [CrossRef] [PubMed]
30. World Health Organization. *WHO Guidelines for Assessing Quality of Herbal Medicines with Reference to Contaminants and Residues*; WHO Press: Geneva, Switzerland, 2007.
31. Matos, A.L.; Bruno, D.F.; Ambrósio, A.F.; Santos, P.F. The Benefits of Flavonoids in Diabetic Retinopathy. *Nutrients* **2020**, *12*, 3169. [CrossRef] [PubMed]
32. Ghorbani, A. Mechanisms of Antidiabetic Effects of Flavonoid Rutin. *Biomed. Pharmacother.* **2017**, *96*, 305–312. [CrossRef]
33. Gülçin, İ. Antioxidant Activity of Food Constituents: An Overview. *Arch. Toxicol.* **2012**, *86*, 345–391. [CrossRef]
34. Nakamura, K.; Ogasawara, Y.; Endou, K.; Fujimori, S.; Koyama, M.; Akano, H. Phenolic Compounds Responsible for the Superoxide Dismutase-like Activity in High-Brix Apple Vinegar. *J. Agric. Food Chem.* **2010**, *58*, 10124–10132. [CrossRef] [PubMed]

35. Vinayagam, R.; Jayachandran, M.; Xu, B. Antidiabetic Effects of Simple Phenolic Acids: A Comprehensive Review. *Phytother. Res.* **2016**, *30*, 184–199. [CrossRef]

36. Moldovan, L.; Gaspar, A.; Toma, L.; Craciunescu, O.; Saviuc, C. Comparison of Polyphenolic Content and Antioxidant Capacity of Five Romanian Traditional Medicinal Plants. *Rev. Chim.* **2011**, *62*, 299–303.

37. Nankar, R.; Prabhakar, P.K.; Doble, M. Hybrid Drug Combination: Combination of Ferulic Acid and Metformin as Anti-Diabetic Therapy. *Phytomedicine* **2017**, *37*, 10–13. [CrossRef] [PubMed]

38. Yan, Y.; Zhou, X.; Guo, K.; Zhou, F.; Yang, H. Use of Chlorogenic Acid against Diabetes Mellitus and Its Complications. *J. Immunol. Res.* **2020**, *2020*, 9680508. [CrossRef] [PubMed]

39. Mocan, A.; Babotă, M.; Pop, A.; Fizeșan, I.; Diuzheva, A.; Locatelli, M.; Carradori, S.; Campestre, C.; Menghini, L.; Sisea, C.R.; et al. Chemical Constituents and Biologic Activities of Sage Species: A Comparison between *Salvia officinalis* L., *S. glutinosa* L. and *S. transsylvanica* (Schur Ex Griseb. & Schenk) Schur. *Antioxidants* **2020**, *9*, 480. [CrossRef]

40. Viapiana, A.; Wesolowski, M. The Phenolic Contents and Antioxidant Activities of Infusions of *Sambucus nigra* L. *Plant Foods Hum. Nutr.* **2017**, *72*, 82–87. [CrossRef] [PubMed]

41. Geszprych, A.; Przybył, J.L.; Kuczerenko, A.; Węglarz, Z. Diversity of Wormwood (*Artemisia absinthium* L.) Growing Wild in Poland in Respect of the Content and Composition of Essential Oil and Phenolic Compounds. *Acta Hortic.* **2011**, *925*, 123–130. [CrossRef]

42. Chawla, R.; Thakur, P.; Chowdhry, A.; Jaiswal, S.; Sharma, A.; Goel, R.; Sharma, J.; Priyadarshi, S.S.; Kumar, V.; Sharma, R.K.; et al. Evidence Based Herbal Drug Standardization Approach in Coping with Challenges of Holistic Management of Diabetes: A Dreadful Lifestyle Disorder of 21st Century. *J. Diabetes Metab. Disord.* **2013**, *12*. [CrossRef]

43. Faraone, I.; Rai, D.K.; Russo, D.; Chiummiento, L.; Fernandez, E.; Choudhary, A.; Milella, L. Antioxidant, Antidiabetic, and Anticholinesterase Activities and Phytochemical Profile of Azorella Glabra Wedd. *Plants* **2019**, *8*, 265. [CrossRef]

44. Pokorny, J.; Yanishlieva, N.; Gordon, M.H. *Antioxidants in Food: Practical Applications*; CRC Press: Boca Raton, FL, USA, 2001.

45. Ghorbani, A.; Esmaeilizadeh, M. Pharmacological Properties of Salvia Officinalis and Its Components. *J. Tradit. Complement. Med.* **2017**, *7*, 433–440. [CrossRef]

46. Zengin, G.; Aktumsek, A.; Ceylan, R.; Uysal, S.; Mocan, A.; Guler, G.O.; Mahomoodally, M.F.; Glamočlija, J.; Ćirić, A.; Soković, M. Shedding Light on the Biological and Chemical Fingerprints of Three Achillea Species (*A. biebersteinii*, *A. millefolium* and *A. teretifolia*). *Food Funct.* **2017**, *8*, 1152–1165. [CrossRef]

47. Jiménez-Aspee, F.; Theoduloz, C.; Soriano, M.D.P.C.; Ugalde-Arbizu, M.; Alberto, M.R.; Zampini, I.C.; Isla, M.I.; Simirigiotis, M.J.; Schmeda-Hirschmann, G. The Native Fruit Geoffroea Decorticans from Arid Northern Chile: Phenolic Composition, Antioxidant Activities and In Vitro Inhibition of Pro-Inflammatory and Metabolic Syndrome-Associated Enzymes. *Molecules* **2017**, *22*, 1565. [CrossRef] [PubMed]

48. Oboh, G.; Agunloye, O.M.; Adefegha, S.A.; Akinyemi, A.J.; Ademiluyi, A.O. Caffeic and Chlorogenic Acids Inhibit Key Enzymes Linked to Type 2 Diabetes (in vitro): A Comparative Study. *J. Basic Clin. Physiol. Pharmacol.* **2015**, *26*, 165–170. [CrossRef] [PubMed]

49. Li, Y.Q.; Zhou, F.C.; Gao, F.; Bian, J.S.; Shan, F. Comparative Evaluation of Quercetin, Isoquercetin and Rutin as Inhibitors of α-Glucosidase. *J. Agric. Food Chem.* **2009**, *57*, 11463–11468. [CrossRef]

50. Adisakwattana, S.; Ruengsamran, T.; Kampa, P.; Sompong, W. In Vitro Inhibitory Effects of Plant-Based Foods and Their Combinations on Intestinal α-Glucosidase and Pancreatic α-Amylase. *BMC Complement. Altern. Med.* **2012**, *12*, 1–8. [CrossRef] [PubMed]

51. Van Hung, P. Phenolic Compounds of Cereals and Their Antioxidant Capacity. *Crit. Rev. Food Sci. Nutr.* **2016**, *56*, 25–35. [CrossRef] [PubMed]

52. Leopoldini, M.; Russo, N.; Toscano, M. The Molecular Basis of Working Mechanism of Natural Polyphenolic Antioxidants. *Food Chem.* **2011**, *125*, 288–306. [CrossRef]

53. Diaz, P.; Jeong, S.C.; Lee, S.; Khoo, C.; Koyyalamudi, S.R. Antioxidant and Anti-Inflammatory Activities of Selected Medicinal Plants and Fungi Containing Phenolic and Flavonoid Compounds. *Chin. Med.* **2012**, *7*, 26. [CrossRef]

54. Berker, K.I.; Güçlü, K.; Tor, İ.; Apak, R. Comparative Evaluation of Fe(III) Reducing Power-Based Antioxidant Capacity Assays in the Presence of Phenanthroline, Batho-Phenanthroline, Tripyridyltriazine (FRAP), and Ferricyanide Reagents. *Talanta* **2007**, *72*, 1157–1165. [CrossRef]

55. Kling, B.; Bücherl, D.; Palatzky, P.; Matysik, F.-M.; Decker, M.; Wegener, J.; Heilmann, J. Flavonoids, Flavonoid Metabolites, and Phenolic Acids Inhibit Oxidative Stress in the Neuronal Cell Line HT-22 Monitored by ECIS and MTT Assay: A Comparative Study. *J. Nat. Prod.* **2014**, *77*, 446–454. [CrossRef]

56. Wojdyło, A.; Oszmiański, J.; Czemerys, R. Antioxidant Activity and Phenolic Compounds in 32 Selected Herbs. *Food Chem.* **2007**, *105*, 940–949. [CrossRef]

57. Mahdi, S.; Azzi, R.; Lahfa, F.B. Evaluation of in Vitro α-Amylase and α-Glucosidase Inhibitory Potential and Hemolytic Effect of Phenolic Enriched Fractions of the Aerial Part of *Salvia officinalis* L. *Diabetes Metab. Syndr. Clin. Res. Rev.* **2020**, *14*, 689–694. Available online: https://pubmed.ncbi.nlm.nih.gov/32442919/ (accessed on 1 December 2020). [CrossRef] [PubMed]

58. Józefczuk, J.; Malikowska, K.; Glapa, A.; Stawińska-Witoszyńska, B.; Nowak, J.K.; Bajerska, J.; Lisowska, A.; Walkowiak, J. Mulberry Leaf Extract Decreases Digestion and Absorption of Starch in Healthy Subjects—A Randomized, Placebo-Controlled, Crossover Study. *Adv. Med. Sci.* **2017**, *62*, 302–306. [CrossRef] [PubMed]

59. Meng, Q.; Qi, X.; Fu, Y.; Chen, Q.; Cheng, P.; Yu, X.; Sun, X.; Wu, J.; Li, W.; Zhang, Q.; et al. Flavonoids Extracted from Mulberry (*Morus alba* L.) Leaf Improve Skeletal Muscle Mitochondrial Function by Activating AMPK in Type 2 Diabetes. *J. Ethnopharmacol.* **2020**, *248*, 112326. [CrossRef] [PubMed]
60. Patra, P.; Das, D.; Behera, B.; Maiti, T.K.; Islam, S.S. Structure Elucidation of an Immunoenhancing Pectic Polysaccharide Isolated from Aqueous Extract of Pods of Green Bean (*Phaseolus vulgaris* L.). *Carbohydr. Polym.* **2012**, *87*, 2169–2175. [CrossRef]
61. Loizzo, M.R.; Saab, A.M.; Tundis, R.; Menichini, F.; Bonesi, M.; Piccolo, V.; Statti, G.A.; de Cindio, B.; Houghton, P.J.; Menichini, F. In Vitro Inhibitory Activities of Plants Used in Lebanon Traditional Medicine against Angiotensin Converting Enzyme (ACE) and Digestive Enzymes Related to Diabetes. *J. Ethnopharmacol.* **2008**, *119*, 109–116. [CrossRef]
62. Mota, A.H.; Andrade, J.M.; Rodrigues, M.J.; Custódio, L.; Bronze, M.R.; Duarte, N.; Baby, A.; Rocha, J.; Gaspar, M.M.; Simões, S.; et al. Synchronous Insight of In Vitro and In Vivo Biological Activities of *Sambucus nigra* L. Extracts for Industrial Uses. *Ind. Crop. Prod.* **2020**, *154*, 112709. [CrossRef]
63. Ho, G.T.T.; Kase, E.T.; Wangensteen, H.; Barsett, H. Effect of Phenolic Compounds from Elderflowers on Glucose- and Fatty Acid Uptake in Human Myotubes and HepG2-Cells. *Molecules* **2017**, *22*, 90. [CrossRef]

Article

Effect of Spaceflight on Tomato Seed Quality and Biochemical Characteristics of Mature Plants

Elena Dzhos [1], Nadezhda Golubkina [1,*], Marina Antoshkina [1], Irina Kondratyeva [1], Andrew Koshevarov [1], Anton Shkaplerov [2], Tatiana Zavarykina [3,4], Galina Nechitailo [3] and Gianluca Caruso [5]

[1] Federal Scientific Center of Vegetable Production, Selectsionnaya 14, 143072 Moscow, Russia; elenadzhos@mail.ru (E.D.); limont_m@mail.ru (M.A.); tomatvniissok@mail.ru (I.K.); zato@inbox.ru (A.K.)
[2] Yuri Gagarin Cosmonaut Training Center, Star City, 141160 Moscow, Russia; shkaplerov@mail.ru
[3] Emanuel Institute of Biochemical Physics RAS, Kosigina 4, 119334 Moscow, Russia; tpalievskaya@yandex.ru (T.Z.); spacemal@mail.ru (G.N.)
[4] Federal Scientific Center "Nemchinovka", Agrokhimikov 6, Novoivanovskoe, 143026 Moscow, Russia
[5] Department of Agricultural Sciences, University of Naples Federico II, Portici, 80055 Naples, Italy; gcaruso@unina.it
* Correspondence: segolubkina45@gmail.com

Citation: Dzhos, E.; Golubkina, N.; Antoshkina, M.; Kondratyeva, I.; Koshevarov, A.; Shkaplerov, A.; Zavarykina, T.; Nechitailo, G.; Caruso, G. Effect of Spaceflight on Tomato Seed Quality and Biochemical Characteristics of Mature Plants. *Horticulturae* **2021**, *7*, 89. https://doi.org/10.3390/horticulturae7050089

Academic Editor: Leo Sabatino

Received: 1 April 2021
Accepted: 20 April 2021
Published: 22 April 2021

Publisher's Note: MDPI stays neutral with regard to jurisdictional claims in published maps and institutional affiliations.

Abstract: Intensive space exploration includes profound investigations on the effect of weightlessness and cosmic radiation on plant growth and development. Tomato seeds are often used in such experiments though up to date the results have given rather vague information about biochemical changes in mature plants grown from seeds subjected to spaceflight. The effect of half a year of storage in the International Space Station (ISS) on tomato seeds (cultivar Podmoskovny ranny) was studied by analyzing the biochemical characteristics and mineral content of mature plants grown from these seeds both in greenhouse and field conditions. A significant increase was recorded in ascorbic acid, polyphenol and carotenoid contents, and total antioxidant activity (AOA), with higher changes in the field conditions compared to greenhouse. Contrary to control plants, the ones derived from space-stored seeds demonstrated a significant decrease in root AOA. The latter plants also showed a higher yield, but lower content of fruit dry matter, sugars, total dissolved solids and organic acids. The fruits of plants derived from space-stored seeds demonstrated decreased levels of Fe, Cu and taste index. The described results reflect the existence of oxidative stress in mature tomato plants as a long-term consequence of the effect of spaceflight on seed quality, whereas the higher yield may be attributed to genetic modifications.

Keywords: space-stored seeds; *Solanum lycopersicum* L.; weightlessness; cosmic radiation; antioxidants

1. Introduction

International space stations provide unique conditions for the investigation of the effects of space radiation and microgravity on plant growth and development [1–5]. Such experiments can serve as the basis for subsequent cultivation of vegetable crops in space during long-distance spaceflights, for understanding the mechanisms of physiological changes in plants and for evaluating the prospects of quick plant selection from space seeds [1,6–8].

Previous investigations of the effects of space radiation and microgravity on plant growth and development achieved on Russian and American space stations revealed insignificant changes in plant morphology, which suggests good prospects of plant cultivation in space [9,10]. Another direction of investigations includes studies of the effect of space on the quality of both the seeds and the plants grown from these seeds on the Earth. According to Liu et al. [11], this direction opens high prospects of quick plant selection aimed to increase yield, tolerance to diseases and vegetation period shortening. A series of such investigations were completed on different agricultural crops, and a beneficial

effect of spaceflight was demonstrated on rice seeds [11]. On the other hand, the results obtained in recent times are often controversial. Indeed, investigations on rocket seeds [12] showed that spaceflight reduced seed germination vigor and increased aging sensitivity but did not compromise seed viability and the development of normal seedlings. An investigation on tomato seeds after 6 years of spaceflight indicated that the tested plants exhibited higher variability in yield than the control ones, and some of the tested plants were infertile; moreover, various differences in cell walls, chloroplasts and mitochondria were observed. The results obtained point out significant changes occurring at the molecular level in tomato plants [10,13]. Experiments carried out in Ukraine [14] demonstrated that spaceflight conditions during a 6-year exposure of tomato seeds increased tomato productivity, whereas the plants were more resistant to viruses and had higher polyphenol concentration than those of the stationary control. In the early experiment of Kahn and Stoffella [15], the authors showed that tomato seeds could survive in space for several years without adverse effects on germination, emergence and fruit yield. Investigations of the effect of 15-day spaceflight revealed acceleration of alfalfa seed germination and inhibition of the root growth due to chromosomal damage and abnormal mitosis induced by cosmic radiation [16]. Other results revealed reduced germination, lethality, sterility and accelerated senescence [17–19].

The chronic exposure to low doses of ionizing radiation also led to significant differences in the expression of radical scavenging enzymes and DNA-repair genes and an increase in the activity of several antioxidant enzymes [20].

Long-term exposure to microgravity inside spaceships also resulted in the important discovery that these conditions are associated with accelerated aging of humans and plants [4,21]. The conditions of space stations reduced seed vigor and viability, which are connected with oxidation of the most important molecules: proteins, lipids and nucleic acids [22,23]. The best shielding for crop seed transport during long-distance space travel was the seed storage inside spaceships [12].

The aim of the present study was to conduct a quality evaluation of mature tomato plants grown from seeds exposed to half a year of spaceflight on the International Space Station (ISS).

2. Materials and Methods

Tomato seeds (*Solanum lycopersicum* L., cultivar Podmoskovny ranny of dwarf type) were obtained in 1992 and exposed in the cosmic station Mir (from 1992 to 1998). After returning to the Earth, 8 generations of plants were grown both in greenhouse and open field [13], and only the seeds of the 8th tomato generation were used in the present research. Seeds were transported to the International Space Station (ISS) on 19 December 2017 by the cosmonaut Shraplerov through the transportable manned spacecraft MS-07. Space-treated seeds were stored for six months inside the ISS, at the average temperature of 22–23 °C. The seed samples were brought back to the Earth on 3 June 2018. Control seeds of cultivar Podmoskovny ranny had been kept at room temperature in the laboratory of the Emanuel Institute of Biochemical Physics since 1998.

2.1. Growing Conditions and Experimental Protocol

Seeds of tomato (cultivar Podmoskovny ranny) kept for half a year in the ISS and seeds of control plants were used in the present investigation.

A research was carried out in 2020 on plants grown in (a) unheated film-covered greenhouse and (b) in open field, at the experimental fields of Federal Scientific Center of Vegetable Production (Moscow region, 55°39.51′ N, 37°12.23′ E), on sod-podzolic clay-loam soil, pH 6.8, 2.1% organic matter, 1.1 $g \cdot kg^{-1}$ N, 0.045 $g \cdot kg^{-1}$ P_2O_5, 0.357 $g \cdot kg^{-1}$ K_2O. The mean values of temperature (°C) and relative humidity (%) were the following: 16.1 and 71.8 in May, 21.0 and 73.0 in June, 23.8 and 74.9 in July, 19.0 and 76.9 in August and 14.8 and 86.0 in September.

Upon the 22 April sowing, the control seeds showed 33% germination and space-treated seeds reached 61%. The seedlings were transferred into cassettes (5 × 5 cv) with a peat mix substrate containing mineral fertilizers and pH 6.5–7. Most of the seedlings emerged on 4 May and were transplanted in greenhouse on 28 May and in open field on 16 June, when they had 6–7 true leaves, with 3 plants per m^2, and each treatment was replicated thrice. Fertilization was practiced by supplying 30 kg·ha^{-1} N (ammonium sulfate), 60 kg·ha^{-1} P$_2$O$_5$ (superphosphate) and 100 kg·ha^{-1} (potassium sulfate) prior to planting and 50 kg·ha^{-1} N (ammonium nitrate) during the crop cycle in two applications, two and five weeks after transplant, respectively.

At harvest, started in early August, the following determinations were performed in all plots: plant height; weight of tomato fruits; number of marketable trusses per plant.

2.2. Resistance to Phytophtorosis

The evaluation of plant resistance to phytophtorosis was carried out during the natural development of the disease against a severe infection background, according to the guidelines for tomato selection relevant to phytophtorosis resistance [24].

2.3. Sample Preparation

Ten samples of healthy red-ripe stage fruit were used, harvested in August. Each sample consisted of at least three tomatoes from the second to fourth trusses, with a minimum total weight for sample of 250 g.

After harvesting, leaves, fruits and roots were separated and weighed; roots were washed with water and dried with filter paper. Samples were homogenized, and fresh homogenates were used for the determination of ascorbic acid, nitrates and total dissolved solids (TDS). Some of the samples were dried at 70 °C to constant weight and used for the determination of total polyphenol content (TP), total antioxidant activity (AOA) and mineral composition.

2.4. Dry Matter

The dry matter was assessed gravimetrically by drying the samples in an oven at 70 °C until constant weight.

2.5. Ascorbic Acid

The ascorbic acid content was determined by visual titration of plant extracts in 6% trichloracetic acid with Tillman's reagent [25]. Three grams of fresh tomato fruits were homogenized in a porcelain mortar with 5 mL of 6% trichloracetic acid and quantitatively transferred to a measuring cylinder. The volume was brought to 60 mL using trichloracetic acid, and the mixture was filtered through filter paper 15 min later. The concentration of ascorbic acid was determined from the amount of Tillman's reagent that went into titration of the sample.

2.6. Preparation of Ethanolic Extracts

Half a gram of dry homogenized tomato fruit or root powder was extracted with 20 mL of 70% ethanol at 80 °C over 1 h. The mixture was cooled and quantitatively transferred to a volumetric flask, and the volume was adjusted to 25 mL. The mixture was filtered through filter paper and used further for the determination of polyphenols and total antioxidant activity.

2.7. Total Polyphenols (TP)

Polyphenols were determined spectrophotometrically based on the Folin–Ciocalteu colorimetric method according to Golubkina et al. [26]. The concentration of polyphenols was calculated according to the absorption of the reaction mixture at 730 nm using 0.02% gallic acid as an external standard. The results were expressed in mg of gallic acid equivalent per g of dry weight (mg GAE g^{-1} d.w.).

2.8. Antioxidant Activity (AOA)

The antioxidant activity was evaluated via titration of 0.01 N $KMnO_4$ solution with ethanolic extracts of dry samples [26].

2.9. Total Dissolved Solids (TDS)

TDS were determined in water extracts using TDS-3 conductometer (HM Digital, Inc., Seoul, Korea) and expressed in $mg\ kg^{-1}$ d.w.

2.10. Nitrates

Nitrates were assessed using ion-selective electrode on ionomer Expert-001 (Econix Inc., Moscow, Russia).

2.11. Monosaccharides (SS)

The monosaccharides were determined using the ferricyanide colorimetric method based on the reaction of monosaccharides with potassium ferricyanide [27]. The total sugars were analogically determined after acidic hydrolysis of water extracts with 20% hydrochloric acid. Fructose was used as an external standard.

2.12. Titratable Acidity (TA)

TA was determined potentiometrically by titrating a 50 mL diluted (1:5) sample with 0.1 N NaOH to pH 8.1 on ionomer Expert 001 (Econix Inc., Russia) and was expressed as percentage of citric acid.

2.13. Taste Index (TI)

TI was determined according to Navez et al. [28] from the total sugar content (TS) and TA values using the formula

$$TI = TA + TS/(20 \times TA). \tag{1}$$

2.14. Carotenoid Content

Determination of carotenoid content was achieved according to Golubkina et al. [26]. First, 0.5 g of homogenized sample was ground in a mortar with ceramic powder and extracted with small portions of acetone until color disappearance. The combined extract was diluted with 9 mL of hexane and washed 4–5 times with distilled water to remove traces of acetone. The residual extract was quantitatively transferred to a volumetric flask, and the volume was adjusted to 10 mL. The resulting extract was mixed, filtered through a small portion of anhydrous Na_2SO_4 and subjected to the analysis. The separation of carotenoids was achieved using quantitative thin-layer chromatography on Whatman 3A chromatographic paper in two chromatographic systems: (1) hexane to separate β-carotene and (2) hexane–acetone, 10:0.5, for separation of lycopene and lutein. Appropriate zones of carotenoid compounds were cut out and filled with 3 mL of hexane. The determination of carotenoid content in tomato fruit was performed using appropriate specific absorption $E^{1\%}_{1cm}$ for β-carotene (2580 at $\lambda = 450$ nm), lycopene (3470 at $\lambda = 474$ nm) and lutein (2560; $\lambda = 447$ nm). The internal standards were β-carotene, lutein and lycopene from Sigma Inc. (Kawasaki, Japan).

2.15. Statistical Analysis

Data were processed by analysis of variance, and mean separations were performed through the Duncan multiple range test, with reference to 0.05 probability level, using SPSS software version 21 (IBM, Armonk, NY, USA). Data expressed as percentages were subjected to angular transformation before processing.

3. Results and Discussion

Tomato plants grown both from control and space-stored seeds had a shorter crop cycle in greenhouse (90–94 days) compared to open field (115–120 days) (Table 1), due to the higher temperatures recorded in the first environment. The differences in crop cycle length between plants from space-stored and control seeds were 5 and 4 days in open field and greenhouse respectively.

Table 1. Phenological progress of tomato plants grown in greenhouse and in open field, from control and spaceflight-exposed seeds, expressed as days from sowing referring to 50% of plants that reached each stage.

	Control Seeds		Space-Stored Seeds	
	Greenhouse	Field	Greenhouse	Field
Two-leaf stage	13	13	12	12
Flowering	52	57	49	56
Fruit ripening	94	115	90	120

3.1. Yield, Dry Matter Content, TDS and Nitrates

Higher height, fruit yield and marketability were recorded for tomato plants grown from space-stored seeds compared to the control ones, with higher values in the more favorable growth conditions in greenhouse (Table 2). The plants derived from space-stored seeds also showed better tolerance to diseases in the field conditions compared to control plants, which was in accordance with the results of the previous investigation [14].

Table 2. Plant height, fruit yield, dry weight and disease occurrence of tomato grown in greenhouse and in open field, from control and spaceflight-exposed seeds.

	Control Seeds		Space-Stored Seeds	
	Greenhouse	Field	Greenhouse	Field
Plant height (cm)	65 b	57 c	77 a	78 a
Yield (t ha^{-1})	39.0 b	34.3 c	42.0 a	36.9 b
Fruit weight (g)	58 b	52 c	63 a	56 b
Number of trusses per plant	5 a	4 b	5 a	4 b
Marketability (%)	89 ab	85 c	92 a	87 bc
Diseases (number of points)	3.0 a	2.5 b	3.0 a	3.0 a
Dry matter (%)	11.5 ± 1.0 a	7.1 ± 0.6 bc	8.2 ± 0.6 b	6.4 ± 0.5 c

In each row, the values with the same letters do not differ statistically according to Duncan test at $p < 0.05$.

At the same time, a significant decrease in fruit dry matter content was found in the fruits obtained from plants derived from space-stored seeds compared to control ones. Notably, in greenhouse conditions, the fruit dry matter associated with the seed space treatment was 1.4 times lower than that of control plant fruits, while the corresponding difference in the open-field-grown plants was only 1.12-fold.

At the same time, the total dissolved solids (TDS) did not differ statistically between fruits produced from control and those obtained from space-treated seeds, though the latter merely showed a slight decreasing tendency. TDS values detected by portable conductometer reflect both the amount of soluble solids and the organic acid content. As can be seen in Figure 1, higher TDS values were recorded in greenhouse conditions, which is connected with a higher nutrient uptake rate.

A similar phenomenon was observed for nitrate accumulation, though with greater differences between the control fruits and those produced by the plants derived from space-stored seeds (Figure 2). The latter results are not surprising, taking into account that all nitrate derivatives are highly soluble in water.

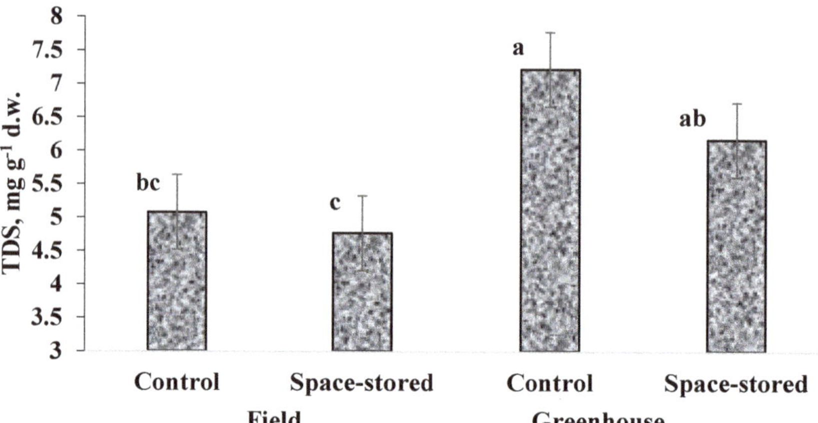

Figure 1. TDS content in tomato fruits grown from control and spaceflight-exposed seeds, in open field and in greenhouse. Values with the same letters do not differ statistically according to Duncan test at $p < 0.05$.

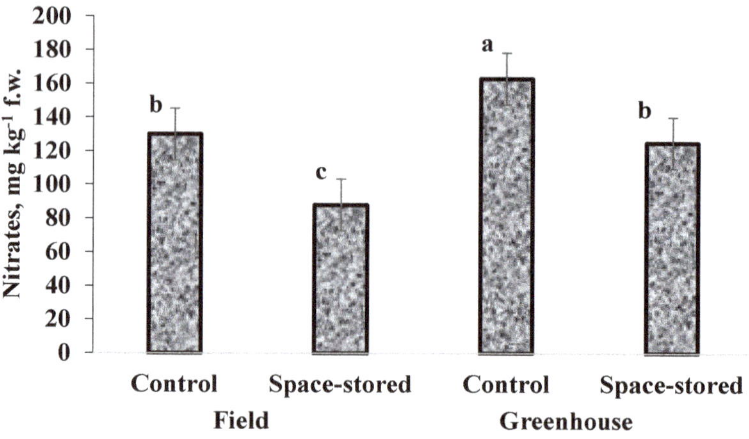

Figure 2. Nitrate accumulation in tomato fruits grown from control and spaceflight-exposed seeds, in open field and in greenhouse. Values with the same letters do not differ statistically according to Duncan test at $p < 0.05$.

3.2. Antioxidant Status

Secondary metabolites and antioxidants in particular are known to be involved in the process of stress adaptation [29]. Tomato fruits are rich in antioxidants, including carotenoids, polyphenols and ascorbic acid, demonstrating high biological activity, thus reducing the risk of cancer and cardiovascular diseases [30,31]; improving immunity [32]; and showing neuroprotective, anti-inflammatory and antimicrobial properties [33]. The most common carotenoids of tomato fruits are β-carotene, lutein and lycopene, which are synthesized at the highest levels in red varieties [34]. The data shown in Figure 3 indicate that the β-carotene/lycopene/lutein ratio in ordinary growing conditions of cultivar Podmoskovny reached 1.00:5.02:1.01 in open field and 1.00:7.50:1.03 in greenhouse, with a significantly higher total carotenoid content in the former case (152.7 mg 100 g^{-1} d.w. vs. 127.4 mg 100 g^{-1} d.w. with $p < 0.01$ significance). The corresponding carotenoid ratio in fruits of plants derived from space-stored seeds was equal to 1.00:3.90:1.04 in open field conditions and 1.00:7.10:0.94 in greenhouse. Taking into account that lycopene is a precursor in the biosynthesis of β-carotene and lutein, it may be inferred either that the

reduction of lycopene biosynthesis took place in greenhouse or that the transformation of lycopene to β-carotene and lutein in these conditions was reduced both in control plants and in plants derived from space-stored seeds. Furthermore, the significant increase in total carotenoid content due to the spaceflight effect on tomato seed quality proves the existence of oxidative stress in plants derived from space-stored seeds, which is reflected by the increase in the total carotenoid content by 15% in tomato fruits produced in open field and by 28% in greenhouse (Figure 3).

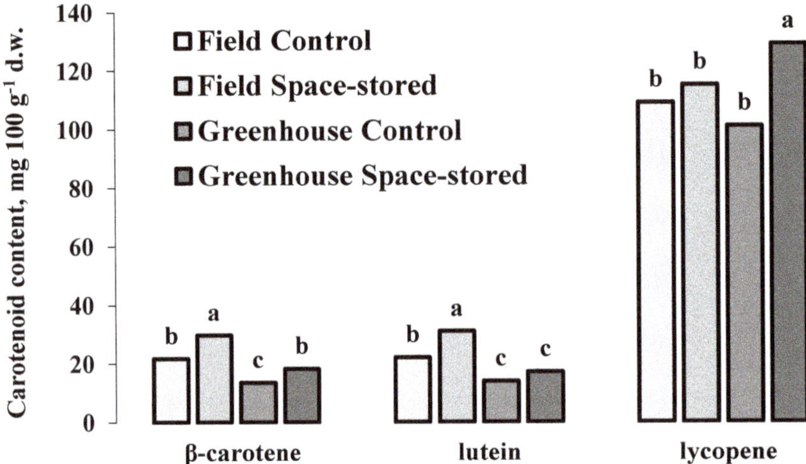

Figure 3. Carotenoid profile of tomato fruits obtained from control plants and plants derived from space-stored seeds grown in open field and in greenhouse. For each carotenoid, values with the same letters do not differ statistically according to Duncan test at $p < 0.05$.

Ascorbic acid (AA) plays an important role in plant antioxidant defense, being a key substrate for the detoxification of reactive oxygen species [35]. Overproduction of reactive oxygen species (ROS) in plants under stress conditions is reduced by the production of enzymatic and nonenzymatic antioxidants. In this respect, AA is one of the universal nonenzymatic antioxidants having substantial potential for scavenging ROS and also modulating a number of fundamental functions in plants both under stress and unstressed conditions. The data of ascorbic acid content in control tomato fruits and tomato fruits derived from space-stored seeds (Table 3) are given on a dry weight basis in order to make an adequate comparison between the treatments having different levels of dry matter. In this respect, the results indicate rather small differences in tomato fruit AA content between control plants and plants derived from space-stored seeds, which reached 6.3% in open field and 20% in greenhouse. In the latter conditions, a similar increase in total antioxidant activity was recorded (21.1%), whereas no AOA differences arose in open field.

Phenolics in tomato fruits, represented by chlorogenic acid and quercetin [34], are important antioxidants both for plant integrity and human health [33] and the major contributors to antioxidant activity in tomatoes [36]. In the present investigation, total phenolics are of special interest as the differences in their content between control plants and plants derived from space-stored seeds reached the highest values of 26.7% in greenhouse and 36.3% in open field. Contrary, no differences in phenolic levels were observed between tomato fruits grown in greenhouse and open field. Interestingly, among the antioxidants studied, polyphenols proved to be the most sensitive to long-term consequences of spaceflight.

Table 3. Antioxidant compounds and activity of tomato fruits obtained from control and spaceflight-exposed seeds, in greenhouse and in open field.

Parameter	Control Seeds		Space-Stored Seeds	
	Greenhouse	Field	Greenhouse	Field
Fruits				
AA (mg 100 g^{-1} d.w.)	399 ± 30 c	537 ± 40 ab	479 ± 35 b	571 ± 42 a
AOA (mg GAE g^{-1} d.w.)	18.0 ± 1 b	22.5 ± 1 a	21.8 ± 1 a	22.5 ± 1 a
TP (mg GAE g^{-1} d.w.)	13.4 ± 1 b	13.5 ± 0.9 b	17.1 ± 1 a	18.4 ± 1 a
Roots				
AOA (mg GAE g^{-1} d.w.)	10.7 ± 0.8 a	12.6 ± 1.0 a	6.5 ± 0.3 b	8.7 ± 0.4 c
TP (mg GAE g^{-1} d.w.)	7.0 ± 0.5 a	6.8 ± 0.4 a	6.5 ± 0.3 ab	6.0 ± 0.3 b

AA: ascorbic acid; AOA: antioxidant activity; TP: total phenolics. In each row, the values with the same letters do not differ statistically according to Duncan test at $p < 0.05$.

The comparison between the antioxidant status of tomato fruits and roots revealed the opposite tendency in antioxidant distribution, especially pronounced in AOA values. Indeed, while root TP decreased in plants derived from space-stored seeds, reaching only 13.3% in open field and 7.7% in greenhouse, root AOA values decreased by 44.8% and 64.6%, respectively (Table 3). The latter phenomenon indirectly indicates the possibility of antioxidant redistribution in plants derived from space-stored seeds, with the AOA showing a decrease in roots and increase in fruits.

3.3. Monosaccharides, Organic Acids, Taste

Organic acids together with sugars are the main soluble components of ripe fruits and have a major effect on taste, being responsible for sourness and contributing to the flavor; it is well known that sugar and organic acid contents are positively correlated in tomato fruit [37]. Contrary to soluble carbohydrates, which are translocated into the fruits as products of photosynthesis, organic acids are synthesized predominantly in fruits from imported sugars. Fruit acidity, measured by titratable acidity and/or pH, is an important component of fruit organoleptic quality [38], and it is connected to the presence of organic acids, with malic and citric acids being the main acids found in most ripe fruits [39]. Both genetic and environmental variations affect organic acid accumulation in tomato fruits. Notably, nitrate accumulation may stimulate organic acid biosynthesis [40], and in this respect, decreased levels of nitrates and TDS result in decreased TA values.

Though recent investigations indicated a significant role of total antioxidant activity (AOA), total phenolics (TP) and ascorbic acid (AA) content in tomato fruit taste [40,41], only one equation between sugar and organic acids content is presently used for appropriate evaluation [28,34,42]. Monosaccharides (glucose and fructose, present at equimolar ratios) are known to dominate in tomato fruits with a negligible amount of disaccharides [43]. Differences in titratable acidity (TA) between control fruits and fruits derived from space-stored seeds were as much as 35.1% in open field and 50% in greenhouse conditions. Citric acid is known to prevail in tomato fruit organic acids [33], whereas malic and oxalic acids contribute to a lesser extent to the titratable acidity value [44]. As can be inferred from the above data, both TDS and TA demonstrated similar differences between control plants and plants derived from space-stored seeds; i.e., they were higher in greenhouse fruits than in the open field ones. Furthermore, according to taste maturity index (TM = TS:TA) [28], fruit of control plants and plants developed from space-stored seeds did not differ at the stage of maturity, with the TM index being equal to 9.00–9.47 for greenhouse plants and 7.56–7.60 for those grown in open field. In this respect, the taste index of fruit from control plants and plants derived from space-stored seeds, calculated according to Navez et al. [28], revealed higher values in greenhouse and lower in open field (Table 4).

Table 4. Monosaccharide and organic acid contents, and taste index (TI) of tomato fruits obtained from control and spaceflight-exposed seeds, in greenhouse and in open field.

Parameter	Control Seeds		Space-Stored Seeds	
	Greenhouse	Field	Greenhouse	Field
Brix (% f.w.)	7.1 ± 0.4 a	3.8 ± 0.2 b	4.2 ± 0.2 b	2.8 ± 0.1 c
TS (% f.w.)	7.1 ± 0.3 a	3.8 ± 0.2 c	4.5 ± 0.3 b	2.8 ± 0.2 d
TA (% f.w.)	0.75 ± 0.04 a	0.50 ± 0.03 b	0.50 ± 0.03 b	0.37 ± 0.02 c
TI	1.22	0.88	0.92	0.78
TM	9.47	7.60	9.00	7.57

TS: total sugar; TA: titratable acidity; TI: taste index; TM: taste maturity index. In each row, the values with the same letters do not differ statistically according to Duncan test at $p < 0.05$.

The recorded monosaccharide trends were similar to TDS changes (Figure 1), as monosaccharides are one of the main components of water-soluble extracts in tomato fruits. Notably, the results indicate a significant decrease in fruit sugar content of plants derived from space-stored seeds, as much as 35.7% and 57.8% in open field and greenhouse, respectively (Table 4), which is in contradiction with the hypothesis of oxidant stress development. According to literature reports [45], carbohydrates also participate in plant antioxidant defense, and their amount usually increases as a consequence of stress conditions. The opposite differences recorded in the present research for fruits of plants derived from space-stored seeds were statistically significant for values calculated on a dry weight basis, i.e., 12.4% decrease for fruits grown in greenhouse and 22.1% decrease for fruits grown in open field.

3.4. Elemental Composition

Changes in fruit elemental composition were revealed only for Fe and Cu content, whose concentrations in fruits derived by space-stored seeds decreased by 1.5 and 1.24 times compared to control plants in greenhouse conditions and by 1.4 and 1.3 times in open field (Table 5). No differences were detected for Mn and Zn content in fruit between control plants and plants derived from space-stored seeds.

Table 5. Elemental composition of tomato fruits obtained from control and spaceflight-exposed seeds, in greenhouse and in open field (mg kg^{-1} d.w.).

Growing Environment	Seed Origin	Zn	Mn	Fe	Cu
Greenhouse	control	7.0 ± 0.5 ab	5.5 ± 0.4 bc	45.8 ± 3.7 a	3.1 ± 0.2 a
	space-stored	6.2 ± 0.4 b	4.8 ± 0.3 c	29.8 ± 2.0 b	2.5 ± 0.1 b
Open field	control	8.7 ± 0.7 a	6.8 ± 0.5 a	39.8 ± 3.1 a	3.0 ± 0.2 a
	space-stored	7.8 ± 0.7 a	6.0 ± 0.5 ab	28.4 ± 0.2 b	2.3 ± 0.1 b

Within each column, the values with the same letters do not differ statistically according to Duncan test at $p < 0.05$.

Iron is an essential micronutrient for almost all living organisms, playing a critical role in DNA synthesis, photosynthesis and respiration [46]. In plants, iron is involved in chlorophyll synthesis, and it is essential for maintaining chloroplast structure and function [47]. As far as Cu is concerned, this microelement is known to serve as an essential cofactor in plant proteins, performing pivotal functions in plant cells by participating in electron transport [48]. Based on the above reports, reduced levels of Cu and Fe may be connected to decreased levels of monosaccharides in tomatoes from space-stored seeds [48].

3.5. Relationships between the Analyzed Parameters

From the results of biochemical and mineral analyses, important characteristics of plants grown from space-kept seeds were revealed (Figure 4). The high levels of antioxidant parameters and the decrease in Fe and Cu accumulation are in agreement with the existence of significant oxidative stress. The decrease in sugar, dry matter and organic acid contents

is in accordance with lower taste index values in fruits derived from space-stored seeds. The controversial aspect lies in the fact that despite the reduced accumulation of Fe, Cu, dry matter and carbohydrates, the plants grown from space-kept seeds demonstrated higher yield, fruit weight and plant height.

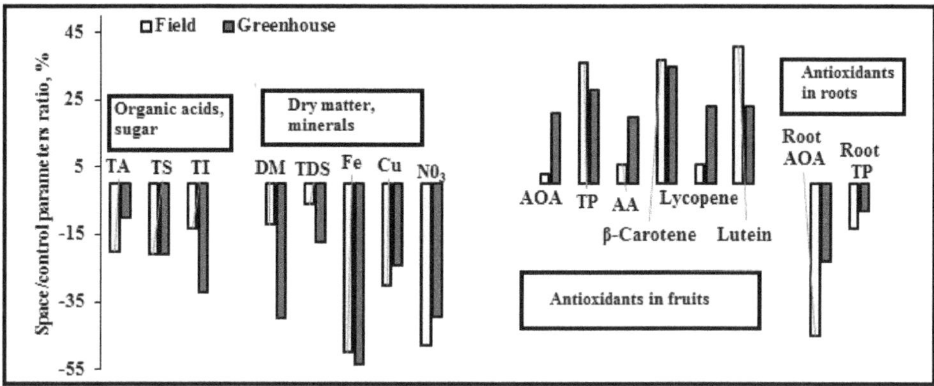

Figure 4. Differences in biochemical parameters and mineral content between control plants and plants derived from spaceflight-treated seeds grown in open field and in greenhouse.

On the other hand, the calculation of dry fruit weight indicated extremely small differences, or a lack of difference, between control plants and plants derived from space-stored seeds. In this respect, the fruit weight of control plants and plants grown from space-kept seeds reached 6.67 ± 0.7 g and 5.17 ± 0.5 g d.w., respectively, in greenhouse and 3.69 ± 0.3 g and 3.56 ± 0.3 g d.w., respectively, in open field. The latter results indicate the increased dilution in fruits derived from space-stored seeds, as a long-term consequence of the effect of spaceflight on seed quality.

4. Conclusions and Future Challenges

Detailed biochemical and mineral characteristics of tomato fruits grown from space-kept seeds revealed the existence of significant oxidative stress in the plants, which was reflected in metabolic antioxidant content changes, a decrease in fruit quality and an increase in fruit yield. The phenomenon relevant to the recorded lowering of biological indicator values resulting from the space-kept seed utilization is dramatically important, though further studies are necessary to evaluate the associated mechanisms and modifications, particularly regarding antioxidant enzyme activity and carbohydrate profile in fruits.

Author Contributions: Conceptualization, E.D., G.N. and G.C.; data curation, T.Z.; formal analysis, T.Z. and G.N.; investigation, N.G., M.A., A.K. and I.K.; methodology, N.G., G.C. and E.D.; supervision, A.S. and G.C.; validation, G.C.; draft manuscript writing, N.G., M.A. and E.D.; manuscript revision and final editing, E.D., N.G. and G.C. All authors have read and agreed to the published version of the manuscript.

Funding: This research did not receive any grants from public, commercial or not-for-profit agencies.

Institutional Review Board Statement: Ethical review and approval were waived for this study, due to the involvement in the authorship of the Istitutions providing the seeds and managing the present research which did not require unusual materials of methods.

Informed Consent Statement: Informed consent was obtained from all subjects involved in the study.

Data Availability Statement: Not applicable.

Conflicts of Interest: The authors declare no conflict of interest.

References

1. Visscher, A.M.; Seal, C.E.; Newton, R.J.; Frances, A.L.; Pritchard, H.W. Dry seeds and environmental extremes: Consequences for seed lifespan and germination. *Funct. Plant Biol.* **2016**, *43*, 656–668. [CrossRef] [PubMed]
2. Vandenbrink, J.P.; Kiss, J.Z. Space, the final frontier: A critical review of recent experiments performed in microgravity. *Plant Sci.* **2016**, *243*, 115–119. [CrossRef] [PubMed]
3. Paul, A.L.; Wheeler, R.M.; Levine, H.G.; Ferl, R.J. Fundamental plant biology enabled by the space shuttle. *Am. J. Bot.* **2013**, *100*, 226–234. [CrossRef] [PubMed]
4. Kordyum, E.; Chapman, D.; Brykova, V. Plant cell development and aging may accelerate in microgravity. *Acta Astronaut.* **2019**, *157*, 157–161. [CrossRef]
5. Musgrave, M.E. Seeds in space. *Seed Sci. Res.* **2007**, *12*, 1–17. [CrossRef]
6. Cannon, K.M.; Britt, D.T. Feeding one million people on Mars. *New Space* **2019**, *7*, 245–254. [CrossRef]
7. Wheeler, R.M. Agriculture for space: People and places paving the way. *Open Agric.* **2017**, *2*, 14–32. [CrossRef]
8. Paul, A.L.; Sng, N.J.; Zupanska, A.K.; Krishnamurthy, A.; Schultz, E.R.; Ferl, R.J. Genetic dissection of the Arabidopsis spaceflight transcriptome: Are some responses dispensable for the physiological adaptation of plants to spaceflight? *PLoS ONE* **2017**, *12*, e0180186. [CrossRef] [PubMed]
9. Poulet, L.; Fontaine, J.-P.; Dussap, C.-G. Plant's response to space environment: A comprehensive review including mechanistic modelling for future space gardeners. *Bot. Lett.* **2016**, *163*, 337–347. [CrossRef]
10. Lu, J.; Xue, H.; Pan, Y.; Kan, S.; Liu, M.; Nechitailo, G.S. Effect of spaceflight duration of subcellular morphologies and defense enzyme activities in earth-grown tomato seedlings propagated from space-flown seeds. *Russ. J. Phys. Chem. B* **2009**, *3*, 981–986. [CrossRef]
11. Jiyuan, L.; Zhenye, Q.; Yongcheng, S.; Tianjun, Q.; Jun, H. Seed growth experiments after space flight: The Chinese experience. In *Selected Papers on Remote Sensing, Space Science and Information Technology*; Seminars of the United Nations Programmer on Space Applications; United Nations: New York, NY, USA, 1999; Volume 10, p. 71.
12. Chandler, J.O.; Haas, F.B.; Khan, S.; Bowden, L.; Ignatz, M.; Enfissi, E.M.A.; Gawthrop, F.; Griffiths, A.; Fraser, P.D.; Rensing, S.A.; et al. Rocket Science: The effect of spaceflight on germination physiology, ageing, and transcriptome of *Eruca sativa* seeds. *Life* **2020**, *10*, 49. [CrossRef]
13. Nechitailo, G.; Jinying, L.; Huai, X.; Yi, P.R.; Chongqin, T.; Liu, M. Influence of long term exposure to space flight on tomato seeds. *Adv. Space Res.* **2005**, *36*, 1329–1333. [CrossRef]
14. Mishchenko, L.T.; Dunich, A.A.; Danilova, O.I. Impact of a real microgravity on the productivity of tomato plants and resistance to viruses. In Proceedings of the Life in Space for Life on Earth, Aberdeen, UK, 18–22 June 2012. ESA SP-706.
15. Kahn, B.A.; Stoffella, P.J. No evidence of adverse effects on germination, emergence, and fruit yield due to space exposure of tomato seeds. *J. Am. Soc. Hortic. Sci.* **1996**, *121*, 414–418. [CrossRef] [PubMed]
16. Ren, W.B.; Zhang, Y.; Deng, B.; Guo, H.; Cheng, L.; Liu, Y. Effect of space flight factors on alfalfa seeds. *Afr. J. Biotechnol.* **2010**, *9*, 7273–7279. [CrossRef]
17. De Micco, V.; Arena, C.; Pignalosa, D.; Durante, M. Effects of Sparsely and Densely Ionizing Radiation on Plants. *Radiat. Environ. Biophys.* **2011**, *50*, 1–19. [CrossRef]
18. Arena, C.; De Micco, V.; De Maio, A. Growth alteration and leaf biochemical responses in *Phaseolus Vulgaris* exposed to different doses of ionizing radiation. *Plant Biol.* **2014**, *16*, 194–202. [CrossRef]
19. Arena, C.; De Micco, V.; Aronne, G.; Pugliese, M.G.; Virzo, A.; DeMaio, A. Response of *Phaseolus vulgaris* L. plants to low-LET ionizing radiation: Growth and oxidative stress. *Acta Astronaut.* **2014**, *91*, 107–114. [CrossRef]
20. Zaka, R.; Vandecasteele, C.M.; Misset, M.T. Effects of low chronic doses of ionizing radiation on antioxidant enzymes and G6PDH activities in *Stipa capillata* (Poaceae). *J. Exp. Bot.* **2002**, *53*, 1979–1987. [CrossRef]
21. Arena, C.; De Micco, V.; Macaevac, E.; Quintens, R. Space radiation effects on plant and mammalian cells. *Acta Astronaut.* **2014**, *104*, 419–431. [CrossRef]
22. Fleming, M.B.; Patterson, E.L.; Reeves, P.A.; Richards, C.M.; Gaines, T.A.; Walters, C. Exploring the fate of mRNA in aging seeds: Protection, destruction, or slow decay? *J. Exp. Bot.* **2018**, *69*, 4309–4321. [CrossRef] [PubMed]
23. Fleming, M.B.; Hill, L.M.; Walters, C. The kinetics of ageing in dry-stored seeds: A comparison of viability loss and RNA degradation in unique legacy seed collections. *Ann. Bot.* **2019**, *123*, 1133–1146. [CrossRef] [PubMed]
24. Alpatiev, A.V.; Skvortsova, R.V.; Gurkina, L.K. Guidelines for the selection of tomato for phytophtorosis resistance. In *Greenhouse and Field Conditions*; VNIISSOK: Moscow, Russia, 1986; pp. 36–45. (In Russian)
25. AOAC Association Official Analytical Chemists. *The Official Methods of Analysis of AOAC International*; 22 Vitamin C; AOAC: Rockville, MD, USA, 2012.
26. Golubkina, N.A.; Kekina, H.G.; Molchanova, A.V.; Antoshkina, M.S.; Nadezhkin, S.M.; Soldatenko, A.V. *Plants Antioxidants and Methods of Their Determination*; Infra-M: Moscow, Russia, 2020. [CrossRef]
27. Swamy, P.M. *Laboratory Manual on Biotechnology*; Rastogi: New Delhi, India, 2008; p. 617.
28. Navez, B.; Letard, M.; Graselly, D.; Jost, J. Les critéres de qualité de la tomate. *Infos-Ctifl* **1999**, *155*, 41–47.
29. Sumalan, R.M.; Ciulca, S.I.; Poiana, M.A.; Moigradean, D.; Radulov, I.; Negrea, M.; Crisan, M.E.; Copolovici, L.; Sumalan, R.L. The Antioxidant profile evaluation of some tomato landraces with soil salinity tolerance correlated with high nutraceutical and functional value. *Agronomy* **2020**, *10*, 500. [CrossRef]

30. Riccioni, G. Carotenoids and cardiovascular disease. *Curr. Atheroscler. Rep.* **2009**, *11*, 434–439. [CrossRef] [PubMed]

31. Keikel, M.; Schumacher, M.; Dicato, M.; Diederich, M. Antioxidant and anti- proliferative properties of lycopene. *Free Radic. Res.* **2011**, *45*, 925–940. [CrossRef]

32. Du, J.; Cullen, J.J.; Buettner, G.R. Ascorbic acid: Chemistry, biology and the treatment of cancer. *Biochim. Biophys. Acta* **2012**, *1826*, 443–457. [CrossRef]

33. Del Río, D.; Rodríguez-Mateos, A.; Spencer, J.P.E.; Tognolini, M.; Borges, G.; Crozier, A. Dietary (poly)phenolics in human health: Structures, bioavailability, and evidence of protective effects against chronic diseases. *Antioxid. Redox Signal.* **2013**, *18*, 1818–1892. [CrossRef]

34. Siddiqui, M.W.; Ayala-Zavala, J.F.; Dhua, R.S. Genotypic variation in tomatoes affecting processing and antioxidant properties. *Crit. Rev. Food Sci. Nutr.* **2015**, *55*, 1819–1835. [CrossRef]

35. Akram, N.A.; Shafiq, F.; Ashraf, M. Ascorbic Acid-A Potential Oxidant Scavenger and Its Role in Plant Development and Abiotic Stress Tolerance. *Front. Plant Sci.* **2017**, *8*, 613. [CrossRef]

36. Toor, R.K.; Savage, G.P. Antioxidant activities in different fractions of tomatoes. *Food Res. Int.* **2005**, *38*, 487–494. [CrossRef]

37. Batista-Silva, W.; Naschimento, V.L.; Medeiros, D.B.; Nunes-Besi, A.; Ribeiro, D.M.; Zsoson, A.; Araujo, W.L. Modification of organic acids profile during fruit development and ripening: Correlation of causation? *Front. Plant Sci.* **2018**, *9*, 1689. [CrossRef] [PubMed]

38. Aoun, A.B.; Lechiheb, B.; Benyahya, L.; Ferchichi, A. Evaluation of fruit quality traits of traditional varieties of tomato (*Solanum lycopersicum*) grown in Tunisia. *Afr. J. Food Sci.* **2013**, *7*, 350–354. [CrossRef]

39. Davies, J.N.; Hobson, G.E. The constituents of tomato fruit—The influence of environment, nutrition, and genotype. *Crit. Rev. Food Sci. Technol.* **1981**, *15*, 205–280. [CrossRef]

40. Etienne, A.; Génard, M.; Lobit, P.; Mbeguié-A-Mbéguié, D.; Bugaud, C. What controls fleshy fruit acidity? A review of malate and citrate accumulation in fruit cells. *J. Exp. Bot.* **2013**, *64*, 1451–1469. [CrossRef] [PubMed]

41. Figàs, M.R.; Prohens, J.; Raigón, M.D.; Fita, A.; García-Martínez, M.D.; Casanova, C.; Borràs, D.; Plazas, M.; Andújar, I.; Soler, S. Characterization of composition traits related to organoleptic and functional quality for the differentiation, selection and enhancement of local varieties of tomato from different cultivar groups. *Food Chem.* **2015**, *187*, 517–524. [CrossRef]

42. Causse, M.; Friguet, C.; Coiret, C.; Lépicier, M.; Navez, B.; Lee, M.; Holthuysen, L.; Sinesio, F.; Moneta, E.; Grandillo, S. Consumer preferences for fresh tomato at the European scale: A common segmentation on taste and firmness. *J. Food Sci.* **2010**, *75*, S531–S541. [CrossRef] [PubMed]

43. Beckles, D.M. Factors affecting the postharvest soluble solids and sugar content of tomato (*Solanum lycopersicum* L.) fruit. *Postharvest Biol. Technol.* **2012**, *63*, 129–140. [CrossRef]

44. Fernández-Ruiz, V.; Sánchez-Mata, M.C.; Cámara, M.; Torija, M.E.; Chaya, C.; Galiana-Balaguer, L.; Roselló, S.; Nuez, F. Internal quality characterization of fresh tomato fruits. *HortScience* **2004**, *39*, 339–345. [CrossRef]

45. Keunen, E.; Peshev, D.; Vangronsveld, J.; ENDE, W.V.D.; Cuylers, A. Plant sugars are crucial players in the oxidative challenge during abiotic stress: Extending the traditional concept. *Plant Cell Environ.* **2013**, *36*, 1242–1255. [CrossRef]

46. Rout, G.R.; Sahoo, S. Role of iron in plants growth and metabolism. *Rev. Agric. Sci.* **2015**, *3*, 1–24. [CrossRef]

47. Sakya, A.; Sulandjari, T. Foliar iron application on growth and yield of tomato. *IOP Conf. Ser. Earth Environ. Sci.* **2019**, *250*, 012001. [CrossRef]

48. Yruela, I. Copper in plants. *Braz. J. Plant Physiol.* **2005**, *17*, 145–156. [CrossRef]

Article

Development of Frozen Pulps and Powders from Carrot and Tomato by-Products: Impact of Processing and Storage Time on Bioactive and Biological Properties

Helena Araújo-Rodrigues [1], Diva Santos [1], Débora A. Campos [1], Modesta Ratinho [2], Ivo M. Rodrigues [3] and Manuela E. Pintado [2,*]

[1] Centro de Biotecnologia e Química Fina—Laboratório Associado, Escola Superior de Biotecnologia, Universidade Católica Portuguesa/Porto, Rua Diogo Botelho, 1327, 4169-005 Porto, Portugal; hrodrigues@porto.ucp.pt (H.A.-R.); djesus@porto.ucp.pt (D.S.); dcampos@porto.ucp.pt (D.A.C.)
[2] Vitacress S.A., Quinta dos Cativos, 7630-033 Odemira, Portugal; modesta.ratinho@vitacress.com
[3] Departamento de Ciências Agrárias e Tecnologias, Escola Superior Agrária, Instituto Politécnico de Coimbra, 3045-601 Coimbra, Portugal; ivorod@esac.pt
* Correspondence: mpintado@ucp.pt; Tel.: +351-226196200

Citation: Araújo-Rodrigues, H.; Santos, D.; Campos, D.A.; Ratinho, M.; M. Rodrigues, I.; E. Pintado, M. Development of Frozen Pulps and Powders from Carrot and Tomato by-Products: Impact of Processing and Storage Time on Bioactive and Biological Properties. *Horticulturae* **2021**, *7*, 185. https://doi.org/10.3390/horticulturae7070185

Academic Editors: Leo Sabatino and Christophe El-Nakhel

Received: 24 May 2021
Accepted: 29 June 2021
Published: 5 July 2021

Abstract: Vegetables and fruits have an interesting nutritional profile, rich in bioactive metabolites, holding a high antioxidant potential and health associated benefits. However, their functional properties, the shorter shelf-life due to their high-water content, and their seasonality nature lead to extensive food losses and waste. The valorization of vegetables and fruits by-products through the development of value-added products and the application of preservation methods is of utmost importance to prevent food losses and waste. In this study, based on a circular economy approach, pulps and powders of baby carrot and cherry tomato by-products were prepared. Freezing, hot air drying and storage time impact on antioxidant activity and bioactive compounds were studied. Microbiological quality and pulps viscosity were also monitored for 6 months. During the freezing storage, TPC and antioxidant capacity by ABTS and ORAC assays decreased. The antioxidant capacity by DPPH method and carotenoid content increased during the first months of freezing, but then decreased. The drying process negatively affected the antioxidant capacity as well as carotenoid and polyphenolic content compared with the fresh vegetables. Both processing methodologies positively impacted the vitamin E content. During drying storage, there were no key variations in antioxidant capacity and bioactive content.

Keywords: tomato and carrot by-products; freezing and drying impact; antioxidant capacity; polyphenolics; carotenoids; vitamin E

1. Introduction

According to the 2019 report of the Food and Agriculture Organization of the United Nations, approximately 21.6% of the worldwide fruits and vegetables were lost in 2016 [1]. The seasonal nature of vegetables and fruits and their short shelf-life contribute to these massive losses and waste [2–4]. However, a significant fraction of food losses also results from vegetables and fruits that do not comply with commercial standards but maintain relevant nutritional and organoleptic profiles [5]. In this context, the valorisation of non-standard vegetables and fruits and other by-products from these industries through the development of new value-added products is of utmost importance to combat food losses and waste.

Clinical and epidemiological studies have described many health-related benefits associated with a diet rich in vegetables and fruits on chronic, cardiovascular, neurological, inflammatory and some cancer diseases [3,6–8]. Their interesting nutritional profile results from the presence of some bioactive metabolites such as, carotenoids, flavonoids, phenolic acids, tocopherols, alkaloids and chlorophyll derivatives, which contribute to the strong

vegetables and fruits antioxidant capacity [6–8] and are also partially responsible for their organoleptic attributes. Tomato and carrot are some of the most popular and consumed fresh and cooked vegetables worldwide [2,7]. High levels of antioxidant capacity are reported in tomato and carrot derived from, for instance, carotenoids such as lycopene or beta-carotene, polyphenols such as anthocyanins or flavonoids and vitamin E [2–4,7]. Thus, the transformation of these by-products into ingredients with extended shelf-life is an opportunity to reduce food losses and acquire added-value ingredients for the food industry.

Several parameters such as plant genetics, growing environment, ripening stage and post-harvest conditions influence the physical and chemical profile of plant foods [9,10]. Moreover, food-processing approaches may include several modifications in their nutritional composition, being recognized as one of the most critical parameters in the food bioactive composition and concentration. Processing may promote the destruction or chemical modification of the natural phytochemicals, and consequent reduction in the antioxidant capacity [8,10,11]. However, the perishability of tomato and carrot, resultant from their high-water content (>80%), coupled to their seasonality nature, requires the application of processing methods to maximize the resources and ensure the availability of raw materials in the off-seasons [2,3,6,10].

Several processing techniques are available to guarantee the shelf-life extension as well as the quality and safety of plant food products [6,8,11]. The freezing is typically applied to preserve the fresh organoleptic characteristics and most of the nutrients [6,9,10,12]. Nevertheless, the freezing process promotes cell membrane and physical integrity damages that can release antioxidant compounds [10]. Although freezing slows the kinetics of some cellular events such as chemical and enzymatic reactions, these are not completely stopped. Accordingly, these reactions during freezing can induce bioactive compounds degradation and consequent decrease in total antioxidant activity [6,10]. Some authors have also suggested that beyond this negative effect, the frozen storage may favor the release of bound phenolic acids and anthocyanins, promoting a positive impact on functional properties [10].

In contrast, the dried fruits and vegetables have gained extensive popularity due to their transversal use in the food industry and long shelf-life resultant from their low water content [2–4,7]. Additionally, the dried fruits and vegetables can be easily produced, stored and transported with reduced packing costs [3,7]. Different drying methods can be employed to preserve vegetables, for instance, freeze, hot air, microwave, vacuum and infrared drying [2,3]. Freeze-drying is one of the most effective drying techniques in the preservation of nutritional and typical sensorial attributes, but it is an expensive and time-demand approach [2–4], rarely feasible for industry companies. Even if the temperature can lead to losses in bioactive compounds, the hot-air drying approach is extensively used as a cost-friendly and rapid approach to generate highly stable products with a long shelf-life [3].

In this study, pulps and powders of baby carrot and cherry tomato by-products were prepared and the impact of freezing, hot air drying and storage time on antioxidant capacity and bioactive compounds (polyphenols, carotenoids and Vitamin E profile) were investigated. In addition, other important parameters, such as pulps viscosity and microbiological quality, were also explored.

2. Materials and Methods

2.1. Chemicals

Sodium hypochlorite was supplied by Honeywell Riedel-de Haën AG, Seelze, Germany (11–15% available chlorine). The microbiological media used were plate count agar (PCA; Biokar), violet red bile glucose agar (VRBGA; VWR), rose bengal chloramphenicol agar (RBCA; Biokar), trypticase soy Agar (TSA; VWR) and *Bacillus Cereus* selective agar (BCSA; Biokar).

ABTS diammonium salt (2,2′-azino-bis(3-ethylbenzothiazoline-6-sulphonic acid)), 2,2-diphenyl-1-picrylhydrazyl, sodium carbonate, 2,2′-azo-bis-(2-methylpropionamidine)-dihydro chloride (≥97%; AAPH), fluorescein, 6-hydroxy-2,5,7,8-tetramethylbroman-2-carboxylic acid

($\geq$97%; Trolox) and butylated hydroxytoluene ($\geq$99%, BHT) were purchased from Sigma-Aldrich (Sintra, Portugal), while Folin-Ciocalteu from Merck (Algés, Portugal). Additionally, maltodextrin, absolute ethanol and ascorbic acid were also purchased from Sigma-Aldrich.

HPLC-grade methanol ($\geq$99.9%) was also supplied by Honeywell Riedel-de Haën AG, while the other solvents used for HPLC analysis (acetonitrile, dichloromethane, formic acid, hexane and 2-propanol) were also purchased from Sigma-Aldrich.

Standards of β- and α-carotene ($\geq$95%), α-tocopherol ($\geq$95%), β-tocopherol ($\geq$95%), γ-tocopherol ($\geq$95%), δ-tocopherol ($\geq$95%), gallic acid ($\geq$99%), rutin ($\geq$95%), ferulic acid ($\geq$98%), isoferulic acid ($\geq$98%), trans-ferulic acid ($\geq$99%), chlorogenic acid ($\geq$99%), p-coumaric acid ($\geq$98%) quercetin-7-O-glucoside ($\geq$98%), sinapic acid ($\geq$98%), myricitin ($\geq$99%) and caffeic acid ($\geq$98%) were supplied from Sigma-Aldrich, while protocatechuic acid ($\geq$99%), lutein ($\geq$95%), lycopene ($\geq$98%), ferulic ($\geq$98%), isoferulic ($\geq$98%), naringenin-7-glycoside ($\geq$99%), naringenin ($\geq$99%), lutein ($\geq$99%) and zeaxanthin ($\geq$98%) from Extrasynthese (Genay Cedex, France).

2.2. Processing

Cherry tomato (*Solanum lycopersicum* var. *cerasiforme*) and baby carrot (*Daucus carota* subsp. *sativus*), which did not comply with size and shape commercial standards, were kindly provided by Vitacress Portugal SA (Odemira, Portugal). The vegetables were washed and disinfected with sodium hypochlorite (150 ppm) for 15 min, in a proportion of water: vegetable of 5 L kg^{-1}. After disinfection, the by-products were rewashed with abundant water and centrifuged to remove washing water. The by-products were submitted to two different processing methods (freezing and drying) and, in both cases, they were stored for 6 months.

2.2.1. Freezing

Based on a circular economy approach, tomato pulps were prepared by grinding whole tomatoes (including seeds and peels) for 2 min, using a vertical slicer blender (Hällde VCB-61). Due to the lower water content, carrots (also including peels but excluding stem) were ground in the same conditions using a carrot: water ratio of 4:1. After processing, the tomato and carrot pulps were frozen at -20 °C, and the sampling times were fresh, M0, M1, M2, M3, M4, M5 and M6. The timepoint M0 corresponded to 24 h after the freezing storage, and M1, M2, M3, M4, M5 and M6 corresponded to month one, two, three, four, five and six of the freezing storage, respectively.

2.2.2. Drying

The carrot and tomato drying process was carried out using a hot air cabinet dryer, with a temperature of 50 °C and an air velocity of 1 m s^{-1}. By-products were cut into slices before drying. Tomatoes were manually cut in halves while carrots were laminated in a vegetable cutting machine (Hallde RG-100, Sweden), with a 5 mm thick disc. The dry by-products were also stored for 6 months under vacuum conditions. Due to the greater stability of dried vegetables, the sampling times were every two months. The sampling points were DM0, DM2, DM4 and DM6, where DM0 corresponded to day 0 of storage and DM2, DM4 and DM6 corresponded to months two, four and six of the storage period, respectively. Regarding microbiological analyses, only DM0 and DM6 sampling points were evaluated.

2.3. Microbiological Evaluation

2.3.1. Fresh and Frozen Pulps

For microbiological analysis, 1 g of carrots or tomatoes non-washed, washed, or in pulp form was added to 9 mL of sterile 0.1% peptone solution. The mixture was homogenized in a stomacher (Seward, Worthing, UK) for 1 min and then, the dilutions prepared were spread in the different microbiological media. Total aerobic bacteria (TAB) were enumerated after incubation for 72 h at 30 °C in PCA media according to the method

of ISO 4833-1, while *Enterobacteriaceae* were grown in VRBGA at 37 °C for 24 h. Yeasts and molds were incubated on supplemented media RBCAat 30 °C for 3 to 5 days. The *Bacillus cereus* group was grown and counted on mannitol egg polymyxin agar at 30 °C for 24 h. All analyses were performed in duplicate and the average values were expressed in log (CFU g^{-1}).

2.3.2. Dried Powders

Regarding microbiological analyses of dry by-products, for the enumeration of TAB, the test sample was prepared by adding 10 g of dry by-product sample to 90 mL of buffered peptone water in a stomacher bag. A tenfold dilution series was done in duplicate, 1 mL of the dilutions was spread-plated onto PCA and incubated as described previously. The enumeration of aerobic and anaerobic spore-formers was performed according to Ijabadeniyi et al. [13]. Firstly, the test sample was also prepared in the same way. Then, the stomacher bag was held at 75 °C for 20 min. A dilution series was done and after that 0.1 mL of the sample was pipetted into Petri dishes and pour-plated into TSA. Both aerobic and anaerobic plates were incubated at 35 °C for 48 h. The anaerobic plates were placed into anaerobic jars and incubated under anaerobic conditions. All analyses were performed in duplicate and the average values were expressed in log (CFU g^{-1}).

2.4. Monitorization of Pulps Viscosity

The pulps viscosity was measured in mPa s^{-1} using a rotational springless viscometer B-one plus (Lamy Rheology Instruments, Champagne au Mont d'Or, France), using an R-2 rotor disc (with a viscosity range between 200 and 240 M mPa s^{-1}) at a speed of 250 rpm for 30 s. All measurements were carried out five times for each sample at room temperature.

2.5. Preparation of Tomato and Carrot Extracts

Extracts for the evaluation of total phenolic compounds, polyphenolic profile and antioxidant activity were prepared by adding 5 g of each homogenized fraction (fresh, frozen pulp or dry by-product) to 50 mL of 80% methanol (methanol: water). Homogenization was carried out in an Ultra-turrax® (T18 IKA, Wilmington, USA) at 12,000 rpm for 30 s. The mixture was left under continuous stirring (300 rpm) at room temperature for 2 h. Afterward, each resulting sample was centrifuged at 5000 rpm at 4 °C for 10 min, and the supernatant was filtered. The methanol was then removed from the extracts using a rotary evaporator (R-210, Buchi, Switzerland) at 40 °C and 175 bar. The resultant fraction was resuspended in deionized water with 2% of maltodextrin and lyophilized for further analysis. All extracts were performed in duplicate.

2.6. Antioxidant Activity

2.6.1. 2,2′-azino-bis(3-ethylbenzothiazoline-6-sulphonic acid (ABTS) Assay

2,2′-azino-bis(3-ethylbenzothiazoline-6-sulphonic acid (ABTS) scavenging assay was carried out accordingly to the Gonçalves et al. established method [14], with slight modifications. The stock solution was prepared by the reaction between ABTS$^{\bullet+}$ (7 mM) with potassium persulfate (2.45 mM) in ultra-pure (UP) water, stirring for 16 h in the dark at room temperature. Daily, the ABTS$^{\bullet+}$ working solution was prepared by the filtration of stock solution with a 0.45 μm syringe filter, and the absorbance at 734 nm was adjusted to 0.70 ± 0.02 with UP water for the control sample (20 μL UP water + 180 μL ABTS$^{\bullet+}$ working solution). The reaction was performed in a 96-well microplate, in a final volume of 200 μL, where 20 and 180 μL of extract sample and ABTS$^{\bullet+}$ working solution were mixed, respectively. Moreover, a blank with UP water as well as a standard curve with Trolox were performed with the same solvent and working solution proportion, where standard concentrations ranged between 25 and 175 μM. After that, the mixture was allowed to react for 5 min in the dark, and the absorbance was immediately recorded at 734 nm, using a multidetection plate reader (Synergy H1, Vermont, USA) operated using the Gen5 Biotek software version 3.04. The scavenging activity was expressed as a % of reduction in

absorbance for the control. All extracts were analyzed in triplicate, and after the calculation of regression equations, the Trolox concentration was expressed in mg Trolox equivalent (TE) 100 g^{-1} of dry matter (DM).

2.6.2. Oxygen Radical Absorbance Capacity (ORAC) Assay

Oxygen radical absorbance capacity (ORAC) assay was carried out as described by Coscueta et al. [15], with few modifications. The reaction was performed in black polystyrene 96-well microplates (Nunc, Denmark), with 75 mM of phosphate buffer (pH 7.4) in a final volume of 200 µL. Fluorescein (with a final concentration in the well of 70 nM; 120 µL) and antioxidant extract (20 µL) were placed in the well and pre-incubated for 10 min at 37 °C. After this period, APPH solution was added (with a final concentration in the well of 12 mM; 60 µL). Then, the incubation occurred immediately during 80 min, also in a multidetection plate reader. The excitation and emission wavelengths were set to 485 and 538 nm, respectively. The fluorescence was read in intervals of 1 min, and after each time point, the microplate was shaken. A calibration standard curve (with concentrations ranging between 1 and 8 µM of Trolox) and a blank (fluorescein + AAPH), where phosphate buffer was added instead of the antioxidant, were also added. The curves of antioxidants were normalized to the curves of blank of the same assay, and the area under the fluorescence decay curve (AUC) was calculated, with the trapezoidal method. The final values of AUC were obtained by subtracting the AUC of the blank to all results, and regression equations of antioxidant concentration in the function of AUC were calculated. All extracts were analyzed in duplicate, and ORAC values were expressed in mg TE 100 g^{-1} DM.

2.6.3. 2-diphenyl-1-picrylhydrazyl (DPPH) Assay

The 2-diphenyl-1-picrylhydrazyl (DPPH) assay was performed using the method described by Schaich et al. [16], with some alterations. The stock solution (600 µM) was prepared by dissolving 24 mg of DPPH in 100 mL of methanol. The stock solution was stored in the dark at −20 °C until the day of use. Daily, the working solution (60 µM) was prepared by adjusting the absorbance with methanol to 0.600 ± 0.100 at 515 nm. The assay was also carried out in a 96-well microplate, in a final volume of 200 µL. The reaction mixture corresponds to 175 µL of DPPH working solution and 25 µL of each extract or trolox (standard curve ranging between 25 and 175 µM) or UP water (negative control). The mixture was incubated for 30 min at 25 °C in the dark. Then, the absorbance at 515 nm was also measured with a multidetection plate reader. The scavenging activity was expressed as % of reduction in absorbance concerning the control and the regression equations between net DPPH scavenging and Trolox concentration were calculated. The Trolox concentration was expressed in mg TE 100 g^{-1} of DM. All extracts were analyzed in triplicate.

2.7. Total Phenolic Content (TPC)

To quantify total phenolic content (TPC), the Folin–Ciocalteau colorimetric method was used as described by Ainsworth et al. [17], with slight variations. The reaction was performed in a 96-well microplate, with a final volume of 230 µL. Firstly, 30 µL of extract and 100 µL of Folin-Ciocalteu reagent (20% *v/v*) were mixed and then, 100 µL of sodium carbonate (7.4% *m/v*) were added and shaken. The mixture was incubated at room temperature in the dark and the absorbance was measured after 1 h at 750 nm, also in a multidetection plate reader. Gallic acid was used as a standard calibration curve (0.015–0.225 mg/mL), and the TPC was expressed as milligrams of gallic acid equivalent per 100 g of dry matter (mg GAE 100 g^{-1} DM).

2.8. Identification and Quantification of Phenolics by High-Performance Liquid Chromatography (HPLC)

The phenolic profile in the extracts from different sampling points under study was evaluated using a chromatographic system with Waters separation module (e2695), with UV–VIS detector (PDA 190–600 nm), as described by Campos et al. [18], with some modifi-

cations. The high-performance liquid chromatography (HPLC) separation was performed with a C18 reverse-phase column coupled to a pre-column (Symmetry® C18, Waters, Milford, MA, USA; 100 Å, 5 µm, 4.6 mm × 150 mm). Two mobile phases were used: phase A containing water (92.5%), methanol (5%) and formic acid (2.5%), while phase B consisting of methanol (92.5%), water (5%) and formic acid (2.5%). The injection volume was 50 µL, and HPLC run was carried out for 59 min in a continuous flow of 0.5 mL min^{-1}, under the following conditions: gradient elution started in 100% of mobile phase A and ended in 55% of mobile phase B, after 55 min; between 50 and 55 min, the mobile phase A returns to 100% and remains in that percentage for 4 min (up to 59 min). Detection was performed using a diode array detector (Waters, Milford, MA, USA) at wavelengths between 200 and 600 nm, measured at 2 nm intervals. The peaks were identified at wavelengths of 280 nm (catechins or procyanidins), 320 nm (phenolic acids) and 350 nm (flavonoids). The various compounds identified were analyzed by comparing the retention time and spectra with pure standards. All the samples replicates were injected in duplicate.

2.9. Monitorization of Carotenoid and Vitamin E Content

2.9.1. Carotenoids and Vitamin E Extraction

Total carotenoids and vitamin E isomers (α-tocopherol, β-tocopherol, γ-tocopherol and δ-tocopherol) were extracted according to the method described by Prates et al. and Oliveira et al. [19,20], with some modifications. Briefly, 0.1 g of each homogenized sample (fresh, frozen pulp or dry by-product), 3 mL of 100% cold ethanol and 0.026 g of ascorbic acid (9 g L^{-1}) were added to the screw teflon-lined cap tube and mixed. For saponification, 380 µL of KOH (5 mol L^{-1} prepared in a solution of 55% absolute ethanol and 45% UP water), freshly prepared each week, were added to the mixture. The mixture was immediately vortexed to avoid agglomeration and remained in a shaking water bath at 85 °C and 200 rpm for 10 min. The mixture was then cooled and maintained on ice during the extraction procedure. Firstly, 3 mL of NaCl (1 mol L^{-1}) were added, and the samples were gently mixed. Secondly, 3 mL of 25 µg mL^{-1} BHT solution in n-hexane was added. The mixture was vigorously vortexed and centrifuged at 5000 rpm for 15 min (at 4 °C),the resultant upper n-hexane layer was transferred to a new screw teflon-lined cap tube. Finally, the residue was re-extracted under the same conditions, and the second n-hexane layer was also recovered and merged with the first one. For each time point, three independent extractions were carried out.

2.9.2. Identification and Quantification of Carotenoids by HPLC

Carotenoid profile was identified and quantified by determination of the absorbance at 454 nm with a UV mini 1240 spectrophotometer (Shimadzu, Tokyo, Japan), according to the method described by Oliveira et al. [19], with slight alterations. The mobile phase used was acetonitrile, methanol, dichloromethane, hexane and ammonium acetate (55:22:11.5:11.5:0.02 *v/v/v/v/w*). HPLC run was carried out under isocratic conditions at 1 mL min^{-1} flow rate for 20 min at 30 °C. The injection volume was 50 µL. The different carotenoids identified at 454 nm, namely, lutein, lycopene, β-carotene and α-carotene, were identified by comparing the retention time with pure standards and quantified using a pure standard calibration curve and were expressed as mg 100 g^{-1} DM of by-product. All the samples' replicates were injected in duplicate.

2.9.3. Identification and Quantification of Vitamin E by HPLC

The α-tocopherol, β-tocopherol, γ-tocopherol and δ-tocopherol were monitored following the HPLC method described by Savlin et al. [21], with some modifications. A Beckman System Gold® coupled to a Waters™ 474 Scanning Fluorescence Detector (with an excitation and emission wavelengths of 290 and 320 nm, respectively) and a Diode Array Detector (DAD) 168 Detector (210 nm) with a Varian ProStar Model 410 AutoSampler were used, recurring to a normal-phase silica column (Kromasil 60-5-SIL, 250 mm, 4.6 mm ID, 5 µm particle size). The mobile phase used corresponded to 1% *v/v* isopropanol in

n-hexane. The running time was 20 min at 1 mL min^{-1} flow rate, and the injection volume was 20 µl. Calibration curves of pure standards were prepared and used to identify and quantify α-tocopherol, β-tocopherol, γ-tocopherol and δ-tocopherol in samples, and their concentration was expressed as mg 100 g^{-1} DM of carrot and tomato by-product. All the samples' replicates were injected in duplicate.

2.10. Statistical Analysis

Statistical analysis was carried out using the SPSS statistical package 27.0 via a one-way analysis of variance (ANOVA), at a degree of significance of $p < 0.05$, to the viscosity, antioxidant capacity, TPC, phenolics, carotenoids and vitamin E content, except for microbiological counts, where only two independent analyses were performed. Firstly, data were compared statistically using ANOVA to understand the significance and confirm a normal distribution of the data. Finally, post hoc multiple comparisons were performed using Turkey's test ($p < 0.05$).

3. Results and Discussion

Based on a circular economy approach, cherry tomato and baby carrot that do not comply with commercial standards were used as by-products to prepare value-added pulps and powders. The impact of two processing methods on several bioactive properties was evaluated. Regarding pulps, their storage was for 6 months in freezing form, and all analyses were performed monthly. By-products from the same batch were also subjected to hot air drying and stored under vacuum conditions, being analyzed every two months. The microbial counts of dried by-products were only monitored in the initial and final phases of storage.

3.1. Monitorization of Microbial Counts

The microbiological analysis of fresh by-products suggested a high microbial load in baby carrot and cherry tomato as present in Table 1. The initial microbial load was considerably high in both vegetables for all microbial groups studied, especially in carrots due to the greater soil microbial contamination [22]. Microbiological control plays an important role in the prevention of foodborne diseases. Therefore, some decontamination agents, for instance, sodium hypochlorite and peroxyacetic acid, are often applied during the washing process to guarantee the safety of fresh vegetables [22]. The initial microbial load of fresh tomato and carrot was greatly reduced after washing with water and disinfection with sodium hypochlorite solution (Table 1).

According to the Portuguese national microbiological guidelines for ready-to-eat foods, samples are classified as satisfactory, for example, when the microbial counts of TAB, *Enterobacteriaceae*, yeasts and *Bacillus cereus* spp. are lower than 6, 5, 5 and 3 log (CFU g^{-1}) [23]. Although the study did not focus on ready-to-eat products (in the future, after demonstrating their potential, these by-products can be used in food formulations or for cooking), before washing and disinfection, these products were above these limits and classified as unsatisfactory, and after this process, tomato and carrot were within satisfactory limits. This fact corroborates the importance and effectiveness of the wash and disinfection process adopted.

During the 6 months of frozen storage, the microbial load stabilized, and there were no substantial variations on microbial counts during the entire period as represented in Table 1. Generally, after 6 months of storage at room temperature of the dry by-products, the microbial counts in the tomato and carrot powders decreased slightly for the microbial groups studied, except for the aerobic and anaerobic spore-formers group where a slight increase of microbial counts was observed in the carrots powders. These results are in agreement with those observed for dried spices described by Ijabadeniyi et al. [13].

Table 1. Microbial counts ($n = 2$) present in cherry tomato and baby carrot by-products before and after the washing process, as well as during the 6 months of pulps freezing and drying storage. All microbial count results (mean ± standard deviation) were expressed in $\log(\text{CFU g}^{-1})$.

Vegetable	Sample	TAB	*Enterobacteriacea*	Yeasts and Molds	*Bacillus cereus* spp.	Aerobic Spore-Forms	Anaerobic Spore-Forms
Tomato	Non washed	8.14 ± 0.15	6.28 ± 0.14	6.16 ± 0.12	4.63 ± 0.14	-	-
	Washed	4.32 ± 0.14	3.82 ± 0.07	3.77 ± 0.07	2.57 ± 0.12	-	-
	Fresh pulp	4.54 ± 0.11	3.91 ± 0.08	3.89 ± 0.15	2.65 ± 0.17	-	-
	M0	4.55 ± 0.15	3.98 ± 0.07	3.91 ± 0.10	2.57 ± 0.14	-	-
	M1	4.52 ± 0.07	4.03 ± 0.06	3.85 ± 0.14	2.53 ± 0.14	-	-
	M2	4.52 ± 0.15	4.00 ± 0.01	3.92 ± 0.11	2.56 ± 0.04	-	-
	M3	4.56 ± 0.04	4.02 ± 0.11	3.87 ± 0.21	2.51 ± 0.09	-	-
	M4	4.68 ± 0.10	4.13 ± 0.05	3.92 ± 0.20	2.63 ± 0.11	-	-
	M5	4.61 ± 0.13	4.07 ± 0.07	3.89 ± 0.18	2.56 ± 0.09	-	-
	M6	4.39 ± 0.01	3.80 ± 0.38	3.87 ± 0.27	2.62 ± 0.06	-	-
	DM0	2.39 ± 0.12	2.74 ± 1.04	-	-	3.00 ± 0.00	3.00 ± 0.00
	DM6	2.15 ± 0.21	2.15 ± 0.21	-	-	2.00 ± 0.00	2.39 ± 0.12
Carrot	Non washed	9.18 ± 0.10	8.03 ± 0.10	5.96 ± 0.10	4.77 ± 0.08	-	-
	Washed	5.29 ± 0.10	4.63 ± 0.13	3.65 ± 0.16	2.90 ± 0.09	-	-
	Fresh pulp	5.36 ± 0.09	4.82 ± 0.10	3.66 ± 0.11	2.93 ± 0.05	-	-
	M0	5.34 ± 0.15	4.85 ± 0.13	3.69 ± 0.03	2.74 ± 0.15	-	-
	M1	5.27 ± 0.07	4.82 ± 0.10	3.81 ± 0.08	2.69 ± 0.17	-	-
	M2	5.26 ± 0.06	4.86 ± 0.07	3.74 ± 0.29	2.70 ± 0.05	-	-
	M3	5.23 ± 0.01	4.83 ± 0.09	3.73 ± 0.03	2.67 ± 0.09	-	-
	M4	5.31 ± 0.37	4.54 ± 0.21	3.76 ± 0.10	2.82 ± 0.16	-	-
	M5	5.28 ± 0.02	4.81 ± 0.21	3.72 ± 0.12	2.65 ± 0.16	-	-
	M6	5.22 ± 0.06	4.82 ± 0.24	3.73 ± 0.03	2.79 ± 0.06	-	-
	DM0	4.49 ± 0.88	4.22 ± 1.10	-	-	2.67 ± 0.95	2.36 ± 0.51
	DM6	4.61 ± 0.88	4.16 ± 1.00	-	-	2.85 ± 0.00	2.50 ± 0.28

M0 = 24 h after the freezing storage of pulps. M1, M2, M3, M4, M5, M6 = month one, two, three, four, five and six of the freezing storage of pulps, respectively. DM0 = day 0 of the powders' storage. DM6 = month six of the storage period of powders. TAB = total aerobic bacteria.

3.2. Monitorization of Pulps Viscosity

The viscosity was measured during the freezing period in pulps and, in both cases, a significant decrease in viscosity after freezing was visible as suggested in Table 2. The viscosity decreases from 486.98 ± 1.40 mPa s^{-1} in the fresh pulps to 349.28 ± 2.92 mPa s^{-1} in M6 of storage in baby carrot pulp. Although the decline in viscosity was more evident in the first months of freezing, a significant decrease in carrot pulp viscosity was registered throughout the storage period, except from M5 to M6 where there were no significant changes in viscosity values. Relating to cherry tomato, the variation in viscosity was also more evident in the first months of frozen storage, but the results suggest significant variations in viscosity values in all months analyzed. The viscosity values ranged between 195.88 ± 3.73 mPa s^{-1} in the fresh pulp to 100.34 ± 1.39 mPa s^{-1} in the final month of storage. The ice crystals produced during the frozen process may promote damage to the integrity of the cellular membranes. Consequently, the resultant loss of water from the intracellular compartments affects physio-chemical and physical properties such as viscosity [10,12]. Generally, lower textural properties are reported in frozen plant foods

compared to the corresponding fresh vegetables and fruits [10], aligned with the decline in tomato and carrot viscosity during the storage period verified (Table 2). As suggested by the work of Lisiewska et al. [24], the freezing storage at $-20\,^{\circ}\text{C}$ of tomato cubes resulted in a significant reduction of texture.

Table 2. Pulps' viscosity ($n = 5$) from fresh to freezing storage of baby carrot and cherry tomato pulps for 6 months were expressed in mPa s^{-1}. Columns corresponding to the same vegetable with different superscript letters are significantly different ($p < 0.05$).

Vegetable	Sample	Viscosity (mPa s^{-1})
Tomato	Fresh pulp	195.88 ± 3.73 [a]
	M0	153.36 ± 1.27 [b]
	M1	142.42 ± 1.27 [c]
	M2	133.98 ± 2.15 [d]
	M3	122.04 ± 0.97 [e]
	M4	110.98 ± 3.66 [f]
	M5	105.08 ± 0.74 [g]
	M6	100.34 ± 1.39 [h]
Carrot	Fresh pulp	486.98 ± 1.40 [a]
	M0	399.98 ± 4.35 [b]
	M1	390.58 ± 1.78 [c]
	M2	380.58 ± 3.43 [d]
	M3	368.94 ± 2.2 [e]
	M4	358.62 ± 2.41 [f]
	M5	351.28 ± 2.37 [g]
	M6	349.28 ± 2.92 [g]

M0 = 24 h after the freezing storage of pulps. M1, M2, M3, M4, M5, M6 = month one, two, three, four, five and six of the freezing storage of pulps, respectively.

3.3. Variation of TPC and Antioxidant Activity

TPC was evaluated during 6 months for both storage types (frozen and dried) as represented in Figure 1. The compounds extraction was performed through methanolic extractions. The TPC results of fresh baby carrot and cherry tomato pulps and after 24 h of freezing did not suggest significant variations in TPC. However, in both cases, a decrease was verified after one month of frozen storage, and a sequential reduction in TPC values was registered during the 6 months of storage. The TPC values in fresh carrot pulp ranged between 145.21 ± 2.74 and 50.09 ± 2.99 mg GAE 100 g^{-1} on a dry basis after 6 months of freezing, while in tomato fresh pulp the TPC were 408.89 ± 12.11 and 277.24 ± 11.29 mg GAE 100 g^{-1} DM after all period of frozen storage. In the case of tomato, after 4 months of freezing, the TPC values indicated a tendency to stabilize.

Bouzari et al. [25] also studied the TPC in fresh carrots and during 3 months of storage. The results suggested a TPC concentration of 108 mg GAE 100 g^{-1} DM in fresh carrots and a slight improvement in TPC after 24 h of freezing with values around 133 mg GAE 100 g^{-1} DM. During 3 months of freezing storage, the TPC concentration decreased to approximately 112 mg GAE 100 g^{-1} DM. These values are in the same concentration order and aligned with the results of the present study. Leja et al., work [26] also reported similar TPC values for fresh carrots, namely 29.3 mg 100 g^{-1} of fresh weight (based on the dry weight of carrot in the present study: 7.96%). TPC of red tomato through the preparation of acetone: water extracts was described by Georgé et al. [27], being around 268 mg GAE 100 g^{-1} DM in fresh tomato. Phenolic composition and antioxidant activity of some tomato varieties

and by-products (composed only by peel and seeds) were studied by Perea-Domínguez et al. [28], through ethanolic extracts. The TPC results of grape, saladette and tomato by-product were 904, 1108 and 1340 mg GAE 100 g^{-1} DM, respectively. The reported TPC results of Georgé et al., are slightly lower, while the TPC results of Perea-Domínguez were higher than the TPC values of the present study. These differences may be due to the different tomato species studied, or even due to the different extraction methodologies (different solvents, different times of residency and different proportions of sample and solvent) employed, leading to different extraction profiles of phenolic compounds and consequent antioxidant activity [18]. In the present study, the whole vegetable by-product was used for processing, including seeds and peels in tomatoes and including peels but excluding stem in carrots. This fact may also contribute to the variations found in the literature, but especially the conditions where they were produced. There are some conflicting results in the literature regarding the effect of the freezing process and frozen storage upon TPC and antioxidant activity [10,12]. As previously appointed, the cell breakages promoted during freezing can lead to the decompartmentalization of antioxidants. Their posterior interaction with oxidative enzymes can lead to a reduction in antioxidant concentrations compared with fresh forms. However, some authors' results have also indicated an improvement in functional properties during the frozen storage due to the possible release of bound phenolic acids and anthocyanins [10]. The conflicting reported results may result from the several pre-treatments used after the freezing and frozen storage, the frozen systems and conditions used as well as other intrinsic characteristics of vegetables and fruits [10,12].

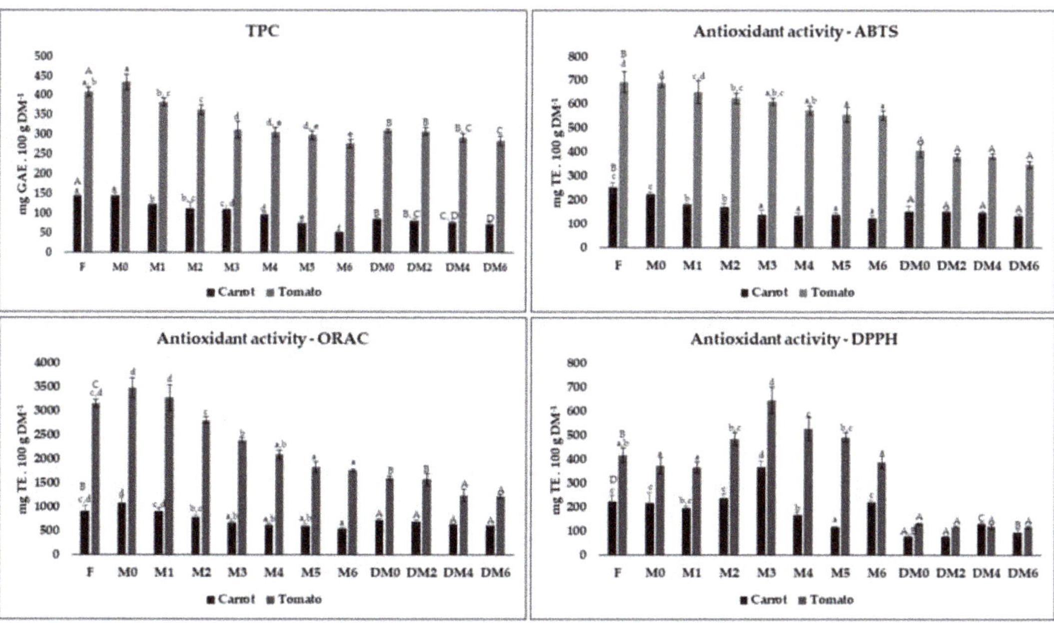

Figure 1. Total phenolic compounds (TPC; Folin–Ciocalteu method) and antioxidant capacity by ABTS, ORAC and DPPH assays (*n* = 4) of fresh carrot and tomato by-products and during the pulps freezing and drying storage. All antioxidant results (mean ± standard deviation) were expressed in mg Trolox equivalent (TE) 100 g^{-1} on a dry basis, while TPC results were represented in mg gallic acid equivalent (GAE) 100 g^{-1}. Bars of the same color and corresponding to the same storage method with different superscript letters are significantly different (*p* < 0.05). Lowercase letters were used for fresh and frozen pulps forms while uppercase letters were used for fresh and dry powders forms. M0 = 24 h after the freezing storage of pulps. M1, M2, M3, M4, M5, M6 = month one, two, three, four, five and six of the freezing storage of pulps, respectively. DM0 = day 0 of the powders storage. DM2, DM4, DM6 = month two, four and six of the storage period of powders.

Concerning drying, the TPC values varied between 85.49 ± 3.48 and 73.47 ± 4.28 mg GAE 100 g^{-1} in carrot and 310.33 ± 10.38 and 283.64 ± 11.84 mg GAE 100 g^{-1} DM in tomato, during the dried by-products storage. The results showed a significant variation in TPC after the drying process comparatively with fresh pulps, but there were no key variations during the several months of storage analyzed. Previous studies corroborate that drying decreases phenolic content in tomato [29] and carrot [7]. The grounding process promotes cell membrane damage, similar to what happens during the freezing process, making these compounds more unstable. This higher instability coupled with the use of temperature in the drying process can explain this reduction [30]. Hung et al. also evaluated the TPC present in carrot and tomato after drying processing [7]. In this study, carrot and tomato (without seeds) were dried in an oven at 55 °C and the results were compared with the freeze-drying method. Ethanolic extracts were prepared, and the results suggested a significant decrease in TPC of vegetables resultant from heat-drying. The TPC of carrot was approximate, 2.16 mg GAE 100 g^{-1}, while TPC of tomato was around 3.31 mg GAE 100 g^{-1} of samples. The results also suggested that, as expected, the TPC values of freeze-drying vegetables were higher than the vegetable resultant from heat-drying vegetables. However, the TPC values were lower than those of the present study but reported as fresh weight. In the case of our study, cherry tomatoes possess a dry weight of around 10.71%. Other justifications for TPC variations are the type of extraction used, the vegetable variety used and still, Hung et al., did not use air circulation during heat-drying, which may lead to losses of bioactive compounds [7]. Kamiloglu et al. [31] work reported the effects of tomato home drying at 70 °C for 36 h by preparing 75% methanolic extracts. The TPC values of fresh tomato and after the drying process were approximately 274 and 157 mg GAE 100 g^{-1} DM, respectively. In this study, a significant decrease was also suggested after the drying process, aligned with the present study results. Despite the results being in the same order of magnitude, they were slightly inferior to our study. Tomato variety and the use of a conventional oven could be conditioning factors.

The antioxidant capacity of methanolic extracts was evaluated by ABTS, ORAC and DPPH assays, being the results in fresh, during freezing and drying storage also present in Figure 1. ABTS and ORAC assays were used to monitor antioxidant capacity variations due to hydrophilic and amphipathic compounds, while the DPPH assay was used to evaluate the contribution of lipophilic compounds to the total antioxidant activity [16].

The ABTS results suggested a significant decrease in total antioxidant activity after the first month of freezing in both pulps, but no significant variations were verified in the case of carrot after M3 of freezing and M4 and M5 in the case of tomato. In tomato pulps the ABTS capacity values ranging between 694.07 ± 45.00 and 558.73 ± 29.06 mg TE 100 g^{-1} on a dry basis in fresh and M5 of freezing storage, respectively. Regarding carrot, the results of fresh pulps suggested an antioxidant capacity of 252.74 ± 21.83 mg TE 100 g^{-1} DM, decreasing until around 140.82 ± 16.76 mg TE 100 g^{-1} DM in M3 of storage. Danesi et al. [32] also studied the impact of freezing (−18 °C) and storage during 6 months of different vegetables, including tomato and carrot. In tomato, no significant variations were registered, while in carrot, a decrease in antioxidant capacity was verified. In this study, the results also suggested that tomato possess a higher antioxidant capacity than carrot and that the impact of freezing in antioxidant activity largely depends on the type of vegetable.

The ABTS antioxidant capacity of dry by-products significantly diminished compared with fresh vegetables. However, no significant differences in the antioxidant capacity were registered during the storage time, varying between 350.15 ± 14.37 and 407.56 ± 25.93 mg TE 100 g^{-1} DM in tomato and 134.69 ± 8.68 and 153.13 ± 25.89 mg TE 100 g^{-1} DM in carrot. As previously indicated, Kamiloglu et al. [31] work reported the effects of home drying at 70 °C for 36 h. The antioxidant capacity of fresh was 432 mg TE 100 g^{-1} DM, decreasing significantly until 46 mg TE 100 g^{-1} DM. The results of the antioxidant capacity of fresh tomatoes were similar to those of this study. However, the reduction in antioxidant activity was more pronounced in the reported study, possibly for the reasons previously

mentioned. Previous studies also suggested a significant decrease in antioxidant capacity using hot air drying using temperatures of 50 °C and 100 °C [33].

The results of the ORAC assay indicated an increase in antioxidant capacity in the first period of storage. In the case of tomato, a significant increase until M1 of storage was visible. On the other hand, in carrot a significant rise was registered after 24 h of freezing and in M1 no significant differences were registered compared with the antioxidant capacity of fresh carrot. The antioxidant capacity of fresh tomato and after one month of freezing was 3165.18 ± 77.48 and 3285.77 ± 271.25 mg TE 100 g^{-1} DM, respectively. Regarding carrot, the antioxidant capacity increase from 907.98 ± 121.69 in fresh pulp to 1091.90 ± 113.28 mg TE 100 g^{-1} DM after 24 h of freezing, being 914.06 ± 59.88 mg TE 100 g^{-1} DM in M1 of storage. After these periods, the ORAC results also indicated a decrease in antioxidant capacity during freezing storage, with a tendency to stabilize from M4 of storage in the case of tomato and M3 in the case of carrot. The antioxidant capacity of tomato and carrot after 6 months of frozen storage was 1771.66 ± 31.25 and 557.71 ± 11.19 mg TE 100 g^{-1} DM, respectively. During ORAC results analysis, a significant decline in the antioxidant capacity was shown compared with fresh forms in both vegetables. However, in the case of carrot no significant variations were verified during all storage period, ranging the antioxidant capacity between 613.53 ± 5.81 to 722.93 ± 28.29 mg TE 100 g^{-1} DM. Concerning tomato results, key modifications were not detected during the first 2 months (1581.76 ± 124.90 and 1610.74 ± 46.51 mg TE 100 g^{-1} DM), but after this period, a significant decrease was registered after 4 months but then stabilized until M6 of storage (1229.74 ± 38.52 mg TE 100 g^{-1} DM).

In the evaluation of antioxidant capacity through the DPPH method, the antioxidant capacity of tomato and carrot improved during the first months of freezing, but then it reduced, being registered significant variations during the 6 months (Figure 1). The antioxidant capacity of fresh tomato was 418.79 ± 30.92 mg TE 100 g^{-1} DM, increasing until 648.06 ± 55.38 mg TE 100 g^{-1} DM in M3 and decreasing until the last month of freezing (388.53 ± 27.18 mg TE 100 g^{-1} DM). In carrot, the antioxidant capacity increased from fresh until M3 of freezing (from 223.21 ± 26.76 to 367.26 ± 28.92 mg TE 100 g^{-1} DM). Subsequently, the antioxidant capacity was reduced until 115.69 ± 4.49 mg TE 100 g^{-1} DM in M5 and improved until 219.43 ± 11.49 mg TE 100 g^{-1} DM in M6 of freezing, with significant variations. This could be explained by the loss of cellular membrane integrity promoted by freezing, which promotes bioactive compounds release (especially lipophilic) and increases antioxidant activity. However, after this phase, chemical and enzymatic reactions induce bioactive compounds degradation and antioxidant activity decrease [10].

As in the ABTS and ORAC methods, a significant decrease in antioxidant capacity was verified in dry tomato and carrot compared with fresh from the DPPH analysis. Although the results did not suggest variations during tomato storage (varying antioxidant capacity between 117.78 ± 4.99 to 130.44 ± 3.51 mg TE 100 g^{-1} DM), some alterations were registered in carrot storage from M4. The antioxidant capacity of dry carrot was 76.28 mg TE 100 g^{-1} DM in the first two months of storage, improving until 131.41 ± 5.22 mg TE 100 g^{-1} DM in M4 and decreasing until 96.87 ± 9.83 mg TE 100 g^{-1} DM in M6.

Generally, the reported studies suggested a decrease in TPC and antioxidant capacity by ABTS and ORAC assays during freezing and a significant negative impact in these properties and in the DPPH antioxidant capacity, aligned with the present study results. The variation between the different antioxidant assays tested results from the different reaction mechanisms underlying. In ABTS and ORAC assay differences, the radical possesses a higher molecular weight than the molecule of transfer mechanisms of hydrogen atom used in ORAC assay. Consequently, a steric block of the active centers of ABTS can occur, decreasing their reaction rate. The small molecules of atom transfer of ORAC assay, make this method considered more accurate as suggested by Campos et al. [18]. In contrast, some researchers also reported that ORAC assay is as a more relevant technique due to the use of a biological radical source. However, ORAC assay is a more time-demanding and expensive

approach [34]. In the case of DPPH, organic solvents are required, being used mainly to evaluate the contribution of lipophilic compounds to the total antioxidant activity [16].

3.4. Phenolics Composition

Quantitative and qualitative determination of polyphenolic compounds present in cherry tomato and baby carrot by-products was performed by HPLC. The qualitative polyphenolic profile of fresh tomato and during frozen and dry storage are depicted in Figure 2, while their quantitative profile in Table 3, respectively. The quantitative analysis of tomato suggested the main presence of compounds belonging to the hydroxycinnamic acid family such as chlorogenic acid, sinapic acid and some derivatives of this group. Moreover, some flavonols, for instance, rutin, myricetin and quercetin derivative as well as the naringenin belonging to the flavanones groups were detected (Figure 2).

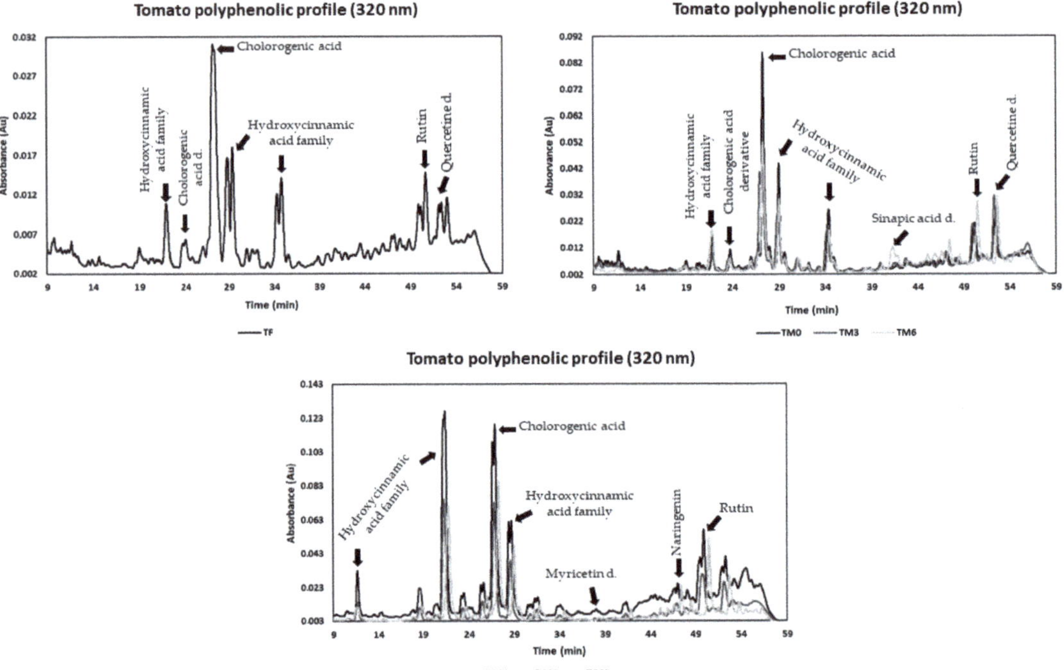

Figure 2. Chromatograms of identified free phenolic compounds of (**a**). fresh tomato by-products and during (**b**). the freezing pulps and (**c**). drying storage of powders at a wavelength of 320 nm. TF = fresh tomato. TM0 = 24 h after the freezing storage of tomato pulp. TM3 and TM6 = month three and six of the freezing storage of tomato pulps, respectively. DM0 = day 0 of the tomato powders storage. DM2 and DM6 = month two and six of the storage period of tomato powders.

Chlorogenic acid was detected and quantified in fresh, frozen and dry tomato, with concentrations varying from 15.9 ± 0.1 to 32.4 ± 0.7 mg 100 g^{-1} DM during frozen storage and from 34.2 ± 0.2 to 38.0 ± 0.3 mg 100 g^{-1} DM. During freezing storage, a significant increase was registered after 24 h of freezing but, then, a significant decrease was verified in M1. After this period, no significant variations were found during storage. Concerning the drying approach, a significant improvement in chlorogenic acid concentration was verified but no variations during the storage period were registered. The rutin concentration varied from 8.0 ± 2.2 to 13.2 ± 1.3 mg 100 g^{-1} DM during the freezing storage. In this case, a significant increase until the final stage of freezing was registered. During drying storage, also a substantial improvement in rutin concentration with no variations during the time was detected. Sinapic acid, quercetin and naringenin derivatives were only detected on

fresh tomato and during freezing storage, while naringenin at the dried tomato. Naringenin was the most incident phenolic compound in dry tomato, with concentrations ranging from 1132.2 ± 16.5 and 1311.5 ± 29.3 mg 100 g^{-1} DM. Regarding fresh and frozen tomato, a naringenin derivative was the most prevalent compound, with concentrations varied between 314.5 ± 93.8 and 556.0 ± 7.4 mg 100 g^{-1} DM.

Table 3. Identification and quantification of phenolic compounds (n = 4) present in fresh tomato by-products and during Table 280. 320 and 350 nm. All results (mean $\pm$ standard deviation) were expressed in mg 100 g^{-1} on a dry basis.

Tomato Sample	Chlorogenic Acid	Sinapic Acid Derivative	Rutin	Quercetin Derivative	Naringenin
Fresh	27.2 ± 2.6 [b,A]	0.46 ± 0.28 [a]	11.4 ± 0.8 [b,c,d,A]	16.9 ± 0.3 [c]	500.0 ± 27.4 [a,b,*]
M0	32.4 ± 0.7 [c]	0.67 ± 0.09 [a]	11.2 ± 0.4 [b,c,d]	27.5 ± 2.0 [d]	501.3 ± 10.7 [a,b,*]
M1	18.4 ± 0.9 [a]	0.64 ± 0.28 [a]	8.4 ± 2.0 [a,b]	13.0 ± 0.9 [a,b]	445.3 ± 5.9 [b,c,*]
M2	16.6 ± 0.5 [a]	0.52 ± 0.16 [a]	9.9 ± 1.4 [a,b,c]	10.8 ± 0.7 [a]	367.1 ± 48.1 [c,e,*]
M3	19.5 ± 2.7 [a]	0.56 ± 0.30 [a]	8.0 ± 2.2 [a]	13.3 ± 1.0 [b]	367.6 ± 52.4 [c,e,*]
M4	17.2 ± 2.4 [a]	0.40 ± 0.12 [a]	8.6 ± 0.3 [a,b]	13.1 ± 0.8 [b]	314.5 ± 93.8 [e,*]
M5	17.2 ± 0.3 [a]	1.50 ± 0.02 [b]	12.5 ± 0.3 [c,d]	17.3 ± 0.4 [c]	556.0 ± 7.4 [a,*]
M6	15.9 ± 0.1 [a]	1.67 ± 0.09 [b]	13.2 ± 1.3 [d]	16.4 ± 0.4 [c]	514.3 ± 3.1 [a,b,*]
DM0	35.6 ± 3.5 [B]	N.D.	30.5 ± 3.1 [B]	N.D.	1203.5 ± 142.3 [A,B]
DM2	34.2 ± 0.2 [B]	N.D.	32.0 ± 1.0 [B]	N.D.	1311.5 ± 29.3 [B]
DM4	35.4 ± 0.1 [B]	N.D.	30.4 ± 0.2 [B]	N.D.	1132.2 ± 16.5 [A]
DM6	38.0 ± 0.3 [B]	N.D.	32.0 ± 0.2 [B]	N.D.	1224.8 ± 22.7 [A,B]

Columns with different superscript letters and corresponding to the same processing and storage approach are significantly different ($p < 0.05$). Lowercase letters were used for fresh and frozen pulps forms while uppercase letters were used for fresh and dry powders forms. N.D. = Not detected. * = derivative. M0 = 24 h after the freezing storage of pulps. M1, M2, M3, M4, M5, M6 = month one, two, three, four, five and six of the freezing storage of pulps, respectively. DM0 = day 0 of the powders storage. DM2, DM4, DM6 = month two, four and six of the storage period of powders.

The polyphenolic profile of tomatoes is consistent with the data found in the literature. Macheix et al. [35] reported the presence of some hydrocinnamic acid family derivatives, while chlorogenic acid, caffeic acid derivative compounds, rutin and naringenin chalcone were reported by Bénard et al. [36] in tomato. Moreover, Georgé et al. reported the presence of some phenolic compounds of the hydrocinnamic acid family such as chlorogenic, caffeic acid and its derivatives as well as rutin and naringenin in red tomato [27]. Regarding Kamiloglu et al., work [31], rutin and chlorogenic acid were also found in fresh and dried tomato and, naringenin was only found in dried tomato. This previous work suggested that the presence of naringenin only in dried tomato may result from the conversion of naringenin chalcone (present only in fresh tomato) into naringenin after hot-air drying at 70 °C for 36 h. This may be the reason why naringenin was only identified in dried tomatoes, while a naringenin derivative in fresh and frozen tomato. In Kamiloglu et al. work [31], the chlorogenic acid concentration was approximately 24 and 15 mg 100 g^{-1} DM in fresh and dried tomato while rutin concentration varied from 12 in fresh tomato to 9 mg 100 g^{-1} DM in dried tomato. These values are aligned with the results of the present study. As previously mentioned, the variations (the concentration of naringenin found in the dry product was also significantly lower) resulted mainly from the drying approach (conventional oven) and also HPLC conditions used.

The baby carrot qualitative polyphenolic profile is illustrated in Figure 3. The main polyphenolic compounds detected belonged to the hydroxycinnamic acid family, for instance, chlorogenic acid, transferulic, isoferulic and trans-ferulic acids, caffeic acid and (4-)p-coumaric acid. Moreover, myricetin from flavonols group was identified while pro-

tocathecuic acid of hydroxybenzoic acid family was found in M1 of frozen storage, as indicated in Table 4, where quantitative polyphenolic concentration was found.

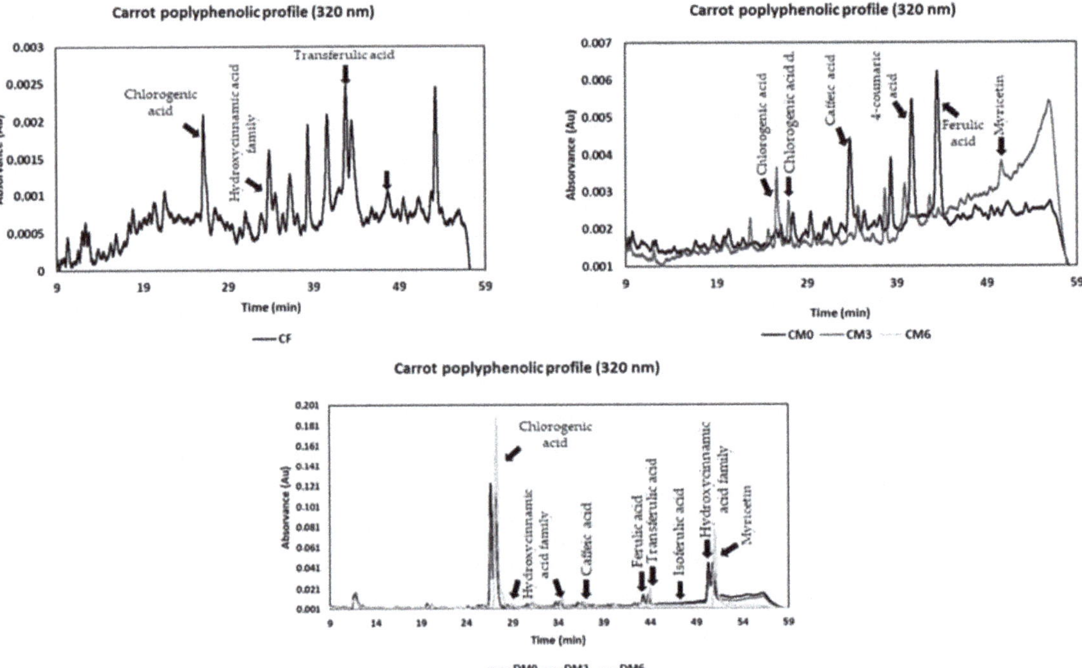

Figure 3. Chromatograms of identified free phenolic compounds of (**a**). fresh carrot by-products and during (**b**). the pulps freezing and (**c**). drying storage of powders at a wavelength of 320 nm. CF = fresh carrot. CM0 = 24 h after the freezing storage of carrot pulp. CM3 and CM6 = month three and six of the freezing storage of carrot pulps, respectively. DM0 = day 0 of the carrot powders storage. DM2 and DM6 = month two and six of the storage period of carrot powders.

The chlorogenic acid was the main phenolic compound detected in fresh, frozen and dry baby carrot by-products. No significant variation in concentration of this hydroxycinnamic acid was verified until M1 of storage (with a concentration of 2.48 ± 0.07 mg 100 g^{-1} DM in fresh carrot). After M1, the results suggested a significant improvement in its concentration, apparently stabilizing at M5 with a concentration value of 62.13 ± 0.41 mg 100 g^{-1} DM. Trans-ferulic acid was only detected in fresh carrot and M0 and M1 (but below the quantification limit). Protocatechuic acid was only quantified in M1 of freezing storage but detected during the total freezing period analyzed, but below the detection limit. The caffeic acid concentration was only detected since M2, with a concentration of 0.25 ± 0.02 mg 100 g^{-1} DM in M5 and 0.41 ± 0.03 and 0.18 ± 0.01 in DM0 and DM2, respectively. These results may suggest that caffeic acid was positively affected for both processing methods although despite being present in residual quantities.

The literature also aligned with the polyphenolic profile detected in the present study. Keser et al., work [2] reported the main presence of phenolic compounds of hydroxycinnamic acid in carrot powders. The chlorogenic acid and a caffeic acid derivative were the most incidents phenolic compounds, corroborating the results of this study. Moreover, other hydroxycinnamic acid family compounds were detected. Previous works also suggested that chlorogenic acid was the most prevalent phenolic compound [37,38], which is formed by the esterification of hydrocinnamic acids, such as caffeic, ferulic, and p-coumaric acids, with (-)-quinic acid. Zhang et al. also described chlorogenic acid as the key hydrox-

ycinnamic acid present in different carrot tissues, representing between 42.2% to 61.8% of phenolic compounds [39], which also supported the present study results.

Table 4. Identification and quantification of phenolic compounds (n = 4) present in fresh carrot by-products and during the freezing and drying storage by HPLC analysis at wavelengths of 280, 320 and 350 nm. All results (mean $\pm$ standard deviation) were expressed in mg 100 g^{-1} on a dry basis.

Carrot Sample	Protocatechuic Acid	Chlorogenic Acid	Caffeic Acid	Trans-Ferulic Acid
Fresh	B.Q.L.	2.48 $\pm$ 0.07 [a,A]	N.D.	3.02 $\pm$ 0.03
M0	B.Q.L.	2.85 $\pm$ 0.23 [a]	N.D.	B.Q.L.
M1	14.28 $\pm$ 1.67	5.36 $\pm$ 2.72 [a]	N.D.	B.Q.L.
M2	B.Q.L.	16.58 $\pm$ 0.43 [b]	B.Q.L.	N.D.
M3	B.Q.L.	39.32 $\pm$ 2.51 [c]	B.Q.L.	N.D.
M4	B.Q.L.	53.19 $\pm$ 1.88 [d]	B.Q.L.	N.D.
M5	B.Q.L.	62.13 $\pm$ 0.41 [e]	0.25 $\pm$ 0.02	N.D.
M6	B.Q.L.	61.51 $\pm$ 0.42 [e]	B.Q.L.	N.D.
DM0	N.D.	56.22 $\pm$ 3.15 [D]	0.41 $\pm$ 0.03	N.D.
DM2	N.D.	40.04 $\pm$ 0.31 [C]	0.18 $\pm$ 0.01	N.D.
DM4	N.D.	37.76 $\pm$ 0.08 [B,C]	B.Q.L.	N.D.
DM6	N.D.	36.18 $\pm$ 0.02 [B]	B.Q.L.	N.D.

Columns with different superscript letters and corresponding to the same processing and storage approach are significantly different (p < 0.05). Lowercase letters were used for fresh and frozen pulps forms while uppercase letters were used for fresh and dry powders forms. B.Q.L. = below the quantification limit. N.D. = not detected. M0 = 24 h after the freezing storage of pulps. M1, M2, M3, M4, M5, M6 = month one, two, three, four, five and six of the freezing storage of pulps, respectively. DM0 = day 0 of the powders' storage. DM2, DM4, DM6 = month two, four and six of the storage period of powders.

The phenolic acids identified in cherry tomato and/or baby carrot by-products as well as their retention time and maximum absorbance wavelength (nm) are present in Table 5. As suggested by TPC values, the HPLC quantifications suggested a higher presence of phenolic compounds in fresh tomato and from both processing methodologies adopted. The differences between HPLC quantifications and the results of TPC by Folin-Ciocalteu assay may result from the detection limit of phenolic compounds. Moreover, in TPC assay, Folin-Ciocalteu reagent can react with some interferent molecules present in extracts such as, proteins and sugars that can result in overestimation of TPC [40].

3.5. Carotenoids and Vitamine E Content

The carotenoid profile at a wavelength of 454 nm was monitored for all samples previously described and is represented in Figure 4. It is well known that tomato's red color is mainly related to its content in lycopene and β-carotene [41]. In accordance, in this work, these carotenoids were the most prominent in both fresh and processed tomato (both tomato pulp and powder). Lycopene was the most predominant carotenoid in tomato, with concentrations between 1250.29 $\pm$ 306.39 (M6) and 2393.11 $\pm$ 112.68 (M2) mg 100 g^{-1} DM in frozen pulps and between 161.45 $\pm$ 3.85 and 187.16 $\pm$ 37.91 mg 100 g^{-1} DM in dry tomato. These values are in accordance with previous studies, where lycopene content ranged between 8 and 18 mg 100 g^{-1} fresh weight [42]. In another study where tomatoes were submitted to high temperatures, lycopene concentration ranged between 70 and 120 mg 100 g^{-1} DM [43]. Similarly in this work, the powders which suffer temperature treatment had lower carotenoids content than the pulps. The higher values comparing to the Hidalgo et al. work [43] could be justified by the lower temperatures that were used in the present work (50 °C) comparing to the temperatures used in that work (around 100 °C).

Table 5. Phenolic acids identified in cherry tomato and/or baby carrot by-products.

Phenolic Acids	Chemical Formula	Retention Time (min)	λ Max (nm)
Hydroxybenzoic acid			
Protocatechuic acid	$C_7H_6O_4$	17.4	294
Hydroxycinnamic acid			
Chlorogenic acid	$C_{16}H_{18}O_9$	27.2	326.0
Caffeic acid	$C_9H_8O_4$	31.0	323.6
p-coumaric acid	$C_9H_8O_3$	39.9	309.3
Ferulic acid	$C_{10}H_{10}O_4$	42.4	322.4
Sinapic acid	$C_{11}H_{12}O_5$	42.7	322.0
Trans-ferulic acid	$C_{10}H_{10}O_4$	42.6	322.4
Iso-ferulic acid	$C_{10}H_{10}O_4$	45.3	322.4
Flavonols			
Myricetin	$C_{15}H_{10}O_8$	38.9	375.3
Rutin	$C_{27}H_{30}O_{16}$	49.8	354.7
Quercetin	$C_{15}H_{10}O_7$	49.9	343.9
Flavanones			
Naringenin	$C_{15}H_{12}O_5$	48.8	283.0

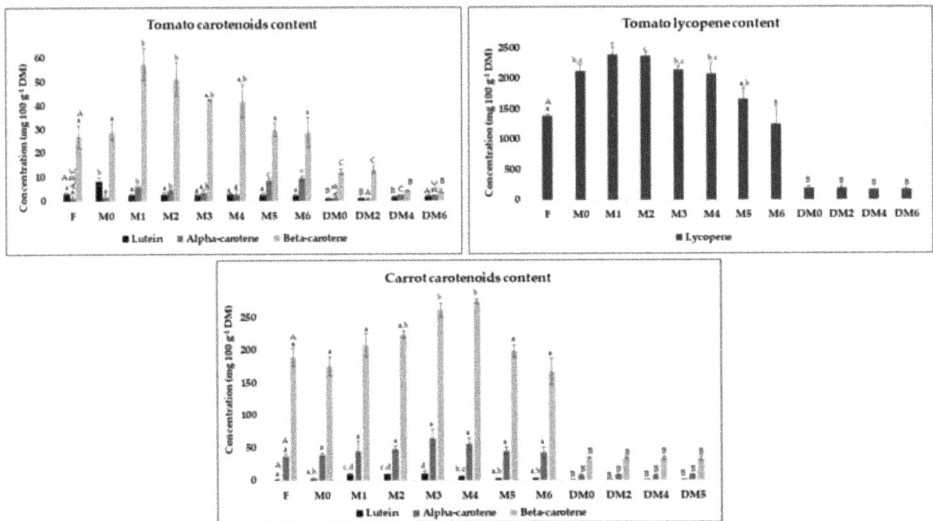

Figure 4. The concentration of carotenoids ($n = 3$) detected in fresh carrot and tomato by-products and during the freezing and drying storage by HPLC-DAD at a wavelength of 454 nm. All results (mean ± standard deviation) were expressed in mg 100 g^{-1} on a dry basis. Bars of the same color and corresponding to the same storage method with different superscript letters are significantly different ($p < 0.05$). Lowercase letters were used for fresh and frozen pulps forms while uppercase letters were used for fresh and dry powders forms. M0 = 24 h after the freezing storage of pulps. M1, M2, M3, M4, M5, M6 = month one, two, three, four, five and six of the freezing storage of pulps, respectively; DM0 = day 0 of the powders storage. DM2, DM4, DM6 = month two, four and six of the storage period of powders.

During freezing, a significant improvement in carotenoids content was verified in the first months of freezing. As previously mentioned, this could be explained by the cells membranes damage that freezing induces upon thawing depending on the size and location of the ice crystals [10]. After this period, the concentration of lutein and β-carotene carotenoids tended to decrease while α-carotene tended to rise, showing the higher concentration in M6 of freezing storage. Lutein, α- and β-carotene concentrations vary between 2.21 ± 0.57 (M6) to 8.10 ± 1.79 (24 h after freezing), 1.16 ± 1.03 (fresh) to 9.47 ± 1.11

(M6) and 27.09 ± 5.07 (fresh) to 57.49 ± 6.76 (M1), respectively. The values of carotenoids content in fresh tomato are aligned with previous studies and comparing to the Saini et al., work [44]; these values are in conformity to a higher stage of ripening as expected since the samples in this work were ripe red tomatoes. The decrease in carotenoids content after the first month of storage is probably related to chemical and enzymatic reactions that cause carotenoids degradation after their release from the membranes due to the ice crystals [10].

In contrast, there were no key variations in lycopene profile during dry storage, but a drastic decrease compared with fresh tomato (from 1502.12 to 187.16 mg 100 g^{-1} DM). This reduction comparatively to fresh by-product was transversal to the other carotenoids identified and quantified for dry tomato. Previous studies corroborate these results, demonstrating that drying decreases carotenoids content in tomato [29]. Moreover, before drying, vegetables were grounded and fresh-cut processed, which is known to promote cell membrane rupture. Bioactive compounds are exposed to the oxidative enzyme systems and are more vulnerable to temperature during processing [30]. Concerning lutein, α- and β-carotene, some alterations were verified during storage. In the case of lutein and α-carotene, there was a significant increase during storage from 0.86 ± 0.43 (DM0) to 1.76 ± 0.46 (DM6) mg 100 g^{-1} DM and 0.90 ± 0.15 (DM0) to 2.33 ± 0.11 mg 100 g^{-1} DM (DM4), respectively. Nevertheless, despite significant differences between the months analyzed, the concentration increase was not meaningful, which can be seen by the value of α-carotene between at DM6, being statistically not different from the previous months of storage. Otherwise, the β-carotene concentration decreased during dried storage, from 12.06 ± 1.10 (DM0) to 3.94 ± 0.79 mg 100 g^{-1} DM (DM6). The conjugated systems of double bonds of the carotenoids justify their vulnerability to degradation during storage by isomerization and degradation [30].

The most incident carotenoid in carrot was β-carotene in all months and storage conditions, followed by α-carotene and lutein, and lycopene was not detected which is in line with previous works on carrot [30]. The results suggested that β-carotene concentration improved from 189.89 ± 13.78 in the fresh form to 275.18 ± 3.49 mg 100 g^{-1} DM in M4 of frozen storage and after this phase, decreased until 167.01 ± 19.77 mg 100 g^{-1} DM in M6 of storage. No significant variations were registered in β-carotene concentration until the M2 of freezing and after M5 of storage. Regarding α-carotene, there was a key increase of its concentration of about half until the concentration of 64.54 ± 13.25 mg 100 g^{-1} DM (M3), reducing posteriorly until 42.06 ± 8.73 mg 100 g^{-1} DM in M6. Lutein was also detected varying between 1.87 ± 0.34 (fresh) to 10.61 ± 2.99 mg 100 g^{-1} DM (M3) and decreasing until 2.58 ± 0.35 mg 100 g^{-1} DM in M6 of storage. As previously discussed for tomato, the increased carotenoid content after the first months of freezing storage could be related to the liberation of bound compounds due to cell membrane rupture. In carrot, this liberation was slower, which could be explained by the stronger structure of carrot comparing to tomato due to higher insoluble fiber content [45]. Carotenoids are located in the chromoplasts which have a double layer membrane located inside the plant cells; thus, carotenoids are released slowly because of the several physical barriers that have to be broken for their release [30]. Previous works also demonstrated that blanching and freezing carrot slices promoted the increase in carotenoids content [46]. Similarly to tomato powders, there was a substantial loss of carotenoids concentration after drying, which is also explained by the combination of the two processing conditions during production of the vegetable powders that affect carotenoids content: (a) fresh-cut processing that liberates carotenoids from the cell wall structures turning them susceptible to oxidative enzyme systems and (b) the drying processing where the temperature promotes degradation of the free forms of these compounds that were liberated during the fresh-cut processing [30]. Previous works also demonstrated that carrot powders have lower β-carotene content than fresh carrots [41]. However, there were no significant changes during storage. The lutein, α- and β-carotene showed concentrations values of 0.66, 7.51 and 34.01, respectively.

Tocopherols (isomers of vitamin E) are lipid-soluble compounds synthesized in plastids that have antioxidant activity and are essential in the human diet [47]. α-tocopherol,

β-tocopherol, γ-tocopherol and δ-tocopherol concentrations were analyzed during the entire storage monitorization (Figure 5). The four vitamins E isomers were detected in tomato while in carrot only α-tocopherol, β-tocopherol and γ-tocopherol were identified and quantified. In both cases, a significant increase after the first month of freezing storage was verified in all isomers detected. Similar to carotenoids results, the ice crystals that are created upon freezing may induce the plastids membrane rupture thus liberating these compounds, which explains the increase in tocopherols content after freezing [30].

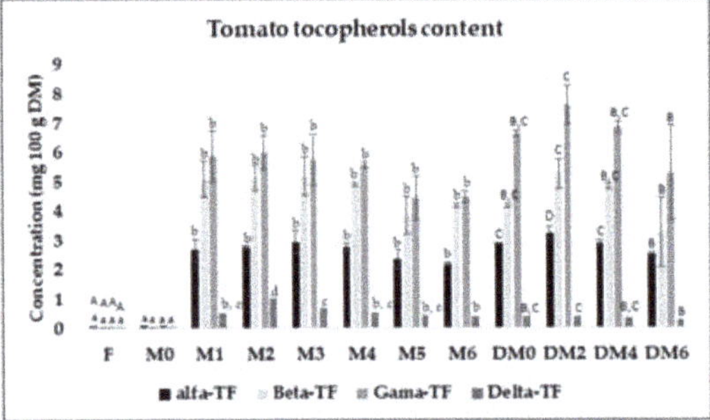

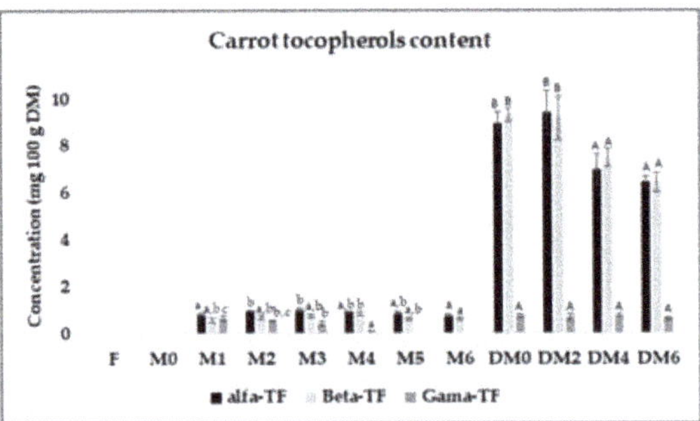

Figure 5. The concentration of α-tocopherol, β-tocopherol, γ-tocopherol and δ-tocopherol (n = 3) detected in fresh carrot and tomato by-products and during the freezing and drying storage by HPLC-DAD. All results (mean ± standard deviation) were expressed in mg 100 g^{-1} on a dry basis. Bars of the same color and corresponding to the same method of storage with different superscript letters are significantly different ($p < 0.05$). Lowercase letters were used for fresh and frozen pulps forms while uppercase letters were used for fresh and dry powders forms. M0 = 24 h after the freezing storage of pulps. M1, M2, M3, M4, M5, M6 = month one, two, three, four, five and six of the freezing storage of pulps, respectively. DM0 = day 0 of the powders' storage. DM2, DM4, DM6 = month two, four and six of the storage period of powders.

β- and γ-tocopherols were the most incident isomers in tomato. In previous works the α-tocopherol was the most incident tocopherol in tomato followed by γ-tocopherol (the other tocopherols were not analyzed) [43], these could be related to different ripening stages of samples [44], the different varieties used and influence of the peel in the sampling as in this work peels were used contrary to the study in comparison [43].

In carrot, α-, β- and γ-tocopherols were found, which is in agreement with previous studies, where also there 3 tocopherols isomers were reported and the values are similar to those found in the pulps of baby carrots [48]. In general, the content of α-tocopherol was similar to the values reported before in literature [30]. The work by Attila et al. [48] showed that different water supply (rain-fed or irrigated) may significantly affect the tocopherols content and profile. Despite that α-tocopherol was always the most incident tocopherol, the second and third tocopherol may be β- or γ-tocopherol depending on the year of production and type of water supply.

For both vegetables, freezing and powders production increased tocopherols content, especially for tomato and carrot powder. The results also indicated that there were no significant alterations in the tocopherols concentration of tomato during all freezing period, which shows the stability of these compounds. In tomato, α-tocopherol concentration was 2.78 mg 100 g^{-1} DM, β-tocopherol concentration 5.19 mg 100 g^{-1} DM and γ-tocopherol concentration 6.00 mg 100 g^{-1} DM. In the case of δ-tocopherol concentration, some variations were verified during the frozen phase analyzed, with a significant increase until 1.03 ± 0.07 mg 100 g^{-1} DM in M2 and a decrease until 0.43 ± 0.03 mg 100 g^{-1} DM in the last month analyzed. In carrot, some slight variations were registered in α-tocopherol, β-tocopherol and γ-tocopherol concentrations during the frozen period. α-tocopherol concentration ranging between 0.77 ± 0.08 and 1.04 ± 0.07 mg 100 g^{-1} DM, β-tocopherol between 0.59 ± 0.10 and 0.87 ± 0.08 mg 100 g^{-1} DM and γ-tocopherol between 0.07 ± 0.03

and 0.68 ± 0.11 mg 100 g^{-1} DM. In general, for both vegetables, there was stability on tocopherols content of pulps during freezing storage.

The results of tocopherols concentrations also showed some variations during dry tomato storage in all isomers but in all cases with a positive impact relative to fresh by-product. This could be related to the release of these compounds during milling. Previous studies have demonstrated that granulometry affects antioxidant activity due to the increased availability of antioxidant compounds [49]. Generally, the tocopherol profile was also slightly improved comparatively with freezing storage. α-tocopherol, β-tocopherol, γ-tocopherol and δ-tocopherol showed the higher concentrations in M2 of dry storage (3.28 ± 0.23, 5.27 ± 0.49, 7.62 ± 0.67 and 0.43 ± 0.03 mg 100 g^{-1} DM, respectively).

In carrot, the tocopherol profile was substantially improved in the powders. The higher fiber content of carrot compared to tomato may explain this higher liberation of tocopherols during powders production comparing to tomato, as these compounds are released from fiber during this processing [21]. The higher tocopherols content found in carrot powders comparing to carrot pulps may be related to the milling step that is absent in pulp production, as previous studies demonstrated that milling promotes the liberation of bioactive compounds from the fibres [49]. No significant variations were registered in γ-tocopherol concentration during storage with concentrations around 0.68 mg 100 g^{-1} DM. The results also suggested no significant differences in α-tocopherol and β-tocopherol concentrations in the first two months of storage (8.95 and 9.32 mg 100 g^{-1} DM, respectively), with a significant decrease in month 4 and subsequent stabilization until M6 (6.97 and 7.50 mg 100 g^{-1} DM, respectively).

4. Conclusions

This work proves that baby carrot and cherry tomato by-products have high nutritional and functional value as ingredients when transformed both in pulps or powders as they are rich in bioactive compounds such as phenolic compounds, carotenoids and tocopherols. This study also demonstrates that the nutritional profile of these by-products is similar in bioactive terms to the profile of carrot and tomato vegetables that comply with commercialization standards. Generally, vitamin E isomers are positively impacted by both storage methods. The concentrations of α-, β-, γ- and δ-tocopherol in tomato and α-, β- and γ-tocopherol in carrot increase significantly during the freezing storage and after the drying process. The drying process negatively affects the carotenoid and polyphenolic content as well as antioxidant capacity, with significant declines after the drying process. However, there were no relevant variations in antioxidant capacity and bioactive content in dry by-products during storage. The freezing storage of pulps negatively impacted TPC and antioxidant capacity evaluated by ABTS and ORAC assays, with sequential decreases during the frozen period. In contrast, the antioxidant capacity assessed by the DPPH method, as well as the carotenoid profile, was positively affected in the first months of freezing but then in both cases, a progressive decline was verified.

Although in the case of freezing process and storage, generally there are sequential losses of antioxidant activity and bioactive compounds and the drying storage maintain the properties, the drying process showed to be more aggressive, suggesting that the nutritional profile of carrot and tomato pulps is more appealing than that of the equivalent powders. Even so, through this study, it is evident that pulps or powders exhibit a good functional character and the dry by-products present guarantee the stability during the selected storage. Overall, this study demonstrated that both processing methods result in value-added products with interesting nutritional profiles rich in bioactive substances and considered safe in terms of microbial counts. Consequently, the recovery of these by-products may have a very interesting positive economic and environmental impact.

Author Contributions: H.A.-R.: investigation, methodology, formal analysis, writing—original draft preparation. D.S.: methodology, formal analysis, writing—review and editing. D.A.C.: supervision, writing—review and editing. M.R.: resources; I.M.R.: methodology, writing—review and editing. M.E.P.: conceptualization, project administration, supervision, writing—review and editing. All authors have read and agreed to the published version of the manuscript.

Funding: This work was supported by Project MOBFOOD POCI-01-0247-FEDER-024524-LISBOA-01-0247-FEDER-024524, cofounded by PORTUGAL2020, Lisb@a2020, COMPETE 2020 and the EU. This work was also supported by National Funds from FCT—Fundação para a Ciência e a Tecnologia through project UIDB/50016/2020.

Institutional Review Board Statement: Not applicable.

Informed Consent Statement: Not applicable.

Conflicts of Interest: The authors declare no conflict of interest.

References

1. Jenkins, W.; Tucker, M.E.; Grim, J. *Routledge Handbook of Religion and Ecology*; Routledge: London, UK, 2016. [CrossRef]
2. Keser, D.; Guclu, G.; Kelebek, H.; Keskin, M.; Soysal, Y.; Sekerli, Y.E.; Selli, S. Characterization of aroma and phenolic composition of carrot (Daucus carota 'Nantes') powders obtained from intermittent microwave drying using GC–MS and LC–MS/MS. *Food Bioprod. Process.* **2020**, *119*, 350–359. [CrossRef]
3. Kamiloglu, S.; Toydemir, G.; Boyacioglu, D.; Beekwilder, J.; Hall, R.; Capanoglu, E. A review on the effect of drying on antioxidant potential of fruits and vegetables. *Crit. Rev. Food Sci. Nutr.* **2016**, *56*, S110–S129. [CrossRef] [PubMed]
4. Kapoor, S.; Aggarwal, P. Drying method affects bioactive compounds and antioxidant activity of carrot. *Int. J. Veg. Sci.* **2014**, *21*, 467–481. [CrossRef]
5. Kyriacou, M.; Rouphael, Y. Towards a new definition of quality for fresh fruits and vegetables. *Sci. Hortic.* **2018**, *234*, 463–469. [CrossRef]
6. Patras, A.; Tiwari, B.; Brunton, N. Influence of blanching and low temperature preservation strategies on antioxidant activity and phytochemical content of carrots, green beans and broccoli. *LWT* **2011**, *44*, 299–306. [CrossRef]
7. Hung, P.V.; Duy, T.L. Effects of drying methods on bioactive compounds of vegetables and correlation between bioactive compounds and their antioxidantsfile. *Int. Food Res. J.* **2012**, *19*, 327–332.
8. Nayak, B.; Liu, R.H.; Tang, J. Effect of processing on phenolic antioxidants of fruits, vegetables, and grains—A review. *Crit. Rev. Food Sci. Nutr.* **2015**, *55*, 887–918. [CrossRef] [PubMed]
9. Ribas-Agustí, A.; Martín-Belloso, O.; Soliva-Fortuny, R.; Elez-Martínez, P. Food processing strategies to enhance phenolic compounds bioaccessibility and bioavailability in plant-based foods. *Crit. Rev. Food Sci. Nutr.* **2018**, *58*, 2531–2548. [CrossRef]
10. Neri, L.; Faieta, M.; Di Mattia, C.; Sacchetti, G.; Mastrocola, D.; Pittia, P. Antioxidant activity in frozen plant foods: Effect of cryoprotectants, freezing process and frozen storage. *Foods* **2020**, *9*, 1886. [CrossRef]
11. Murcia, M.A.; Jiménez, A.M.; Martínez-Tomé, M. Vegetables antioxidant losses during industrial processing and refrigerated storage. *Food Res. Int.* **2009**, *42*, 1046–1052. [CrossRef]
12. Celli, G.B.; Ghanem, A.; Brooks, M.S.-L. Influence of freezing process and frozen storage on the quality of fruits and fruit products. *Food Rev. Int.* **2015**, *32*, 280–304. [CrossRef]
13. Ijabadeniyi, O.A.; Pillay, Y. Microbial safety of low water activity Foods: Study of simulated and durban household samples. *J. Food Qual.* **2017**, 1–7. [CrossRef]
14. Gonçalves, B.; Falco, V.; Moutinho-Pereira, J.; Bacelar, E.; Peixoto, F.; Correia, C. Effects of elevated CO_2 on grapevine (*Vitis vinifera* L.): Volatile composition, phenolic content, and in vitro antioxidant activity of red wine. *J. Agric. Food Chem.* **2009**, *57*, 265–273. [CrossRef]
15. Coscueta, E.R.; Campos, D.; Osório, H.; Nerli, B.B.; Pintado, M. Enzymatic soy protein hydrolysis: A tool for biofunctional food ingredient production. *Food Chem. X* **2019**, *1*, 100006. [CrossRef] [PubMed]
16. Schaich, K.; Tian, X.; Xie, J. Reprint of "Hurdles and pitfalls in measuring antioxidant efficacy: A critical evaluation of ABTS, DPPH, and ORAC assays". *J. Funct. Foods* **2015**, *18*, 782–796. [CrossRef]
17. Ainsworth, E.A.; Gillespie, K.M. Estimation of total phenolic content and other oxidation substrates in plant tissues using Folin-Ciocalteu reagent. *Nat. Protoc.* **2007**, *2*, 875–877. [CrossRef] [PubMed]
18. Campos, D.A.; Ribeiro, T.B.; Teixeira, J.A.; Pastrana, L.; Pintado, M.M. Integral valorization of pineapple (*Ananas comosus* L.) By-products through a green chemistry approach towards Added Value Ingredients. *Foods* **2020**, *9*, 60. [CrossRef] [PubMed]
19. Oliveira, A.; Pintado, M.M.; Almeida, D. Phytochemical composition and antioxidant activity of peach as affected by pasteurization and storage duration. *LWT* **2012**, *49*, 202–207. [CrossRef]
20. Prates, J.A.M.; Quaresma, M.A.G.; Bessa, R.J.B.; Fontes, C.M.A.; Alfaia, C.M.M. Simultaneous HPLC quantification of total cholesterol, tocopherols and β-carotene in Barrosã-PDO veal. *Food Chem.* **2006**, *94*, 469–477. [CrossRef]
21. Slavin, M.; Yu, L. A single extraction and HPLC procedure for simultaneous analysis of phytosterols, tocopherols and lutein in soybeans. *Food Chem.* **2012**, *135*, 2789–2795. [CrossRef]

22. Vandekinderen, I.; Van Camp, J.; Devlieghere, F.; Veramme, K.; Denon, Q.; Ragaert, P.; De Meulenaer, B. Effect of Decontamination Agents on the Microbial Population, Sensorial Quality, and Nutrient Content of Grated Carrots (*Daucus carota* L.). *J. Agric. Food Chem.* **2008**, *56*, 5723–5731. [CrossRef]

23. Saraiva, M.; Correia, C.B.; Cunha, I.C.; Maia, C.; Bonito, C.C.; Furtado, R.; Calhau, A. *Interpretação de Resultados de Ensaios Microbiológicos em Alimentos Prontos Para Consumo e em Superfícies do Ambiente de Preparação e Distribuição Alimentar: Valores-guia*; INSA IP—Instituto Nacional de Saúde Doutor Ricardo Jorge: Lisboa, Portugal, 2019.

24. Lisiewska, Z.; Kmiecik, W. Effect of storage period and temperature on the chemical composition and organoleptic quality of frozen tomato cubes. *Food Chem.* **2000**, *70*, 167–173. [CrossRef]

25. Bouzari, A.; Holstege, D.; Barrett, D.M. Mineral, fiber, and total phenolic retention in eight fruits and vegetables: A comparison of refrigerated and frozen storage. *J. Agric. Food Chem.* **2015**, *63*, 951–956. [CrossRef] [PubMed]

26. Leja, M.; Kamińska, I.; Kramer, M.; Maksylewicz-Kaul, A.; Kammerer, D.; Carle, R.; Baranski, R. The content of phenolic compounds and radical scavenging activity varies with carrot origin and root color. *Plant. Foods Hum. Nutr.* **2013**, *68*, 163–170. [CrossRef] [PubMed]

27. Georgé, S.; Tourniaire, F.; Gautier, H.; Goupy, P.; Rock, E.; Caris-Veyrat, C. Changes in the contents of carotenoids, phenolic compounds and vitamin C during technical processing and lyophilisation of red and yellow tomatoes. *Food Chem.* **2011**, *124*, 1603–1611. [CrossRef]

28. Perea-Domínguez, X.P.; Hernández-Gastelum, L.Z.; Olivas-Olguin, H.R.; Espinosa-Alonso, L.G.; Valdez-Morales, M.; Medina-Godoy, S. Phenolic composition of tomato varieties and an industrial tomato by-product: Free, conjugated and bound phenolics and antioxidant activity. *J. Food Sci. Technol.* **2018**, *55*, 3453–3461. [CrossRef]

29. Smith, J.C.; Biasi, W.V.; Holstege, D.; Mitcham, E.J. Effect of passive drying on ascorbic acid, α-tocopherol, and β-carotene in tomato and mango. *J. Food Sci.* **2018**, *83*, 1412–1421. [CrossRef]

30. Char, C.D. Carrots (*Daucus carota* L.). *Fruit Veg. Phytochem.* **2017**, 969–978.

31. Kamiloglu, S.; Demirci, M.; Selen, S.; Toydemir, G.; Boyacioglu, D.; Capanoglu, E. Home processing of tomatoes (Solanum lycopersicum): Effects on in vitro bioaccessibility of total lycopene, phenolics, flavonoids, and antioxidant capacity. *J. Sci. Food Agric.* **2014**, *94*, 2225–2233. [CrossRef]

32. Danesi, F.; Bordoni, A. Effect of home freezing and Italian style of cooking on antioxidant activity of edible vegetables. *J. Food Sci.* **2008**, *73*, H109–H112. [CrossRef]

33. Nguyen, V.T.; Le, M.D. Influence of various drying conditions on phytochemical compounds and antioxidant activity of carrot peel. *Beverages* **2018**, *4*, 80. [CrossRef]

34. Thaipong, K.; Boonprakob, U.; Crosby, K.; Cisneros-Zevallos, L.; Byrne, D.H. Comparison of ABTS, DPPH, FRAP, and ORAC assays for estimating antioxidant activity from guava fruit extracts. *J. Food Compos. Anal.* **2006**, *19*, 669–675. [CrossRef]

35. Macheix, J.-J.; Fleuriet, A.; Billot, J. *Fruit Phenolics*; CRC Press: Boca Raton, FL, USA, 2018. [CrossRef]

36. Bénard, C.; Gautier, H.; Bourgaud, F.; Grasselly, D.; Navez, B.; Caris-Veyrat, C.; Weiss, M.; Génard, M. Effects of low nitrogen supply on tomato (Solanum lycopersicum) fruit yield and quality with special emphasis on sugars, acids, ascorbate, carotenoids, and phenolic compounds. *J. Agric. Food Chem.* **2009**, *57*, 4112–4123. [CrossRef]

37. Sun, T.; Simon, P.; Tanumihardjo, S.A. Antioxidant phytochemicals and antioxidant capacity of biofortified carrots (*Daucus carota* L.) of various colors. *J. Agric. Food Chem.* **2009**, *57*, 4142–4147. [CrossRef] [PubMed]

38. Arscott, S.A.; Tanumihardjo, S.A. Carrots of many colors provide basic nutrition and bioavailable phytochemicals acting as a functional food. *Compr. Rev. Food Sci. Food Saf.* **2010**, *9*, 223–239. [CrossRef]

39. Zhang, D.; Hamauzu, Y. Phenolic compounds and their antioxidant properties in different tissues of carrots (*Daucus carota* L.). *Int. J. Food Agric. Environ.* **2004**, *2*, 332.

40. Gómez-García, R.; Campos, D.A.; Oliveira, A.; Aguilar, C.N.; Madureira, A.R.; Pintado, M. A chemical valorisation of melon peels towards functional food ingredients: Bioactives profile and antioxidant properties. *Food Chem.* **2021**, *335*, 127579. [CrossRef] [PubMed]

41. Chaudhary, P.; Sharma, A.; Singh, B.; Nagpal, A.K. Bioactivities of phytochemicals present in tomato. *J. Food Sci. Technol.* **2018**, *55*, 2833–2849. [CrossRef]

42. Martí, R.; Roselló, S.; Cebolla-Cornejo, J. Tomato as a source of carotenoids and polyphenols targeted to cancer prevention. *Cancers* **2016**, *8*, 58. [CrossRef]

43. Hidalgo, A.; Di Prima, R.; Fongaro, L.; Cappa, C.; Lucisano, M. Tocols, carotenoids, heat damage and technological quality of diced tomatoes processed in different industrial lines. *LWT* **2017**, *83*, 254–261. [CrossRef]

44. Saini, R.K.; Zamany, A.J.; Keum, Y.S. Ripening improves the content of carotenoid, α-tocopherol, and polyunsaturated fatty acids in tomato (*Solanum lycopersicum* L.) fruits. *3 Biotech* **2017**, *7*, 1–7. [CrossRef]

45. Slavin, J.L.; Lloyd, B. Health benefits of fruits and vegetables. *Adv. Nutr.* **2012**, *1*, 506–516. [CrossRef] [PubMed]

46. Behsnilian, D.; Mayer-Miebach, E. Impact of blanching, freezing and frozen storage on the carotenoid profile of carrot slices (*Daucus carota* L. cv. Nutri Red). *Food Control.* **2017**, *73*, 761–767.

47. Bermudez, L.F.; Del Pozo, T.; Lira, B.S.; De Godoy, F.; Boos, I.; Romanó, C.; Previtali, V.; Almeida, J.; Bréhélin, C.; Asis, R.; et al. A Tomato Tocopherol Binding Protein Sheds Light on Intracellular α-tocopherol Metabolism in Plants. *Plant. Cell Physiol.* **2018**, *59*, 2188–2203. [CrossRef]

48. Ombódi, A.; Daood, H.G.; Helyes, L. Carotenoid and tocopherol composition of an orange-colored carrot as affected by water supply. *HortScience* **2014**, *49*, 729–733.
49. Esposito, F.; Arlotti, G.; Bonifati, A.M.; Napolitano, A.; Vitale, D.; Fogliano, V. Antioxidant activity and dietary Wbre in durum wheat bran by-products. *Food Res. Int.* **2005**, *38*, 1167–1173. [CrossRef]

Article

Influence of Berry Ripening Stages over Phenolics and Volatile Compounds in Aged Aglianico Wine

Alessandro Genovese [1], Boris Basile [1,*], Simona Antonella Lamorte [2], Maria Tiziana Lisanti [1], Giandomenico Corrado [1], Lucia Lecce [3], Daniela Strollo [4], Luigi Moio [1] and Angelita Gambuti [1]

[1] Department of Agricultural Sciences, University of Naples Federico II, 80055 Portici, Italy; alessandro.genovese@unina.it (A.G.); mariatiziana.lisanti@unina.it (M.T.L.); giandomenico.corrado@unina.it (G.C.); luigi.moio@unina.it (L.M.); angelita.gambuti@unina.it (A.G.)

[2] Ministero Delle Politiche Agricole Alimentari e Forestali, PIUE VII—Settore Vitivinicolo, 00187 Roma, Italy; sa.lamorte@politicheagricole.it

[3] Department of Clinical and Experimental Medicine, University of Foggia, 71122 Foggia, Italy; lucia.lecce@unifg.it

[4] Mastroberardino Winery, 83042 Avellino, Italy; strollo@mastroberardino.com

* Correspondence: boris.basile@unina.it

Abstract: The harvest time of grapes is a major determinant of berry composition and of the wine quality, and it is usually established through empirical testing of main biochemical parameters of the berry. In this work, we studied how the ripening stage of Aglianico grapes modulates key secondary metabolites of wines, phenolics and volatile compounds. Specifically, we analyzed and compared four berry ripening stages corresponding to total soluble solids of 18, 20, 22, and 25 °Brix and related aged wines. Wine color intensity, anthocyanins level and total *trans*-resveratrol (free + glycosidic form) increased with grape maturity degree. Wines obtained from late-harvested grapes significantly differed from the others for a higher content of aliphatic alcohols, esters, acetates, α-terpineol and benzyl alcohol. The content of glycosidic terpene compounds, such as nerol, geraniol and α-terpineol, was higher in wines obtained with grapes harvested at 25 °Brix compared to the earlier harvests. Our work indicated that the maturity of the grape is a determining factor in phenolic and volatile compounds of red Aglianico wines. Moreover, extending grape ripening to a sugar concentration higher than 22 °Brix improves the biochemical profile of aged wine in terms of aroma compounds and of phytochemicals with known health-related benefits.

Keywords: *trans*-resveratrol; esters; terpenols; glycosidic precursors; harvest time; *Vitis vinifera*

Citation: Genovese, A.; Basile, B.; Lamorte, S.A.; Lisanti, M.T.; Corrado, G.; Lecce, L.; Strollo, D.; Moio, L.; Gambuti, A. Influence of Berry Ripening Stages over Phenolics and Volatile Compounds in Aged Aglianico Wine. *Horticulturae* **2021**, *7*, 184. https://doi.org/10.3390/horticulturae7070184

Academic Editor: Qiuhong Pan

Received: 19 May 2021
Accepted: 30 June 2021
Published: 5 July 2021

Publisher's Note: MDPI stays neutral with regard to jurisdictional claims in published maps and institutional affiliations.

1. Introduction

Wine quality largely depends on identifying the optimal maturity of the grapes at harvest. During ripening, berry weight, sugars, pH, and acidity are traditionally monitored to choose the best moment for harvest. These parameters are often measured to empirically define the technological maturity of the grapes. However, for some red varieties, harvesting at technological maturity does not guarantee the production of balanced wines, because important characteristics of red wines depend also on polyphenols and volatiles, and these compounds are usually not monitored during ripening. For instance, grape phenolics are responsible for color, astringency, bitterness, and aging behavior of red wines. Although their bioavailability is still questioned [1,2], the interest of the scientific community towards phenolics, in particular *trans*-resveratrol and quercetin, has increased because they are accountable for several health-related effects correlated to wine intake [3]. Moreover, several odor active compounds, belonging to the chemical classes of monoterpens, norisoprenoids, C6 alcohols, and shikimate derivatives, which are present in the volatile fraction of grapes, play a decisive role in the aroma of wine. Most odor compounds as monoterpenes are present both as free volatiles and non-volatile, glycosidically bound forms and represent a potential reservoir of flavor [4,5].

Grape ripening is an important physiological process that influences the composition of the wine, as it is important for the development of its sensory properties. Due to their notable oenological and biological importance, numerous authors have investigated the changes during grape ripening of phenolics [6–10] as well as free and glycosylated bound volatiles [11,12]. The ripening process is also characterized by other important aspects influencing the ability of each class of chemical compounds to be released from grape to must during vinification. They include seed lignification for phenolics, the enzymatic release from non-volatile precursors for volatiles, and, generally, the gradual breakdown of the cellular structure of the skins. Therefore, investigations on the effect of the grape composition at harvest on wine quality should include an understanding of the ability of phenolics and volatiles to be extracted from the solid part of the grape to must during vinification.

During aging, polyphenolic and volatile compounds of red wines undergo numerous transformations. Anthocyanins, the grape polyphenols responsible for color of red wine, are involved in numerous reactions with other wine components, as well as tannins, flavanols, acetaldehyde, vinyl phenol and pyruvic acid, with the consequent development of new pigments and a yellow-orange hue [13–15]. The involvement of tannins in the polymerization reaction with anthocyanins and flavanols and in reactions with proteins and polysaccharides often causes a decrease in wine astringency and bitterness [16]. Changes in the content of aroma substances during wine aging include: the oxidation, isomerization and cyclization reactions of monoterpens to compounds that have higher odor thresholds [17]; the acid and enzyme-catalyzed hydrolysis of glycosidic aroma precursors of grape [18]; and the decrease in acetates and ethyl esters of fatty acids, resulting in an overall decrease in fruity notes [19]. Several studies have aimed at understanding the effect of grape maturity on volatile compounds in grapes and wine [20–29]. For instance, a positive influence of the degree of grape ripening on wine color and flavanol and anthocyanins content of 18 months aged red wines has been reported [30]. Papers detailing the influence of grape maturity on the content of free and glycosidically bound *trans*-resveratrol as well as on free and bound volatiles of red wine are relatively scarce. To our knowledge, no study has reported the influence of grape maturity degree on the content of free and glycosidically bound *trans*-resveratrol as well as on free and bound volatile compounds in Aglianico red wine. *Vitis vinifera* cv. "Aglianico" is one of the most prestigious southern Italian native grapes used to produce aged wine in agreement with related Appellation of Controlled and Guaranteed Origin (DOCG) and Appellation of Controlled Origin (DOC) regulation.

In the present work, we studied chromatic characteristics, anthocyanins, tannins, catechin, epicatechin, quercetin, *trans*-resveratrol, *trans*-polydatin content and free and bound volatiles of aged wines obtained with grapes harvested at four stages according to the soluble sugar content of the berries.

2. Materials and Methods

2.1. Plant material, Location, and Treatments

The grape production was carried out in a commercial grapevine vineyard (380 m a.s.l.) located in Foglianise (Benevento, Campania region, Italy) within the Taburno area, one of the major Italian viticultural areas for wine production. Vines were 8-year old *V. vinifera* L. cv. "Aglianico" plants grafted onto 1103P and trained to a spur-pruned permanent cordon. Vines were spaced 2.20 m × 1.30 m (3497 vines/ha) and rows had a north–south orientation. The trials compared four stages of berry ripening (four harvest times) corresponding to total soluble solids (TSS) in the berries of 18, 20, 22, and 25 °Brix (hereafter these treatments will be indicated as S18, S20, S22, and S25, respectively). For each treatment, the harvest date was determined by monitoring TSS accumulation in the berries every five days starting from veraison on three samples (replicates) of 50 berries (a total of 150 berries per treatment). When the target TSS values defined for each treatment were reached, an additional three samples (each made of 250 berries) were collected and used for berry

composition analyses. In addition, on the same dates, a sample of 100 kg of grapes was harvested and used for vinification.

2.2. Vinification

Vinification was carried out using 100 kg of grapes. Grapes were destemmed and crushed and the must was treated with $K_2S_2O_5$ (60 mg/kg of grapes). Fermentation was carried out at 26 °C using a *S. cerevisiae* D254 active dry yeast inoculum (Lallemand Italy, Castel d'Azzano, Italy) and the cap was immersed twice a day for 10 days. The must was then pressed (at about 8 bar), and 65 L of finished wine was obtained. Upon completion of alcoholic fermentation (residual sugars < 2 g/L), wines were racked with 8 g/hL of $K_2S_2O_5$, bottled and stored for 24 months at 10 °C. Two repetitions per vinification were carried out. All wines were analyzed two years after winemaking.

Physicochemical analysis of grapes. TSS, titratable acidity and pH were measured according to the Official European Method (1990). Berries of each sample were analyzed for phenolic potential (total anthocyanins), potential anthocyanins (A pH 1), extractable anthocyanins (A pH 3.2), extractability and tannins of skins and seeds, according to Glories [31].

2.3. Main Wine Characteristics and Spectrophotometric Analysis

Ethanol, sugar, pH, dry extract, titratable acidity, volatile acidity, total and free SO_2, optical densities (OD) and total polyphenols (Folin-Ciocalteu Index) were measured according to the Official European Methods (1990). Color intensity, hue, percentage of free anthocyanins on the total amount of anthocyanins (dAL %) and percentage of anthocyanins combined with tannins and not bleachable by SO_2 on the total amount of anthocyanins (dTaT %) were evaluated according to the method described by Glories [32]. When the absorbance values were higher than 0.9 Abs units, a dilution with a hydroalcoholic buffer solution (12% *v/v* of ethanol in water) at pH 3.3 (obtained by using tartaric acid and sodium hydroxide) was used, in order to be in the range of validity of the Lambert–Beer law. Total anthocyanins and tannins were determined by spectrophotometric methods on the berry extracts and directly on wines [31].

2.4. Determination of Trans-Resveratrol, Trans-Polydatin, Catechin, Epicatechin and Quercetin

The analysis was performed using an HPLC apparatus (Shimadzu Italy, Milan, Italy), consisting of an SCL-10AVP system controller, two LC-10ADVP pumps, a SPD-M 10 AVP detector, an injection system full Rheodyne model 7725 (Rheodyne, Cotati, CA, USA) equipped with a 20 µL loop. Separation and quantification of catechin, epicatechin, *trans*-resveratrol and quercetin were carried out by HPLC, as previously described [33]. The glucosidic form of *trans*-resveratrol was detected following 24 h of enzymatic hydrolysis with a glycosidase rich enzyme preparation in wine, at pH 4.5 and at a temperature of 45 °C. For each sample of wine, analysis was carried out in triplicate.

2.5. Extraction and Analysis of Free and Bound Aroma Compounds

C18 reversed-phase solid-phase extraction (SPE) was adopted for the extraction of free and glycosidically bound aroma compounds from wines according to the method earlier described [34]. A sample of 50 mL was diluted (1:1 *v/v*) with water after the addition of 2-octanol as internal standard (250 µL 2-octanol at 200 mg/L in methanol) loaded onto 1-g C18 cartridges (Phenomenex, Torrance, CA, USA). Free and bound volatiles were eluted with dichloromethane and methanol, respectively. The dichloromethane fraction was collected in a separating funnel, was dried over Na_2SO_4 and concentrated first in a Kuderna–Danish concentrator (Supelco, Bellefonte, PA, USA) and then under a stream of pure N_2 for gas-chromatographic analysis. The methanol fraction was concentrated to dryness under vacuum at 35 °C, and redissolved in 5 mL of citrate phosphate buffer (pH = 5.0) containing a glycosidase rich enzyme preparation (80 mg). After 16 h of incubation at 40 °C, 2-octanol

was added, and the mixture was extracted with dichloromethane for gas chromatographic analyses. All extractions were carried out in triplicate.

A 1.2 µL aliquot of each concentrated wine extract was injected in splitless mode into a Hewlett-Packard 5890 series II gas chromatograph equipped with a split/splitless injector (Hewlett-Packard, Avondale, PA, USA) and with a flame ionization detector. The column used was a DBWax from J&W Scientific (Folsom, CA, USA), 30 m length × 0.32 mm with 0.5 film thickness. The temperature program was the following: 40 °C for 5 min, 5 °C/min rate to 220 °C, 10 min at maximum temperature. The carrier gas was He flowing at 37 cm/s. Both detector and injector temperatures were maintained at 250 °C. The identity of the volatile compounds was determined through the comparison of their chromatographic retention properties with those of pure references. Comparison of mass spectra stored in the NIST database with those obtained for each compound was performed by GC/MS and consisted of a 6890 series Agilent Technologies gas chromatograph and 5973 Network series mass selective detector (Agilent Technologies) fitted with a 4 electronic impact source. Electron impact mass spectra were recorded with ion-source energy of 70 eV. Quantitative data were obtained by interpolation of relative areas versus the internal standard.

2.6. Chemicals

All chromatographic solvents were HPLC ultra-gradient grade and were purchased from Merck (Darmstadt, Germany). *trans*-Resveratrol and quercetin standards were purchased from Sigma-Aldrich (Milan, Italy), (+)-catechin and (−)-epicatechin standards were purchased from Fluka (Milan, Italy). Because (−)-catechin has recently been found in grape extracts when analyzed with a chiral column [35], the results of the analysis were reported as catechin and epicatechin.

2.7. Statistical Analysis

Analysis of variance and Tukey's test were used to assess the significance of the differences between means. Data elaboration was carried out using XLSTAT version 2014.5.03 (Addinsoft Corp., Paris, France). A correlogram representing the Pearson correlation coefficient ($n = 12$) matrix between grape and wine compounds and chromatic characteristics was built in RStudio (RStudio, Boston, MA, USA) with corrplot package.

3. Results and Discussion

Table 1 shows the anthocyanin and tannins content of grapes harvested at the four different ripening stages, S18, S20, S22, and S25. The tendency observed for total and potential anthocyanins confirms the known trend of these pigments during ripening: a quick increase until a maximum (S20) followed by stabilization or decrease in the over-ripe grape (S25). After prolonging the on-vine berry maturation (S20 to S25), anthocyanin content in the grapes decreased, but these pigments became progressively easier to extract, as suggested by the trends of extractable anthocyanins and of extractability values. A similar result was previously reported and it is considered to be the results of a decrease in cell turgor pressure causing the breakdown of cell walls and the release of anthocyanins from vacuoles [36]. Tannins in berry skins did not significantly vary during ripening, while their level in the seeds declined considerably (Table 1) as also previously reported [8,9].

Alcohol percentage, pH, and dry extract of the wines increased with the maturation stage of the grapes from which it was made. Conversely, total acidity gradually decreased (Table 2). Data reported for wines confirm the soluble solids and acid trends occurring during grape ripening [37,38].

Table 1. Mean values (±standard deviation) of skin and seed tannins, total anthocyanins, potential anthocyanins (A pH 1), extractable anthocyanins (A pH 3.2), and anthocyanin extractability measured in Aglianico grapes harvested at four ripening stages, corresponding to 18 °Brix (S18), 20 °Brix (S20), 22 °Brix (S22) and 25 °Brix (S25).

Ripening Stage	Skin Tannins (g/L of Extract)	Seed Tannins (g/L extract)	Total Anthocyanins (mg/L of Extract)	A pH 1	A pH 3.2	Anthocyanin Extractability (%)
S18	1.19 ± 0.56	2.59 ± 0.92 a	620 ± 9 b	468 ± 6 a	216 ± 2 b	53.9 ± 4.5 a
S20	2.11 ± 0.28	1.57 ± 0.10 ab	742 ± 18 a	431 ± 8 c	237 ± 4 a	45.0 ± 0.2 b
S22	1.77 ± 0.96	0.81 ± 0.49 b	499 ± 4 c	399 ± 4 d	179 ± 3 c	50.0 ± 1.4 a
S25	2.69 ± 0.38	0.76 ± 0.47 b	487 ± 15 c	456 ± 2 b	205 ± 9 b	55.0 ± 1.7 a
Significance	n.s.	*	***	***	**	**

Within each column, means followed by different letters are significantly different according to the Tukey test ($p < 0.05$); n.s., *, **, *** correspond to not significant, and significant at $p \leq 0.05$, 0.01, and 0.001, respectively.

Table 2. Mean value (±standard deviation) of ethanol, sugar, total acidity, pH volatile acidity, free SO_2, total SO_2, and dry extract in wines obtained with Aglianico grapes harvested at four ripening stages, corresponding to 18 °Brix (S18), 20 °Brix (S20), 22 °Brix (S22) and 25 °Brix (S25).

Ripening Stage	Ethanol (v/v %)	Sugar (g/L)	Total Acidity (g/L)	pH	Volatile Acidity (g/L)	Free SO_2 (mg/L)	Total SO_2 (mg/L)	Dry Extract (g/L)
S18	9.0 ± 0.5 d	1.1 ± 0.8	7.2 ± 0.4 a	3.07 ± 0.11 c	0.39 ± 0.27	13.4 ± 3.9	134.0 ± 11.8	22.7 ± 1.1 b
S20	11.1 ± 0.4 c	1.1 ± 1.0	5.7 ± 0.4 b	3.28 ± 0.15 bc	0.57 ± 0.10	14.1 ± 6.7	102.4 ± 12.8	20.9 ± 0.9 b
S22	11.9 ± 0.3 b	1.6 ± 0.3	5.4 ± 0.6 bc	3.34 ± 0.16 b	0.64 ± 0.13	19.2 ± 6.1	110.1 ± 16.1	21.6 ± 1.4 b
S25	14.5 ± 0.6 a	1.8 ± 0.9	4.6 ± 0.6 c	3.63 ± 0.10 a	0.86 ± 0.28	19.8 ± 4.7	107.5 ± 10.1	28.9 ± 0.7 a
Significance	***	n.s.	**	**	n.s.	n.s.	n.s.	***

Within each column, means followed by different letters are significantly different according to the Tukey test ($p < 0.05$); n.s., **, *** correspond to not significant, and significant at $p \leq 0.01$ and 0.001, respectively.

Total polyphenols did not differ between experimental wines, while S22 and S25 wines were characterized by a higher content of anthocyanins than the other treatments (Table 3). These data are not in agreement with the trend of total and potential anthocyanins and of their extractability reported in Table 1. This result may arise from factors affecting the extractability of these pigments during maceration, such as the presence of copigments [39], and the numerous reactions stabilizing the anthocyanins and occurring during aging [40]. Tannins were significantly less in S25 wines compared to the other treatments. This could be due to a predominant effect of seed tannins, but also phenomena determining their precipitation should be considered. Since in both seeds and skins there is some evidence of oxidative processes causing high polymerization of tannins with ripening [6,7], it is possible that an easier precipitation of these high polymerized phenolics occurred during vinification of over-ripe grapes.

Table 3. Mean value (±standard deviation) of total phenolics, total anthocyanins, and tannins in wines obtained with Aglianico grapes harvested at four ripening stages, corresponding to 18 °Brix (S18), 20 °Brix (S20), 22 °Brix (S22) and 25 °Brix (S25).

Ripening Stage	Total Phenolics (Folin-Ciocalteau Index)	Total Anthocyanins (mg/L of Malvidin-3-Monoglucoside)	Tannins (g/L)
S18	69.9 ± 2.3	47.4 ± 2.5 c	2.4 ± 03 a
S20	78.8 ± 6.9	70.1 ± 0.9 b	2.4 ± 0.2 a
S22	72.5 ± 4.3	114.8 ± 5.6 a	2.4 ± 0.1 a
S25	69.0 ± 3.5	125.1 ± 11.7 a	2.0 ± 0.1 b
Significance	n.s.	***	**

Within each column, means followed by different letters on the column are significantly different according to the Tukey test ($p < 0.05$); n.s., **, *** correspond to not significant, and significant at $p \leq 0.01$ and 0.001, respectively.

Values of optical density at 420, 520 and 620 nm and the colorant intensity of wine progressively increased with the degree of grape maturation (Table 4). These results may be explained by considering the higher content of anthocyanins as well as the new pigments derived from the reaction of anthocyanins with other compounds stabilizing flavylium

form of anthocyanins [40], and also the occurrence of the co-pigmentation [41] due to an elevated cofactor/pigment ratio in wines obtained from grapes at a higher ripening degree. It is also well known that wine color depends on ethanol content and pH. Nevertheless, our results disagree with the earlier findings showing a lowering in the magnitude of the hypercromic shift of several wine pigments at the increase of pH and ethanol content of the medium [39]. A higher colorant intensity was detected in that wines less acidic and more alcoholic, suggesting that the effect of pH and ethanol is negligible compared to the anthocyanin and new pigments content in these wines. Concerning the color parameters O.D. 420 nm, O.D. 520 nm and O.D. 620 nm, Pérez-Magarino and Gonzáles-San José [30] reported that aging accentuated the "harvest data effect" owing to the higher values for percentage blue (O.D. 620 nm) in aged wines made from the ripest grapes. The results of our study appear to support these findings, as the contribution of O.D. 620 nm to colorant intensity significantly increased to 12.3% in S25 wines. The shift of the dominant wavelength towards blue is probably correlated with the presence of pigments resulting from the reaction between anthocyanin-pyruvic acid adducts and vinyl flavanols [42]. An important contribution to color is also due to numerous stable pigments deriving from the reaction of native anthocyanins with colorless phenolic compounds because of the action of ethanal [43]. This high reactive aldehyde has a double origin: from yeasts during alcoholic fermentation and from chemical oxidation of wines during aging. In S22 and S25 wines, more sugar was fermented and, consequently, more acetaldehyde was produced [44]. In addition, the higher pH favors the oxidation reactions in wines. It is, therefore, possible that more stable ethyl-bridged pigments were formed in these wines, accounting for the higher values of color intensity and O.D. detected in S22 and S25 wines.

Table 4. Mean value (±standard deviation) of optical densities (O.D. 420 nm, O.D. 520 nm and O.D. 620 nm) colorant intensity, and color hue of wines obtained with Aglianico grapes harvested at four ripening stages, corresponding to 18 °Brix (S18), 20 °Brix (S20), 22 °Brix (S22) and 25 °Brix (S25).

Ripening Stage	O.D. 420 nm	O.D. 520 nm	O.D. 620 nm	Colorant Intensity	Color Hue
S18	1.9 ± 0.05 d	2.01 ± 0.05 c	0.45 ± 0.02 d	4.37 ± 0.12 c	0.94 ± 0.06
S20	2.45 ± 0.04 c	2.91 ± 0.07 b	0.68 ± 0.01 c	6.18 ± 0.12 b	0.84 ± 0.01
S22	3.08 ± 0.04 b	3.59 ± 0.10 a	0.91 ± 0.02 b	7.76 ± 0.16 a	0.86 ± 0.01
S25	3.31 ± 0.07 a	3.70 ± 0.10 a	1.01 ± 0.01 a	8.19 ± 0.19 a	0.89 ± 0.01
Significance	***	***	***	***	n.s.

Within each column, means followed by different letters are significantly different according to the Tukey test ($p < 0.05$); n.s., *** correspond to not significant, and significant at $p \leq 0.001$, respectively.

The content of total *trans*-resveratrol was directly correlated to the grape ripening stage (Table 5). These results are not due to an increase in resveratrol concentration during maturation, because its biosynthesis generally tends to decrease at this developmental stage [45]. However, it is well known that the extraction of *trans*-resveratrol from skins is enhanced by the alcoholic content of the medium. This behavior is less pronounced for the glycosylated form due to its higher hydrophilic character [46]. Thus, both the high level of ethanol produced during maceration and the gradual softening of cell walls during grape ripening may account for values of total *trans*-resveratrol more than three times higher in S25 than in S18 wines.

Table 5. Mean value (±standard deviation) of antioxidant phenolic compounds of wines obtained with Aglianico grapes harvested at four ripening stages, corresponding to 18 °Brix (S18), 20 °Brix (S20), 22 °Brix (S22) and 25 °Brix (S25). n.d.: not determined.

Ripening Stage	Catechin (mg/L)	Epicatechin (mg/L)	*trans-* Polidatin (mg/L)	*trans-* Resveratrol (mg/L)	Quercetin (mg/L)
S18	77.0 ± 2.3 a	48.1 ± 0.2 a	0.7 ± 0.3 b	0.5 ± 0.2 c	6.9 ± 1.1 a
S20	72.7 ± 2.8 a	39.6 ± 0.7 b	0.5 ± 0.6 ab	1.8 ± 0.3 b	7.8 ± 1.9 a
S22	55.7 ± 0.7 b	26.1 ± 0.8 c	2.4 ± 0.7 a	1.0 ± 0.2 c	5.4 ± 0.8 ab
S25	72.3 ± 4.2 a	49.4 ± 0.9 a	n.d.	3.5 ± 0.4 a	3.8 ± 1.8 b
Significance	***	***		***	*

Within each column, means followed by different letters are significantly different according to the Tukey test ($p < 0.05$); *, *** correspond to not significant, and significant at $p \leq 0.05$ and 0.001, respectively.

The level of catechin and epicatechin decreased from the S18 to the S22 wine. Contrasting results are reported in the literature regarding the trend followed by flavan-3-ols during ripening. Kennedy et al. [6] detected a dramatic decrease in flavan-3-ol monomers in Cabernet Sauvignon seeds during grape ripening; whereas, in a more recent study, a definite trend for seeds and skins flavanols of the same grape variety was not detected [47]. Concerning the condition of maceration, it has been reported that the extraction of flavan-3-ols from seeds and skins increases with ethanol [48]. Moreover, the contribution of skins to the flavanol composition of wine is assumed to be prevailing because they are more available for extraction [49]. Therefore, the higher values found in S25 wines may be due to both the advanced senescence of skin cells and the higher level of ethanol of solution. The lower amount of catechin and epicatechin detected in S22 wines compared to S25 wines could be due to the greater involvement of these flavan-3-ols in condensation reactions with anthocyanins and acetaldehyde, as these reactions can be favored at lower pH (3.34 in S22 wine with respect to 3.63 in S25 wine) [50]. However, taking into account the high involvement of these compounds in condensation reactions giving more stable pigments during wine aging [13], it is possible to hypothesize that the higher flavan-3-ols content of Aglianico wines obtained from over-ripened grapes may be responsible for its greater O.D. 620 nm.

A correlation analysis was carried out to highlight the possible relationships between the phenolic parameters of the grape and the main phenolic composition and chromatic characteristics of aged wines (Figure 1). A negative significant correlation was detected for total phenolics and A pH 1 indicating that the evaluation of total and extractable anthocyanins are not useful for predicting the color and anthocyanin content of aged red wine. For skin tannins, after two years of aging, a significant positive correlation with color intensity and O.D 420 nm and O.D. 620 nm was detected instead. This result could be related to the greater extractability of skin tannins and to the role that non-pigmented native phenolics have on all these reactions that stabilize wine color over time [40]. Jensen et al. [51] showed that it is possible to predict the color quality of fresh wines from grape measurements. Pérez-Magariño and González-San José [30] found a correlation between individual native phenolics of Cabernet Sauvignon and Tinto Fino grapes and wines; in our study a positive correlation was detected only between wine tannins and total anthocyanins of grape. This result underlines that, after two years of aging, the whole phenolic composition of wines should be considered for a possible correlation with grape anthocyanins because the reactions of native pigments in the formation of high molecular weight structure involving tannins and stabilizing color over time became dominant.

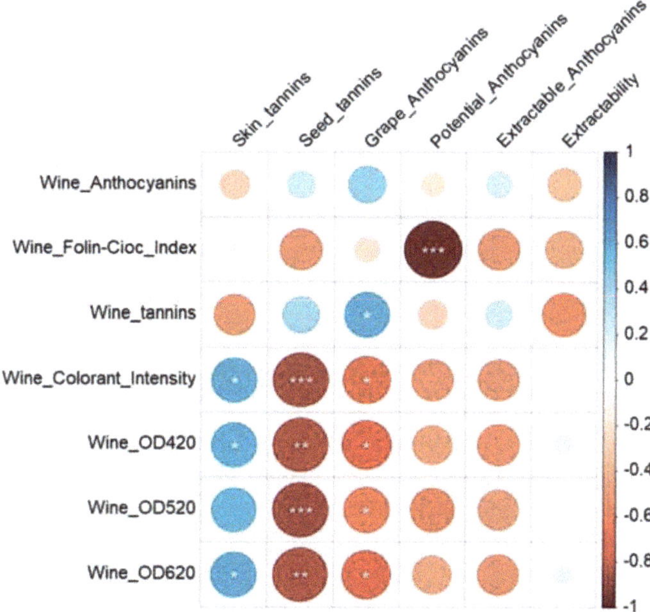

Figure 1. Correlogram representing the Pearson correlation between grape and wine compounds and chromatic characteristics. Asterisks indicate the significance of the Pearson correlation coefficient (*, **, *** correspond to $p \leq 0.05$, 0.01, and 0.001, respectively). Colors indicate different values of the correlation coefficient according to the scale bar reported on the right. The size of the circle is proportional to the correlation coefficient.

A total of 35 volatile compounds were quantified in Aglianico wines (Table 6). The S25 sample showed a significantly higher content of alcohols (1-butanol, 1-pentanol, 3-methyl-1-pentanol, 1-hexanol, (E)-3-hexen-1-ol, (Z)-3-hexen-1-ol, (E)-2-hexen-1-ol, benzyl alcohol, α-terpineol and 7-hydroxy-3,7-dimethyl-1-ol), esters (ethyl butanoate, ethyl hexanoate, ethyl lactate, ethyl octanoate, diethyl succinate), acetates (3-methylbutyl acetate, 2-phenylethyl acetate), thiols (3-methylthio-1-propanol) and amides (N-3-methylbutyl acetamide). Hexanol, (E)-3-hexen-1-ol, (Z)-3-hexen-1-ol, (E)-2-hexen-1-ol and other 1-alkanols are grape constituents and are related to fatty acid metabolism [52,53]. C_6 alcohols are produced during grape ripening from C_{18} fatty acids through the lipoxygenase pathway and the activity of the alcohol dehydrogenase. In contrast, 1-alkanols are obtained from homolytic cleavage of fatty acids when grapes are crushed under oxidative conditions. Our results are in agreement with those reported by Bindon et al. [22] on Cabernet Sauvignon wine. Benzyl alcohol, a derivative of the phenylpropanoid pool in plants [54], showed a higher content in S25 wines. Both a higher concentration and an elevated extraction from the pomace of overripe grape may account for the higher amounts detected.

The level of α-terpineol was almost four times higher in S25 than S18. In an aged wine, this compound may arise from a chemical modification of other more odorous terpenoids [17] and the hydrolysis of non-volatile glycosylated precursors [18]. The higher content detected in S25 wine seems to confirm earlier studies showing that terpenols increase with ripening [11].

Most of the identified esters, responsible for fruity odorous notes, were higher in S25 wine. It is known that the synthesis of esters by yeasts is enhanced when the rate of fermentation decreases [55]. Therefore, the consumption of the reducing sugars during fermentation of experimental musts is also reported to explain the observed behavior (Figure 2).

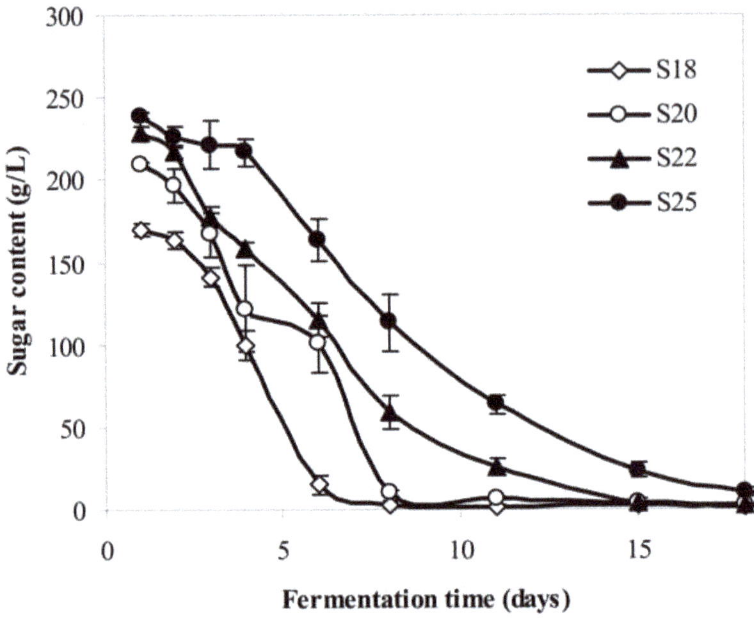

Figure 2. Sugar content during the fermentation of Aglianico grapes harvested at four ripening stages, corresponding to 18 °Brix (S18), 20 °Brix (S20), 22 °Brix (S22) and 25 °Brix (S25).

The fermentation rate was negatively correlated with the grape ripening degree, owing to the rise in the sugar content and to the decrease in the content of yeast's assimilable nitrogen during the last stages of grapes development. As expected, higher amounts of esters are detected in wines obtained through a slower fermentation. Among esters, the amount of 3-methylbutyl acetate, an important wine odorous compound responsible for banana notes, was higher in S25 compared to S18 wines. Similar results have been reported in two studies on the fermentation of Chardonnay juices and Cabernet Sauvignon wine [22,56]. It may be ascribed to a higher activity of alcohols acetyltransferases (AAT) of yeasts in slower fermentation batches. However, the higher values of esters and acetates in S25 wines are likely due also to the higher pH of wines obtained from ripest grapes, which is expected to reduce their hydrolysis during aging. The higher level of esters and acetates in S25 wines could be responsible for an improvement in their fruity notes.

Conversely, ethyl malate is found at the highest level in S18 wines. This ester originates from the reaction of esterification of malic acid by ethanol and the high amounts detected in S18 wine could be explained considering the higher level of malic acid present in the unripe grape (as suggested by the values of total acidity, Table 1).

Concerning the glycosidically bound volatile composition, 21 volatile compounds were identified (Table 7). Glycosidic volatile compounds were present at small concentrations as expected considering that two-year aged wines were analyzed. In some cases, there was an opposite trend in the concentration levels making it difficult to understand a correlation with the grape maturation degree, e.g., some terpene compounds and benzyl alcohol. Numerous factors involved in glycoside decline during fermentation and aging such as pH [57], adsorption by yeast cells and specific hydrolysis of bound aglycons [18].

Table 6. Mean value (±standard deviation) of level of free volatile compounds (µg/L) detected in the wines obtained with Aglianico grapes harvested at four ripening stages, corresponding to 18 °Brix (S18), 20 °Brix (S20), 22 °Brix (S22) and 25 °Brix (S25). n.d.: not detected.

Compounds	Descriptor [1]	S18	S20	S22	S25	Sig.
Alcohols						
2-methyl-1-propanol	wine-like	122.4 ± 15.0	100.3 ± 16.6	104,2 ± 4.4	134.0 ± 17.0	n.s.
1-butanol	medicinal	14.9 ± 3.9 b	16.7 ± 1.6 b	16.1 ± 2.3 b	54.0 ± 10.5 a	***
3+2-methyl-1-butanol	grass	63958 ± 388	53146 ± 633	54892 ± 612	51385 ± 277	n.s.
1-pentanol	mild green	14.1 ± 0.5 b	10.7 ± 0.8 b	9.1 ± 0.7 b	35.3 ± 2.9 a	***
4-methyl-1-pentanol	wine-like	57.4 ± 1.2 c	81.2 ± 4.8 b	96.3 ± 4.5 a	93.3 ± 6.5 ab	***
2-heptanol	herbaceous	n.d.	n.d.	n.d.	19.2 ± 0.5	-
3-methyl-1-pentanol	wine-like, green	93.2 ± 2.5 d	172.5 ± 10.9 c	245.5 ± 18.0 b	292.6 ± 23.3 a	***
1-hexanol	grass, herbaceous	2515 ± 52 c	1842 ± 82 b	1758 ± 67 b	3501 ± 25 a	***
(E)-3-hexen-1-ol	grass, herbaceous	49.4 ± 0.2 a	21.6 ± 0.9 c	17.7 ± 1.2 c	41.7 ± 3.5 b	***
(Z)-3-hexen-1-ol	grass, herbaceous	48.9 ± 1.3 a	25.1 ± 1.9 b	24.4 ± 4.2 b	28.1 ± 4.5 b	***
(E)-2-hexen-1-ol	grass, herbaceous	5.8 ± 0.1 c	12.8 ± 1.3 b	6.3 ± 0.9 c	22.3 ± 3.8 a	***
1-heptanol	apple, fruit	46.0 ± 4.0	41.1 ± 9.9	34.3 ± 1.4	103.2 ± 10.5	n.s.
1-octanol	waxy, citrus	16.8 ± 0.8	23.8 ± 2.1	29.7 ± 0.5	58.7 ± 2.8	n.s.
α-terpineol	sweet, floral	5.9 ± 0.8 c	10.5 ± 1.7 b	11.8 ± 0.3 b	21.7 ± 1.6 a	***
benzyl alcohol	toasty, sweet	21.0 ± 1.5 b	58.3 ± 1.1 b	20.0 ± 2.0 b	99.7 ± 15.3 a	**
2-phenylethanol	rose	38881 ± 3270	40490 ± 4181	41614 ± 2659	35852 ± 4345	n.s.
7-hydroxy-3,7-dimethyl-1-ol		25.5 ± 6.8 bc	29.0 ± 0.1 b	15.8 ± 0.6 c	45.3 ± 6.9 a	*
Esters						
ethyl butanoate	kiwi, fruity	45.7 ± 3.8 b	38.6 ± 1.8 b	38.4 ± 4.1 b	173.1 ± 13.2 a	***
ethyl 2-methylbutanoate	red fruity	72.2 ± 8.5	93.4 ± 0.8	82.4 ± 10.0	76.7 ± 3.7	n.s.
ethyl 3-methylbutanoate	exotic fruit	99.7 ± 3.7 ab	112.1 ± 0.4 a	101.8 ± 11.1 ab	88.3 ± 6.3 b	**
3-methylbutyl acetate	banana	80.2 ± 1.9 c	205.9 ± 2.9 b	184.6 ± 17.5 b	441.2 ± 29.5 a	***
ethyl hexanoate	apple	150 ± 4 b	108 ± 3 b	113 ± 7 b	275 ± 22 a	***
ethyl lactate	butter-scotch	8.6 ± 0.3 c	50.6 ± 8.0 bc	77.0 ± 8.1 ab	117.9 ± 22.6 a	**
ethyl octanoate	pineapple	148 ± 3 ab	111 ± 9 b	114 ± 5 b	174 ± 20 a	***
ethyl decanoate	floral, brandy	6.4 ± 0.1 b	7.7 ± 1.5 b	7.7 ± 0.2 b	39.9 ± 4.0 a	***
diethyl succinate	pleasant	5944 ± 64 c	10058 ± 650 b	11059 ± 675 b	14731 ± 1317 a	***
2-phenylethyl acetate	rose	20.4 ± 0.6 c	56.3 ± 1.6 b	55.0 ± 2.6 b	97.6 ± 16.8 a	***
diethyl malate	fruity	1466 ± 92 a	126 ± 20 b	83 ± 7 b	58 ± 6 b	***
Acids						
acetic acid	vinegar	33.0 ± 1.0 a	23.6 ± 1.6 b	21.0 ± 0.4 b	31.1 ± 3.6 a	**
3+2-methylbutanoic acid	cheese	153 ± 12	141 ± 18	141 ± 19	114 ± 25	n.s.
hexanoic acid	cheesy, rancid	897 ± 15 a	440 ± 55 c	408 ± 11 c	755 ± 22 b	***
octanoic acid	cheese, oily	1200 ± 239	891 ± 86	680 ± 93	992 ± 132	*
decanoic acid	fatty, rancid	99.7 ± 14.9	106.1 ± 18.0	148.0 ± 12.6	181.2 ± 36.3	n.s.
Other						
3-methylthio-1-propanol	potato, garlic	19.5 ± 0.6 ab	11.5 ± 3.1 b	14.5 ± 2.1 b	36.0 ± 6.9 a	***
N-3-methylbutyl acetamide		7.7 ± 0.5 b	19.0 ± 1.1 b	7.6 ± 0.7 b	398.8 ± 61.2 a	***

[1] Descriptors from: Genovese et al. [58]; Genovese et al. [4]; Flavors and Fragrances, Aldrich International Edition, 2011; Within each row, means followed by different letters on the column are significantly different according to the Tukey test ($p < 0.05$); n.s., *, **, *** correspond to not significant, and significant at $p \leq 0.05$, 0.01, and 0.001, respectively.

Table 7. Mean value (±standard deviation) of the level of glycosidic bound compounds (µg/L) of the wines obtained with Aglianico grapes harvested at four ripening stages, corresponding to 18 °Brix (S18), 20 °Brix (S20), 22 °Brix (S22) and 25 °Brix (S25). n.d.: not detected.

Compounds	Descriptor [1]	S18	S20	S22	S25	Sig.
Alcohols						
3-methyl-1-butanol	grass	144 ± 40	80 ± 21	112 ± 11	91 ± 26	n.s.
1-pentanol	mild green	5.1 ± 0.8	4.3 ± 1.4	6.2 ± 0.9	5.2 ± 1.2	n.s.
2-heptanol	herbaceous	7.8 ± 1.7	6.7 ± 1.4	7.6 ± 0.4	6.7 ± 1.1	n.s.
1-hexanol	grass, herbaceous	72.7 ± 18.1	37.7 ± 1.0	75.9 ± 7.0	75.8 ± 14.8	n.s.
(Z)-3-hexen-1-ol	grass, herbaceous	10.1 ± 2.1 a	6.3 ± 1.8 b	8.5 ± 1.2 b	4.8 ± 1.3 c	*
1-heptanol	apple, fruit	12.0 ± 1.5	10.9 ± 1.0	12.0 ± 0.2	11.9 ± 1.2	n.s.
(Z)-5-octen-2-ol		2.3 ± 0.6	2.1 ± 0.4	2.4 ± 0.3	2.3 ± 0.4	n.s.
1-octanol	waxy, citrus	3.6 ± 0.7	2.8 ± 0.6	3.2 ± 0.4	4.1 ± 0.4	n.s.
benzyl alcohol	toasty, sweet	246 ± 44 a	123 ± 20 b	157 ± 22 b	137 ± 25 b	**
2-phenylethanol	rose	163 ± 21 a	89 ± 15 b	121 ± 20 b	102 ± 13 a	*

Table 7. *Cont.*

Compounds	Descriptor [1]	S18	S20	S22	S25	Sig.
Terpene compounds						
α-terpineol	sweet, floral	2.2 ± 0.1 c	4.4 ± 0.1 b	8.8 ± 0.1 b	13.4 ± 2.3 a	**
epoxylinalool (I)		41.0 ± 7.5	30.6 ± 5.9	37.7 ± 6.6	28.4 ± 3.8	n.s.
epoxylinalool (II)		14.2 ± 4.0 a	6.3 ± 1.9 c	11.5 ± 0.2 ab	10.0 ± 1.4 bc	*
nerol	orange flowers	n.d.	n.d.	n.d.	10.0 ± 0.4	-
geraniol	orange flowers	n.d.	n.d.	n.d.	8.2 ± 0.2	-
exo-2-hydroxycineol		5.1 ± 0.5	2.5 ± 0.7	4.5 ± 0.9	4.0 ± 0.1	n.s.
trans-linalool oxide		78.9 ± 15.7 a	54.1 ± 14.6 ab	78.3 ± 9.0 a	48.3 ± 7.7 b	*
cis-linalool oxide		51.4 ± 10.1 a	33.7 ± 9.1 ab	52.9 ± 5.9 a	33.7 ± 4.1 b	*
Aldehydes and ketones						
furfural	sweet, woody, almond	1.9 ± 0.3	2.9 ± 0.4	2.9 ± 0.2	3.0 ± 0.8	n.s.
benzaldehyde	bitter almond	18.8 ± 3.7	18.0 ± 4.8	14.1 ± 2.2	16.6 ± 4.8	n.s.
5-methylfurfural	spice, bitter almond	3.1 ± 0.6	3.2 ± 0.7	4.1 ± 0.8	4.7 ± 0.7	n.s.

[1] Descriptors from: Genovese et al. [58]; Genovese et al. [4]; Flavors and Fragrances, Aldrich International Edition, 2011; within each row, means followed by different letters on the column are significantly different according to the Tukey test ($p < 0.05$); n.s., *, ** correspond to not significant, and significant at $p \leq 0.05$ and 0.01, respectively.

4. Conclusions

This study demonstrated that the choice of the ripening stage when harvesting the grapes is a crucial decision that will be reflected in the phenolic and aromatic composition of Aglianico wines after 20 months of aging in bottles. Aglianico grapes produce wines with more biologically active phenolic compounds (e.g., total *trans*-resveratrol), more stable color (e.g., anthocyanins) and richer in aroma compounds (free and bound) when grapes are harvested at a soluble solids content of 25 °Brix. Our work is a helpful tool to support the decisions of viticulturists and oenologists to modulate the biochemical profile and potentially, the health qualities of aged Aglianico red wines. Our results highlight that taste panel studies will be needed to define the sensory properties (color, taste, mouthfeel) of the chemically different wines that can be obtained by vinification at different maturation stages.

Author Contributions: Conceptualization, A.G. (Alessandro Genovese) and A.G. (Angelita Gambuti); methodology, S.A.L. and M.T.L.; validation, A.G. (Angelita Gambuti) and S.A.L.; formal analysis, L.L. and D.S.; investigation, A.G. (Angelita Gambuti); data curation, S.A.L., M.T.L. and B.B.; writing—original draft preparation, A.G. (Alessandro Genovese); writing—review and editing, A.G. (Angelita Gambuti), A.G. (Alessandro Genovese), B.B. and G.C.; supervision, L.M. and A.G. (Angelita Gambuti); project administration, L.M. and A.G. (Angelita Gambuti); funding acquisition, L.M. All authors have read and agreed to the published version of the manuscript.

Funding: This research received no external funding.

Institutional Review Board Statement: Not applicable.

Informed Consent Statement: Not applicable.

Data Availability Statement: The data presented in this study are contained within the article. Raw data are available on request from the corresponding author.

Conflicts of Interest: The authors declare no conflict of interest.

References

1. Goldberg, D.M.; Yan, J.; Soleas, G.J. Absorption of three wine-related polyphenols in three different matrices by healthy subjects. *Clin. Biochem.* **2003**, *36*, 79–87. [CrossRef]
2. Vitaglione, P.; Sforza, S.; Galaverna, G.; Ghidini, C.; Caporaso, N.; Vescovi, P.P.; Fogliano, V.; Marchelli, R. Bioavailability of trans-resveratrol from red wine in humans. *Mol. Nutr. Food Res.* **2005**, *49*, 495–504. [CrossRef] [PubMed]
3. Marambaud, P.; Zhao, H.; Davies, P. Resveratrol Promotes Clearance of Alzheimer's Disease Amyloid-β Peptides *. *J. Biol. Chem.* **2005**, *280*, 37377–37382. [CrossRef]
4. Genovese, A.; Lamorte, S.A.; Gambuti, A.; Moio, L. Aroma of Aglianico and Uva di Troia grapes by aromatic series. *Food Res. Int.* **2013**, *53*, 15–23. [CrossRef]

5. Genovese, A.; Gambuti, A.; Lamorte, S.A.; Moio, L. An extract procedure for studying the free and glycosilated aroma compounds in grapes. *Food Chem.* **2013**, *136*, 822–834. [CrossRef] [PubMed]
6. Kennedy, J.A.; Matthews, M.A.; Waterhouse, A.L. Changes in grape seed polyphenols during fruit ripening. *Phytochemistry* **2000**, *55*, 77–85. [CrossRef]
7. Di Stefano, R.; Borsa, D.; Ammarino, I.; Gentilizi, N.; Follis, R. Evoluzione della composizione polifenolica di uve da cultivars diverse durante la maturazione. *L'enologo* **2002**, *10*, 81–96.
8. Harbertson, J.F.; Kennedy, J.A.; Adams, D.O. Tannin in Skins and Seeds of Cabernet Sauvignon, Syrah, and Pinot noir Berries during Ripening. *Am. J. Enol. Vitic.* **2002**, *53*, 54–59.
9. Mateus, N.; Marques, S.; Gonçalves, A.C.; Machado, J.M.; Freitas, V.D. Proanthocyanidin Composition of Red Vitis vinifera Varieties from the Douro Valley during Ripening: Influence of Cultivation Altitude. *Am. J. Enol. Vitic.* **2001**, *52*, 115–121.
10. Belviso, S.; Torchio, F.; Novello, V.; Giacosa, S.; de Palma, L.; Segade, S.R.; Gerbi, V.; Rolle, L. Modeling of the evolution of phenolic compounds in berries of "Italia" table grape cultivar using response surface methodology. *J. Food Compos. Anal.* **2017**, *62*, 14–22. [CrossRef]
11. Wilson, B.; Strauss, C.R.; Williams, P.J. Changes in free and glycosidically bound monoterpenes in developing muscat grapes. *J. Agric. Food Chem.* **1984**, *32*, 919–924. [CrossRef]
12. Gomez, E.; Martinez, A.; Laencina, J. Changes in volatile compounds during maturation of some grape varieties. *J. Sci. Food Agric.* **1995**, *67*, 229–233. [CrossRef]
13. Francia-Aricha, E.M.; Guerra, M.T.; Rivas-Gonzalo, J.C.; Santos-Buelga, C. New Anthocyanin Pigments Formed after Condensation with Flavanols. *J. Agric. Food Chem.* **1997**, *45*, 2262–2266. [CrossRef]
14. Remy, S.; Fulcrand, H.; Labarbe, B.; Cheynier, V.; Moutounet, M. First confirmation in red wine of products resulting from direct anthocyanin-tannin reactions. *J. Sci. Food Agric.* **2000**, *80*, 745–751. [CrossRef]
15. Gambuti, A.; Picariello, L.; Rinaldi, A.; Ugliano, M.; Moio, L. Impact of 5-year bottle aging under controlled oxygen exposure on sulfur dioxide and phenolic composition of tannin-rich red wines. *OENO One* **2020**, *54*, 623–636. [CrossRef]
16. McManus, J.P.; Davis, K.G.; Beart, J.E.; Gaffney, S.H.; Lilley, T.H.; Haslam, E. Polyphenol interactions. Part 1. Introduction; some observations on the reversible complexation of polyphenols with proteins and polysaccharides. *J. Chem. Soc. Perkin Trans.* **1985**, *2*, 1429–1438. [CrossRef]
17. Rapp, A.; Marais, J. Shelf life of wine: Changes in aroma substances during storage and ageing of white wines. In *Shel Life Studies of Foods and Beverages*; Elsevier Sciences: Amsterdam, The Netherlands, 1993; ISBN 0167-4501.
18. Gunata, Y.Z.; Bayonove, C.L.; Baumes, R.L.; Cordonnier, R.E. Stability of Free and Bound Fractions of Some Aroma Components of Grapes cv. Muscat during the Wine Processing: Preliminary Results. *Am. J. Enol. Vitic.* **1986**, *37*, 112–114.
19. Chisholm, M.G.; Guiher, L.A.; Zaczkiewicz, S.M. Aroma Characteristics of Aged Vidal blanc Wine. *Am. J. Enol. Vitic.* **1995**, *46*, 56–62.
20. Fang, Y.; Qian, M.C. Quantification of selected aroma-active compounds in Pinot noir wines from different grape maturities. *J. Agric. Food Chem.* **2006**, *54*, 8567–8573. [CrossRef]
21. Gómez-Míguez, M.J.; Gómez-Míguez, M.; Vicario, I.M.; Heredia, F.J. Assessment of colour and aroma in white wines vinifications: Effects of grape maturity and soil type. *J. Food Eng.* **2007**, *79*, 758–764. [CrossRef]
22. Bindon, K.; Varela, C.; Kennedy, J.; Holt, H.; Herderich, M. Relationships between harvest time and wine composition in Vitis vinifera L. cv. Cabernet Sauvignon 1. Grape and wine chemistry. *Food Chem.* **2013**, *138*, 1696–1705. [CrossRef] [PubMed]
23. Vilanova, M.; Genisheva, Z.; Bescansa, L.; Masa, A.; Oliveira, J.M. Changes in free and bound fractions of aroma compounds of four Vitis vinifera cultivars at the last ripening stages. *Phytochemistry* **2012**, *74*, 196–205. [CrossRef] [PubMed]
24. Bowen, A.J.; Reynolds, A.G. Aroma compounds in Ontario Vidal and Riesling icewines. I. Effects of harvest date. *Food Res. Int.* **2015**, *76*, 540–549. [CrossRef] [PubMed]
25. Sherman, E.; Greenwood, D.R.; Villas-Boâs, S.G.; Heymann, H.; Harbertson, J.F. Impact of grape maturity and ethanol concentration on sensory properties of Washington State Merlot wines. *Am. J. Enol. Vitic.* **2017**, *68*, 344–356. [CrossRef]
26. Ferrero-del-Teso, S.; Arias, I.; Escudero, A.; Ferreira, V.; Fernández-Zurbano, P.; Sáenz-Navajas, M.-P. Effect of grape maturity on wine sensory and chemical features: The case of Moristel wines. *LWT* **2020**, *118*, 108848. [CrossRef]
27. Arias-Pérez, I.; Ferrero-Del-Teso, S.; Sáenz-Navajas, M.P.; Fernández-Zurbano, P.; Lacau, B.; Astraín, J.; Barón, C.; Ferreira, V.; Escudero, A. Some clues about the changes in wine aroma composition associated to the maturation of "neutral" grapes. *Food Chem.* **2020**, *320*, 126610. [CrossRef]
28. Wu, Y.; Zhang, W.; Song, S.; Xu, W.; Zhang, C.; Ma, C.; Wang, L.; Wang, S. Evolution of volatile compounds during the development of Muscat grape 'Shine Muscat' (Vitis labrusca× V. vinifera). *Food Chem.* **2020**, *309*, 125778. [CrossRef] [PubMed]
29. Lutskova, V.; Martirosyan, I. Influence of Harvest Date and Grape Variety on Sensory Attributes and Aroma Compounds in Experimental Icewines of Ukraine. *Fermentation* **2021**, *7*, 7. [CrossRef]
30. Pérez-Magariño, S.; González-San José, M.L. Evolution of flavanols, anthocyanins, and their derivatives during the aging of red wines elaborated from grapes harvested at different stages of ripening. *J. Agric. Food Chem.* **2004**, *52*, 1181–1189. [CrossRef]
31. Glories, Y. La maturità fenolica delle uve: Primo parametro da controllare per una corretta vinificazione in rosso. *Vignevini* **1999**, *26*, 46–50.
32. Glories, Y. La couleur des vins rouges. 1° e 2° partie. *Conn. Vigne Vin* **1984**, *18*, 253–271.

Nutritive Value, Polyphenolic Content, and Bioactive Constitution of Green, Red and Flowering Plants

AF389437